Mathematical theory of dislocations and fracture

R.W. LARDNER

Concise, logical, and mathematically rigorous, this introduction
to the theory of dislocations is addressed primarily to students and
researchers in the general areas of mechanics and applied mathematics.
Its scope encompasses those aspects of dislocation theory which are
closely related to the theories of elasticity and macroscopic plasticity,
to modern continuum mechanics, and to the theory of cracks
and fracture. The volume incorporates several new and original pieces
of work, including a development of the theory of dislocation motion
and plastic strain for non-linear materials, a new discussion
of the line tension model, revised calculations of the Peierls resistance,
and a new development of the van der Merwe theory of crystal interfaces.
Mathematical Expositions No. 17.

R.W. LARDNER is professor of mathematics at Simon Fraser
University in Vancouver, British Columbia.

MATHEMATICAL EXPOSITIONS

Editorial Board

H.S.M. COXETER, G.F.D. DUFF, D.A.S. FRASER,
G. de B. ROBINSON (Secretary), P.G. ROONEY

Volumes Published

MATHEMATICAL EXPOSITIONS NO. 17

Mathematical theory of dislocations and fracture

R.W. LARDNER

UNIVERSITY OF TORONTO PRESS

© University of Toronto Press 1974

Printed in Great Britain for

University of Toronto Press, Toronto and Buffalo

ISBN 0-8020-5277-0

CN ISSN 0076-5333

LC 75-190346

AMS 1970 Subject classification 73

CONTENTS

PREFACE

It has seemed to the author for some time that there exists a need for an introductory text on the theory of dislocations addressed primarily to students and researchers in the general areas of mechanics and applied mathematics. The present book represents an attempt to meet this need. It presupposes that the reader has a mathematical background which includes vector field theory and elementary differential equations and in addition has some familiarity with the theory of elasticity. In certain places more advanced mathematical tools are used, such as tensor calculus, contour integration, Fourier transforms, and the Green's functions for certain partial differential equations, but these are generally self-explanatory or concern peripheral topics. Thus a large part of the book, particularly the earlier chapters, should be accessible to advanced undergraduates in the relevant fields, although it is anticipated that the book will be of most interest to students at the postgraduate level.

In this book, we have limited ourselves, broadly speaking, to those aspects of dislocation theory which are closely related to the theories of elasticity and macroscopic plasticity, to modern continuum mechanics, and to the theory of cracks and fracture. We have made no effort to provide a comprehensive treatment of all of the work which has been done on dislocations, if for no other reason than that several excellent books already exist which provide such coverage (see the General Bibliography). In particular, there is very little discussion of experimental techniques or results, although we have tried throughout to keep the reader aware of the applicability of the theory to the case of most physical interest – dislocations in crystals. On the other hand, the fundamentals of the theory of dislocations are covered with some measure of completeness and rigour.

The first chapter contains in its first two sections a brief historical review and some elementary definitions and geometrical properties of dislocations. The remainder of the chapter consists of a qualitative discussion of the physical significance of dislocations and their role in the mechanical behaviour of crystalline materials. The purpose of these sections is to provide some motivation for the theoretical investigations which follow in later chapters, so that they can be reserved for later reading if preferred.

The second chapter consists of two distinct parts, the first being a summary of those portions of the classical theory of elasticity which are needed later, and the second a brief but self-contained account of non-linear elasticity. The latter is used as a basis for the discussion of continuous distributions of dislocations, and so can be omitted until §7.3 is reached. Chapter 3 contains a derivation of the stress-fields and energies of various straight dislocations and, in its final section, of curved dislocation lines.

After this basic material has been set down, the book develops along three more or less independent branches. One of these, Chapter 4, discusses various models that have been proposed to account for the effects on the structure of dislocations of the crystallinity of the media in which they reside. The second branch starts in Chapter 5 with a treatment of continuous linear arrays of dislocations and their applications to crack problems and elastic boundary-value problems. Then in the following chapter these methods and results are applied to the problems of crack formation and growth as they occur in brittle and ductile fracture and in fatigue fracture. The third branch is concerned essentially with continuous distributions of dislocations. In Chapter 7 the kinematic properties of such distributions are described, first within the classical theory of elasticity and then within the framework of the non-linear theory. In Chapter 8 the internal stress problem is posed and solved for linear isotropic materials, and the solution is applied to a number of particular dislocation configurations and to a discussion of the line-tension model. In the final chapter, on dislocation dynamics, we have included derivations of the stress-fields caused by arbitrarily moving dislocations, a discussion of the kinematics of the flow of a continuous distribution, and a partly qualitative discussion of the application of dislocation dynamics to plastic flow.

Throughout we have used tensor-component notation, with the summation convention applied to repeated indices. Rectangular Cartesian coordinates are used almost everywhere, the exceptions being clearly indicated. We have used (x, y, z) and (x_1, x_2, x_3) synonymously and almost interchangeably, and have correspondingly used, for example, σ_{12} and σ_{xy} to denote the same shear stress components. We have used ∂_i to stand for the differential operator $\partial/\partial x_i$, with ∂_{ij} denoting $\partial^2/\partial x_i \partial x_j$, etc. As usual, δ_{ij} denotes the Kronecker delta and ε_{ijk} the permutation symbol in three dimensions. Vectors and tensors are indicated by bold-face type (e.g. $\mathbf{b}$, $\boldsymbol{\sigma}$, with components b_i, σ_{ij}). Finally, we use the standard notation ∂V to denote the surface bounding a volume V in space.

It is a pleasure to thank Professor J.A. Steketee for a large number of constructive comments which have led to improvements in the book. I am also indebted to Mrs Arielle Gerencser for her excellent typing and retyping of the manuscript. I would like to acknowledge research support from the National Research Council and the Defence Research Board of Canada which has contributed towards the production of the book. I also wish to express my thanks for a grant provided by the Publications Fund of the University of Toronto Press.

Simon Fraser University   R.W.L.
July 1971

GENERAL BIBLIOGRAPHY OF TEXTS ON DISLOCATIONS AND RELATED SUBJECTS

Cottrell, A.H., *Dislocations and plastic flow in crystals* (Oxford, 1953)
Friedel, J., *Les Dislocations* (Paris: Gauthier-Villars, 1956); *Dislocations* (Oxford: Pergamon, 1964)
Hirth, J., and Lothe, J.P., *Theory of dislocations* (New York: McGraw-Hill, 1968)
Hull, D., *Introduction to dislocations* (Oxford: Pergamon, 1965)
Kröner, E., *Kontinuumstheorie der Versetzungen und Eigenspannungen* (Berlin: Springer, 1958)
McClintock, F.A., and Argon, A.S., *Mechanical behavior of materials* (Reading, Mass.: Addison-Wesley, 1966)
Nabarro, F.R.N., *Theory of crystal dislocations* (Oxford, 1969)
Noll, W., Toupin, R., and Wang, C.-C., *Continuum theory of inhomogeneities in simple materials* (Berlin: Springer, 1969)
Nye, J.F., *Physical properties of crystals* (Oxford, 1957)
Read, W.T., *Dislocations in crystals* (New York: McGraw-Hill, 1953)
Schmidt, E., and Boas, W., *Kristallplastizität* (Berlin: Springer, 1935); *Plasticity of crystals* (London: Hughes, 1950; Chapman and Hall, 1968)
Van Bueren, H.G., *Imperfections in crystals* (Amsterdam: North-Holland, 1960)
Weertman, J., and Weertman, J.R., *Elementary dislocation theory* (London: Macmillan, 1964)

Mathematical theory of dislocations and fracture

1

INTRODUCTION TO DISLOCATIONS

1.1

HISTORICAL BACKGROUND

A *Volterra dislocations*

Dislocations were first investigated in the early years of the century by Volterra and others[1-3] who were concerned with certain deformations of an elastic body in which the displacement field is not single-valued. The simplest examples of such deformations can be formed in the following way. A hollow circular cylinder is cut through on one side by a half-plane through the axis of the cylinder (*AB* in Fig. 1.1(a)) and the two faces of the cut are given a rigid displacement relative to one another by the application to them of suitable tractions. These tractions are required to be equal and opposite at corresponding points on the two faces. The relative displacement of the faces of the cut can be classified into six basic types corresponding to three independent directions of rigid translation and three independent rotations. The first three of these types are illustrated in Figure 1.1(b)–(d), corresponding respectively to rigid translations normal to the cut, in the plane of the cut and normal to the cylinder axis, and parallel to the cylinder axis. Cases (b) and (c) are called *edge dislocations* and case (d) a *screw dislocation*. The three rotation dislocations are not drawn, and we shall make only occasional mention of them since, unlike the translation dislocations, they appear to play little or no part in the behaviour of real materials.

Because the dislocation is formed by applying equal and opposite tractions to the two faces A_1B_1 and A_2B_2, these may be glued together again in the deformed state, and will then support one another without further deformation. (For the edge dislocation of Figure 1.1(b), a certain amount of extra material has to be inserted, of course, before the gluing can be performed.) In this way, we achieve a state of stress in the cylinder in which the stress and strain are continuous and single-valued, and no external tractions are applied, but the displacement field changes by a certain constant amount in following any path in the cylinder which encircles the axis.

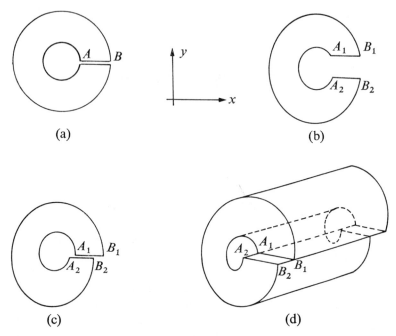

FIGURE 1.1

The change in displacement along such a path, which is equal to the rigid translation imposed in Figure 1.1 in order to form the dislocation, is called the *Burgers vector* of the dislocation and denoted by **b**. To make the definition of **b** precise, we first of all fix a unit vector ξ in some direction along the cylinder axis; then we follow a path around the axis in a right-handed sense with respect to ξ, returning to the starting point. The change in displacement vector, $\mathbf{u}_{\text{final}} - \mathbf{u}_{\text{initial}}$, between the initial and final points on the path, equals the Burgers vector. Clearly, reversing ξ changes the sign of **b**. In Figure 1.1(b)–(d), if $b = |A_1 A_2|$, the Burgers vectors are respectively $(0, -b, 0)$, $(-b, 0, 0)$, and $(0, 0, b)$ when ξ is chosen along the positive z-direction (i.e. out of the paper).

B *Plastic flow in crystals*

In the thirties dislocations were proposed in a different context as the defects in crystalline solids which are responsible for the plastic flow of such materials.[4-7] When single crystals of many materials are stressed above their yield limit they are often observed to undergo plastic deformation which is concentrated as slip on a few parallel planes (Fig. 1.2). The planes on which the deformation occurs are called *slip-planes*. When a

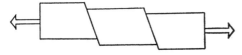

FIGURE 1.2

group of slip-planes become active within the same neighbourhood of the crystal, they are often said to form a *slip-band*. The slip-planes are usually crystal planes on which the atoms are closest packed, regardless of the orientation of the applied stress. The yield stress for such materials is determined by the shear traction on one of the possible slip-planes reaching a critical value, not by the maximum over-all shear stress. Furthermore, the direction of slip within the slip-plane is always crystallographic, no matter what the direction of the shear stress. More complicated slip modes in single crystals can be considered as the simultaneous occurrence of this type of slip on two or more sets of planes.[8]

The common state for metals is not the single crystal but the polycrystal, which consists of a number of single crystals, or grains, whose typical size might be 10^{-2} mm, which join together continuously. In the plastic deformation of a polycrystal, each grain behaves as a separate single crystal, slipping on one or more sets of crystal planes, but with the additional constraint that the slip displacements in adjacent grains have to be continuous across the grain boundary between them. Thus, in detail, the plastic flow of polycrystals is considerably more complex than that of single crystals, although basically it is the same process.

The elementary slip process is illustrated in Figure 1.3 for a simple cubic crystal, where intersections of pairs of lines represent atoms. The two halves of the crystal slide relative to one another on the plane SS', and in this simplest process the amount of slip is one lattice spacing. In practice, slip can and does occur through many spacings on the same slip-plane, leading to visible steps on external surfaces, but such multiple slip corresponds to many repetitions of the above basic process.

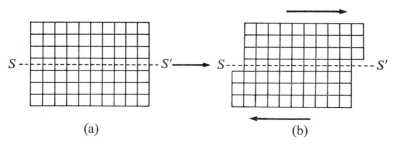

(a)   (b)

FIGURE 1.3

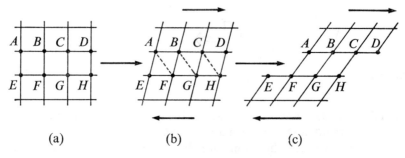

FIGURE 1.4

The simplest hypothesis for the slip mechanism is that slip occurs over the whole plane at once, as illustrated in Figure 1.4. The deformation from (a) to (b) is purely elastic, so that if the shear stress is released the lattice will return to the configuration (a). However, if the shear stress is sufficiently large, a point will be reached when atom A, say, is nearer to F than to its original neighbour E, and at this stage it will change its allegiance. This occurs over the whole slip-plane at the same instant so that the row of atoms $ABCD$ . . . slip forward all together, to give state (c), which has the desired slip-steps.

Unfortunately this model does not fit the observed facts, since it predicts that the deformation from (a) to (b) remains elastic until $AF = AE$, which for a cubic crystal corresponds to an angle of shear of $\tan^{-1}(1/2)$ or about $26\frac{1}{2}°$. In most cases, plastic flow begins at much smaller strains, typically a hundredth or a thousandth of this value. In a similar way, the yield *stress* predicted by this model is too high by the same sort of factor. For example, if Hooke's law were obeyed right up to this maximum strain, the yield shear stress would be $\mu/2$ where μ is the shear modulus. The use of Hooke's law overestimates the yield stress, however, and various refinements have been made to this calculation. Frenkel[9] assumed a sinusoidal stress-strain law which has a periodicity determined by the lattice, and this leads to a shear strength of $\mu/2\pi$. Calculations based on interatomic forces, done by Mackenzie,[10] lower the value further to $\mu/30$. But this is still much higher than the experimental values of yield stress, which are typically about $10^{-3}\mu$ to $10^{-4}\mu$, and in particularly careful experiments can be several orders of magnitude smaller than this.

c *Dislocations and plastic flow*

To overcome this difficulty it was suggested by Taylor,[4] Orowan,[5] and Polanyi[6] that the slip process occurs via the operation of an edge dislocation (see Fig. 1.5). Figure 1.5(a) represents an intermediate stage of the process

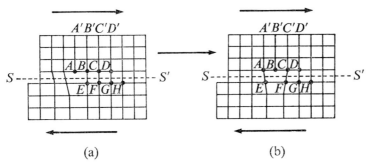

FIGURE 1.5

in which the slip has occurred on part of the slip-plane (between S and A) but not on the remainder. There is a slip-step at S, but not yet at S', and this discrepancy is made up for by an extra half-plane of atoms, AA'. Figure 1.5(b) shows the next stage in the process with the extra half-plane at BB'; it then moves to CC', DD', and so on until it finally emerges on the right-hand side as the slip-step of Figure 1.3(b). The crystalline defect of Figure 1.5(a) corresponds to the edge dislocation of Figure 1.1(c), rotated through 180°, since it can be formed by making a cut from S to A–E in the crystal, sliding SA to the right by one lattice spacing relative to SE, and then re-joining the crystal across the cut. Its Burgers vector has the magnitude of a lattice spacing.

It must be explained in more detail how the motion of the extra half-plane can occur and in particular why it occurs at such low stresses. In Figure 1.6(a)–(d), successive configurations of the atoms on either side of the slip-plane SS' are shown. In (a) the extra half-plane ends at A, and the dominant interatomic bonds are BE, CF, etc. But also, owing to the distortion of the lattice in the neighbourhood of A produced by the defect, atom E is almost

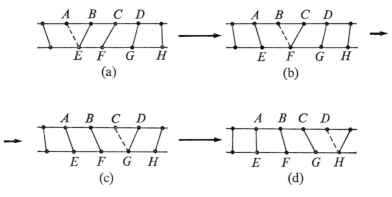

FIGURE 1.6

as near *A* as it is *B*, even when no external stress is applied. The effect of external shear stress is to move *E* even nearer to *A*, and it is clear that only a little extra shear strain is needed to make *EA* less than *EB*. When that happens, *EA* becomes a dominant bond and *B* becomes the extra half-plane (Fig. 1.6(b)). This process then continues down the line of atoms. We shall reserve until later (Chapter 4) the calculation of the shear stress needed to move a dislocation, but we can be confident that it will be much less than the strengths found above on the basis of simultaneous slip over the whole slip-plane.

Returning to the comparison with Volterra dislocations, the edge in Figure 1.5(a) can be regarded as formed in a different way. Instead of cutting along *SA* in the original perfect crystal we could cut along *AA'*, pull the two faces apart by one lattice spacing, and then insert the extra half-plane of atoms. This corresponds to the Volterra dislocation of Figure 1.1(b) rotated through 90°. We can associate with the first method of cutting (along *SA*) the dislocation motion which occurs in the slip process, in the sense that varying the depth of the cut gives a series of positions of the dislocation which occur during slip. Similarly if we vary the depth of the cut *AA'* in the second method, we change the dislocation position in the vertical direction (Fig. 1.5), which corresponds to a type of dislocation motion called *climb*. Although *slip* (or *glide*) of dislocations is a widespread and vitally important phenomenon, climb becomes significant only at high temperatures. This is not surprising since climb requires either the cooperative movement of the

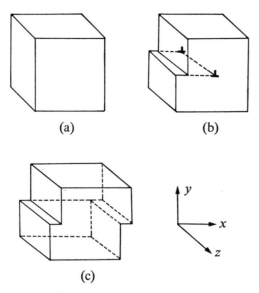

(a)   (b)

(c)

FIGURE **1.7**

whole half-plane of atoms or a change in the number of atoms near the centre of the dislocation, occurring through the diffusion of atoms away from or towards the dislocation.

The slip process occurring via an intermediate edge dislocation is illustrated in Figure 1.7. (Note the symbol ⊥ used for an edge, the long line denoting the slip-plane and the short line denoting the extra half-plane.) Now it is also possible that the process (a)–(c) could proceed through a series of intermediate states such as those drawn in Figure 1.8. In this series the material above the half-plane *ABCD* has slipped to the right by one lattice spacing relative to that below. The boundary of the slipped region is the line *AB*, and, as this moves from the front (*CD*) to the back (*HI*) of the crystal, we get the transition from Figure 1.7(a) to 1.7(c). This process was first suggested by Burgers,[7] and the intermediate states correspond to the Volterra screw dislocation of Figure 1.1(d). The atomic arrangements within the crystal are shown in Figure 1.9, which is a plan view of the crystal in Figure 1.8. The intersections of rigid lines represent atoms below the slip-plane, and circles represent atoms above the slip-plane.

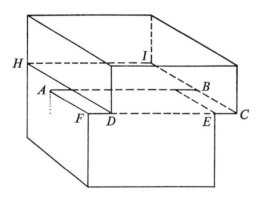

FIGURE 1.8

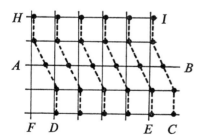

FIGURE 1.9

1.2
THE BURGERS VECTOR OF A DISLOCATION

A *Burgers circuits*

We now give an alternative definition of the Burgers vector, equivalent to that given earlier but more appropriate to the crystal situation, following the original notion of Burgers. First a direction is ascribed to the dislocation line by choosing (arbitrarily) a unit vector ξ along it such that, for the straight dislocations encountered so far, ξ is constant along the line. Then a circuit is drawn in the crystal in a right-hand sense with respect to ξ so as to link the dislocation line once. An example is given in Figure 1.10(a) for an edge dislocation, where ξ has been chosen out of the paper. This circuit is now mapped lattice step by lattice step onto a perfect (i.e. undislocated) reference crystal ($ABC \ldots UV \rightarrow A'B'C' \ldots U'V'$ in Fig. 1.10), and of course the mapped circuit does not close. The additional vector needed to close it is called the *true Burgers vector* ($V'A'$ in the example), and this is always one perfect lattice spacing in length. If we now map the closure failure back onto the real crystal, using the same mapping as before, we obtain the *local Burgers vector* (**b** in Fig. 1.10(a)). This latter quantity has the advantage of having the same orientation as the real crystal at the point of interest, but has the disadvantage that its definition is ambiguous in that it depends on the starting point for the Burgers circuit, assuming for definiteness that the mapping back of the true Burgers vector is carried out at this point. Because of the lattice strains produced by the dislocation itself, the interatomic spacings will vary around the circuit. This difficulty can be reduced by ensuring that the size of the Burgers circuit is sufficient that the strains at points on it are small.

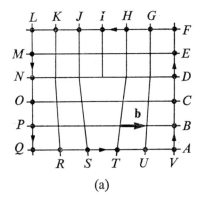

(a)

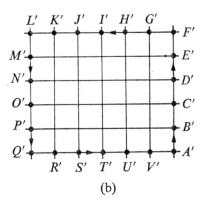

(b)

FIGURE 1.10

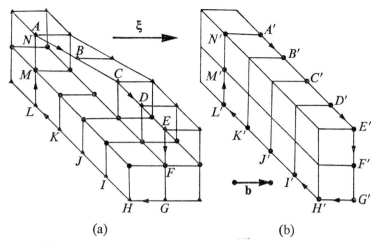

(a)   (b)

FIGURE 1.11

Reversing the sign of ξ, we can use the same circuit in the opposite sense to obtain a Burgers vector $-\mathbf{b}$.

For an edge dislocation, $\mathbf{b}$ is perpendicular to ξ and the two vectors $\mathbf{b}$ and ξ lie in the slip-plane. For a screw dislocation, $\mathbf{b}$ and ξ are parallel. Take ξ as in Figure 1.11(a) and the circuit $ABC \ldots LMN$, which maps onto $A'B'C' \ldots L'M'N'$ in Figure 1.11(b) and gives a closure failure $N'A'$ and hence a (local) Burgers vector $\mathbf{b}$ as shown. If $\mathbf{b}.\xi > 0$, the screw is said to be right-handed, and if $\mathbf{b}.\xi < 0$, it is left-handed.

B *Mixed dislocations*

Another possible intermediate state in the slip process from Figure 1.7(a)–(c) is drawn in Figure 1.12. Here slip has started from the corner D and has

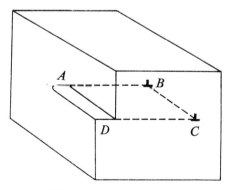

FIGURE 1.12

occurred over a rectangle *ABCD* on the slip-plane. The transition progresses as *BC* moves to the right and *AB* to the back of the material. The boundary of the slipped region now consists of two parts: a screw dislocation line *AB* and an edge dislocation line *BC*. The Burgers vectors of these two segments are identical, provided the line direction ξ is defined continuously along one sense or the other of the dislocation line (i.e. if ξ on *AB* points from *A* to *B*, then on *BC* it points from *B* to *C*, and vice versa).

By extending this idea we can obtain dislocation lines made up of many edge and screw segments, the only condition being that, if ξ is defined continuously along the line, then the Burgers vector is constant along it. If the segments are very small, the dislocation line will appear as a smooth curve on a macroscopic scale. Such a curve, called a *mixed* dislocation, can be resolved into screw and edge components by resolving the Burgers vector into components parallel and perpendicular to ξ at any point:

$$\mathbf{b}_s = (\mathbf{b}\cdot\xi)\xi, \qquad \mathbf{b}_e = \mathbf{b}-(\mathbf{b}\cdot\xi)\xi = \xi\wedge(\mathbf{b}\wedge\xi).$$

Within the theory of elastic continua, dislocation lines can be considered which are arbitrarily smooth curves. For, let Γ be such a curve, which may be a closed curve or an open curve ending on the boundary of the body or extending to infinity, and let *C* be a surface spanning Γ. Then we form a dislocation by making a cut on *C* and deforming the two faces relative to one another by a fixed relative displacement, $\Delta\mathbf{u}$, in such a way that the tractions at corresponding points are equal and opposite. The body can be rejoined across the cut if desired. Calling the two faces arbitrarily C_+ and C_-, let $\mathbf{n}$ be the unit vector normal to *C* pointing from C_+ to C_-, and let ξ define the right-hand sense around Γ with respect to $\mathbf{n}$ (see Fig. 1.13). If $\mathbf{u}_+$ and $\mathbf{u}_-$ are the displacements on C_+ and C_- and $\Delta\mathbf{u} = \mathbf{u}_+-\mathbf{u}_-$ the prescribed relative displacement, then the Burgers vector is $\mathbf{b} = \Delta\mathbf{u}$. This is illustrated in Figure 1.13 when the relative displacements of the two sides are such as to form at the right-hand side of Γ an edge of the same type as in Figure 1.10(a). On the left-hand side there is an opposite edge and at the back and front respectively, right-hand and left-hand screws.

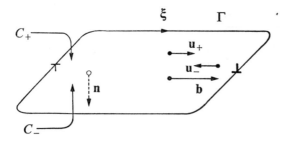

FIGURE 1.13

There is some difficulty associated with the above dislocation, namely that, if Δu is constant over C, there arise singularities of stress and strain on Γ of such a strength as to lead to a divergent energy. There are two ways of removing this difficulty. The first is to suppose that before displacing the two faces relative to one another a small tube of material is removed from around Γ. For straight dislocation lines, this brings us back to the Volterra dislocations of Figure 1.1, which were defined only for hollow cylinders. This is the procedure we shall adopt in Chapter 3 in calculating the stress-fields of dislocations. Perhaps a more satisfactory approach, which will be used in Chapter 8 in the context of continuous distributions of dislocations, is to suppose that Δu does not drop sharply to zero at Γ, but falls off continuously in some small core region around Γ. Thus we no longer have a single dislocation line but a smeared-out filament of dislocation. Of course neither of these approaches precisely describes the situation encountered in a crystal, where the core region of a dislocation cannot properly be treated by any continuum model, but should be dealt with on the basis of inter-atomic forces.

c *Groups of dislocations*

In constructing Burgers circuits, we might draw one large enough to link several dislocation lines. Then the closure failure of the mapping of such a circuit equals the sum of the Burgers vectors of the linked dislocations. This is illustrated for two dislocations in Figure 1.14. The circuit $ABCDA$ linking both is equivalent to the two circuits $ABCA + ACDA$, so its closure failure equals the sum of the two failures, that is $\mathbf{b}_1 + \mathbf{b}_2$. This argument clearly extends to arbitrarily many dislocations.

It is possible for branches to occur in dislocation lines. A simple example

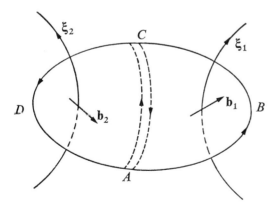

FIGURE 1.14

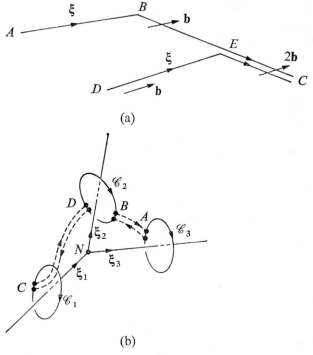

(a)

(b)

FIGURE 1.15

is illustrated in Figure 1.15(a). Here ABE is the boundary on the slip-plane of a region over which an amount $\mathbf{b}$ of slip has occurred, and DEC is the boundary of a region where a second equal amount of slip has occurred. Then EC is an edge dislocation of Burgers vector $2\mathbf{b}$, which at E splits into an edge BE and a screw DE, each with Burgers vector equal to $\mathbf{b}$. Such points are called *nodes*.

In Figure 1.15(b) a general triple node, N, is drawn, at which the three Burgers vectors are $\mathbf{b}_1, \mathbf{b}_2, \mathbf{b}_3$ with the $\boldsymbol{\xi}_i$ as drawn. If $\{l_i\}$ are three Burgers circuits, one for each dislocation, they correspond to closure failures $\{\mathbf{b}_i\}$. Now join the circuits together by the dashed lines, AB and CD, and consider the composite circuit made up of the following sequence, starting from A: l_3, AB, BD following l_2, DC, $-l_1$, CD, DB along l_2, BA. The closure failure for this must be $\mathbf{b}_2 + \mathbf{b}_3 - \mathbf{b}_1$ since the extra parts CD and AB are traced once in each direction. But the circuit can be shrunk to zero without having to cross any dislocation lines; hence it has zero closure failure. Therefore $\mathbf{b}_1 = \mathbf{b}_2 + \mathbf{b}_3$.

This argument extends to an arbitrary node, giving $\sum_i \pm \mathbf{b}_i = 0$, summed over all dislocations at the node, with $+$ if $\boldsymbol{\xi}_i$ points into the node and $-$ if it points away.

1.3
SOURCES OF DISLOCATIONS

A *General remarks*

For a long time after the importance of dislocations in explaining crystal plasticity was realized, the question of their origin remained unanswered. The simplest hypothesis is that an opposite pair of dislocations could be generated spontaneously on the same slip-plane, and then move off in opposite directions, thus spreading the slip from the centre of the crystal outwards. Unfortunately the stress needed to create such dislocation pairs is again much higher than observed yield stresses. It is now known that new-grown crystals contain substantial numbers of dislocations, even before they are strained, typically about 10^5 dislocation lines per square centimetre. Frank[11, 12] first suggested that the presence of dislocations aids the process of growth of a crystal – that indeed it is difficult to understand how crystallization occurs as easily as it does without introducing dislocations. The direct connection between dislocations and crystal growth was later observed by Griffin.[13] A variety of techniques now exist (see, for example, the review article by Pashley[14]) by means of which it is possible to observe the presence of dislocations and to measure their densities.

It is possible, if considerable care is taken, to make crystals with densities of dislocations much lower than 10^5 lines/cm^2. For some materials, thin whiskers can be made which seem in some cases to be actually free of dislocations. It is interesting to note that the yield strengths of these whiskers are very high,[15] approaching $\mu/15$, which is, in fact, higher than some of the estimates of the strength of a perfect crystal quoted in §1.1B.

After plastic straining, the dislocation density is observed to increase, attaining values as high as 10^{11} to 10^{12} lines/cm^2, and the question of the origin of the extra dislocations naturally arises. An important clue is provided by the observation that plastic flow tends to occur in concentrated amounts – that is, the plastic strain is not spread uniformly throughout the material, but occurs as large amounts of slip on relatively few planes. This shows up on the external surface as large slip-steps which are many lattice spacings in height (up to a thousand or more in some cases). Thus a large number of dislocations move on each active slip-plane, whereas we should expect that the original grown-in dislocations would be spread more or less uniformly throughout the volume. There are two possible explanations for this. There could be a cascade process operating in which the motion of one dislocation creates others on the same slip-plane, which in turn begin to move and create yet more, and so on, so that once slip starts on any one plane it will develop to a considerable extent on the same plane. Alternatively, there could be within the material certain sources of dislocations which, once they start operating, produce many dislocations on the same glide-plane. The continued operation of a source which is already operating could plausibly be favoured over the stimulation of new sources.

B *Dynamical sources*

The earliest suggestions made by Frank[16] belong to the first of these categories. When a dislocation emerges from the surface of a crystal a slip-step is created (unless the dislocation Burgers vector happens to be parallel to the surface). Such a step corresponds to an increase in the free surface area of the body and therefore it absorbs a certain amount of the energy of the dislocation. (This effect is analogous to the surface tension effect in liquids: the creation of extra surface area in both solids and liquids requires that an additional surface energy be provided.) Now a fast-moving dislocation has a large amount of energy associated with it both as elastic strain energy and as kinetic energy of the material particles. Upon reaching the surface, part of this energy must go to provide the surface energy of the associated slip-step; but if the velocity of the dislocation is sufficiently high, there will be enough energy left over to create a new dislocation of the opposite sign (and its slip-step) which will then move back across the material, driven by the external stress. Calculations based on energy indicate that a dislocation velocity of the order of half the shear-wave velocity in the material is necessary for this process. Since velocities of this magnitude can only be attained at stresses many times higher than yield, and particularly since the energy calculation assumes that all the incident energy can be transformed into the strain energy of the new dislocation, without any loss, it is now thought that this mechanism occurs in only rather special circumstances, such as shock-loading.

A second mechanism proposed by Frank occurs when a dislocation moving through the crystal meets an obstacle lying across its path, such as a precipitate or impurity particle or the stress-field of a second, immobile, dislocation. This is illustrated in Figure 1.16, in which the slip-plane is the plane of the paper, the obstacle is the shaded circle, the arrow along the dislocation line indicates the ξ direction, and the transverse arrows the velocity. When the dislocation is moving slowly, we get the sequence Figure 1.16(a)–(e). The central part of the dislocation is held up, while the slip still spreads in the outer regions, which link up eventually as the obstacle is surrounded. At stage (c) the parts A and A' of the dislocation line attract one another, and since they are moving only slowly, these two segments come to rest in a position of coincidence at stage (d). In that position they cancel one another out and the dislocation separates into two parts. So the slip progresses beyond the obstacle, leaving a small unslipped region immediately around it. At stage (c), the two segments A and A' have a certain amount of energy (strain energy plus kinetic energy) associated with them and, in order for the process from (c) to (e) to occur, this energy must be dissipated either into heat or into scattered elastic waves. If the dislocation is moving fast, however, the dissipation mechanisms may not be sufficient to arrest the segments A and A' and they will then overshoot stage (d), reaching configurations such as (f) and (g). The overshoot gives a region $\mathscr{R}$ in which slip by two Burgers vectors has occurred, and this slipped region will be propagated by the applied stress just as the original dislocation is moved. Thus $\mathscr{R}$ expands, as an inner loop, which is again held up by the obstacle, and an outer expanding loop. If the ends B and B' approach one another sufficiently fast, the process can occur again to produce a further expanding dislocation loop.

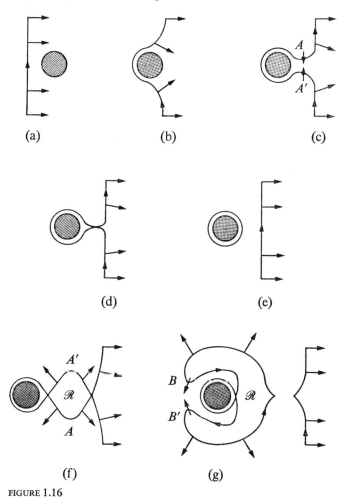

FIGURE 1.16

In order for this process to occur it is necessary that at stage (c) there should be enough energy in the neighbourhood of the colliding segments A and A' to overcome the energy dissipation which must occur during the collision. This illustrates the major difficulty with this explanation of dislocation production in that, like the first one, it is highly favoured by large dislocation velocities and hence by large rates of plastic strain.

A third possible dynamical production mechanism is perhaps even simpler. A fast-moving dislocation has a large energy associated with it, which is concentrated mostly as motion of the atoms near the dislocation line itself. It should therefore be possible for the dislocation to lose this energy (and thus slow down), with the energy being used to create a pair of opposite dislocations. Again this process requires high dislocation velocities.

There is evidence that dynamical cascade processes of the types described above become significant under conditions of very high strain-rates such as those encountered under conditions of shock-loading. However, it is quite clear that they are by no means the whole story. The large increases in dislocation density mentioned earlier are known to occur even at the smallest strain rates, and hence cannot be dependent on the conversion of kinetic energy into dislocation strain energy.

c Double cross-slip[17]

Another mechanism, by means of which a moving screw dislocation can act as a source, is called double cross-slip, and results in a trail of dislocation loops in the wake of the dislocation. The glide-plane for an edge dislocation, or for a mixed dislocation, is uniquely determined by the Burgers vector and the line direction. For a pure screw dislocation, however, these two vectors are parallel, so that any plane through the line of the dislocation can be a possible glide-plane. However, the shear stress needed to move the dislocation will vary from one glide-plane to another, generally being least when the plane is one of closest packing. Thus, although slip is kinematically possible on many planes, it will in fact occur most easily on certain particular planes, which are referred to as *the* glide- or slip-planes. Slip on other planes is still possible, and can often be deliberately caused by so arranging the applied stress that the shear stress on the glide-plane is small while that on some transverse plane is large; this is referred to as *cross-slip*.

Now consider the sequence in Figure 1.17(a)–(e) in which a screw dislocation, moving in its glide-plane, meets an obstacle. The obstacle prevents further slip for some central segment of the screw but, if the stress on the cross-slip plane is high enough, this segment can cross-slip as in (b). As soon as the cross-slip has proceeded beyond the obstacle, slip will again be favoured, as in (c), and eventually the segment may be induced to cross-slip again, back onto its original glide-plane, by the attractive forces between

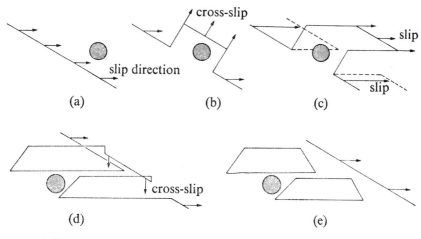

FIGURE 1.17

itself and the two outer segments, which will have continued to move on that plane. The dislocation then continues, leaving behind it a pair of *prismatic loops* (i.e. loops consisting entirely of edge dislocations). (Note that a similar production process for an edge dislocation would involve climb at stages (b) and (d), and hence would not occur under normal circumstances.)

D *Frank-Read sources*

It is now believed that the most important production mechanism in the majority of cases is through the operation of stationary sources, called Frank-Read sources.[18] In order to explain them we need to introduce the idea of the force on a dislocation line, an idea that we shall develop more systematically later (see §3.3). In Figure 1.7, let the body have dimensions L_1, L_2, L_3 in the three coordinate directions, so that the total shear force on the top face in the x-direction is $\sigma_{xy} L_1 L_3$. In the transition from (a) to (c), the dislocation moves a distance L_1 across the body and the top face moves by an amount b in the x-direction, so that the external forces do work equal to $b\sigma_{xy} L_1 L_3$. We can imagine that this is equivalent to a force F on the dislocation line which does work $FL_1 = b\sigma_{xy} L_1 L_3$. Hence the force per unit length, F/L_3 is $b\sigma_{xy}$. For the screw dislocation of Figure 1.8 the same argument applies, giving the same force per unit length, with the difference that the force must now be considered as acting in the $-z$ direction.

It is possible – in fact, common – for dislocation lines in real materials to be anchored down at certain points along their lengths. This could be caused by impurity atoms in the crystal, which often tend to settle on the dislocation line, and having done so then inhibit its motion. Stronger barriers to dislocation motion can be created by larger precipitate particles. Nodes on dislocation lines can have a similar effect. In Figure 1.15(b), for example, the stress-field may be such that there is no force on one of the three lines meeting at N, and this one immobile line will act as an anchor for any motion of either of the other two lines.

Now suppose we have a dislocation segment AB which is firmly pinned at A and B. Under an appropriate component of shear stress, σ, the dislocation will want to move, and since its ends are fixed it will take up some curved configuration (Fig. 1.18). The form of this configuration will depend on the detailed properties of the material. In some cases it will consist of a number of straight lines which follow the lines of closest packing in the crystal and which meet in almost sharp corners, and in other cases a smooth arc will be obtained. The equilibrium configuration of curved dislocation lines can be estimated using the line-tension model, according to which a dislocation line is analogous to a stretched string with a certain equivalent tension T. T is roughly equal to μb^2. Then under a transverse force $b\sigma$ per unit length, the segment becomes a circular arc of radius $r = T/b\sigma$. (In §8.4 we shall discuss the foundations of this model at some length, and it will be clear then that a number of qualifications must be put on its use. Nevertheless, we can be reasonably confident that the circular arc gives at least a rough guide to the actual configuration of the segment.)

If σ is increased, we get a series of arcs of decreasing radius as in Figure 1.19(a), until eventually the minimum radius of $L/2$ is reached (where $L = AB$). This occurs at a shear stress $\sigma_{\text{crit}} = 2T/bL$. Further expansion of the arc

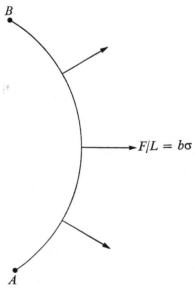

$F/L = b\sigma$

FIGURE 1.18

leads to an increase in radius, and hence can proceed spontaneously, without
further stress increase. The expansion eventually produces a large loop,
together with an equivalent segment AB, which can then proceed to operate
a second time. The complete loop itself produces a stress-field which in part
cancels the original force $b\sigma$ on AB, thus preventing an immediate re-
operation of the source. As the loop expands further, its effect on AB falls off
and eventually is reduced sufficiently to allow the production to proceed.
Thus the source produces a series of concentric loops, all on the same glide-
plane, the rate of production being determined by the speed at which each
loop expands and reduces its stresses at the source. Since the original
suggestion by Frank and Read such series of loops have been observed
directly.[19]

A related and perhaps simpler Frank-Read source involves a long dislocation segment which is fixed at one point. This is shown in Figure
1.20(a) along the z-axis (Oz), fixed at O, and with Burgers vector $\mathbf{b} = (b, 0,
0)$. Under shear stress $\sigma_{xy} > 0$, the dislocation moves in the x-direction,
trailing a length OA of screw dislocation behind it. (In practice the curves
in these figures would not be angular, but would usually be smooth as in
Figure 1.19.) The force on OA moves it in the $-z$ direction, producing a

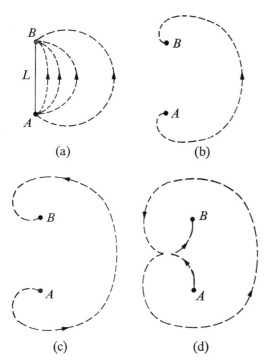

FIGURE 1.19

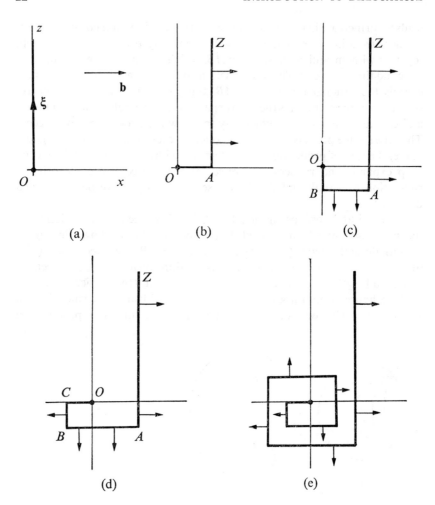

FIGURE 1.20

segment OB of edge, opposite in sign to AZ. OB moves in the $-x$ direction, producing a screw segment OC, and so on, until a spiral of expanding dislocation line results (Fig. 1.20(e)). Upon emerging from a free surface such a spiral would create a slip-step several Burgers vectors in height, as would a number of unconnected dislocations.

1.4

YIELD AND PLASTIC FLOW

In this section we wish to give a preliminary qualitative description of the physical events involved in the phenomenon of plasticity. The discussion will be far from complete, and we shall return later to some of the questions for more detailed investigation.

A *Macroscopic observations*

The basic experimental feature we shall be concerned with is illustrated by Figure 1.21, which shows a schematic stress-strain diagram for the tension and compression of a typical polycrystalline metal. The initial tensile deformation is elastic from O to the *yield point* Y, after which plastic strain begins. The stress continues to increase beyond Y, an occurrence called *strain-hardening*. If at some later point A the stress is reduced to zero, the curve AYO is not retraced, but strain relaxes purely elastically to O'. OO' gives a measure of the plastic strain which has been given to the specimen. If tension is again imposed, the strain-stress path follows the elastic line $O'A$, and further plastic deformation does not occur until A is reached, or nearly so. On the other hand, under compression from O', the elastic part $O'B$ is considerably shorter than $O'A$ and usually shorter than the original elastic range OY. This reduction in the compressive yield stress by plastic flow in tension is called the *Bauschinger effect*.

Under further compression, reverse plastic flow occurs until, at some point C, the total plastic strain is reduced to zero. Relaxation of the compression then gives an elastic deformation back to O. If the stress again becomes tensile, yield occurs at a stress Y' which is close to, but usually greater than, the original yield stress at Y. This is called *cyclic hardening*. Under repeated applications of the stress-strain cycle between fixed plastic strain limits (i.e. with OO' being held constant), cyclic hardening progressively raises the yield stress until eventually, after typically a thousand cycles, *saturation* occurs, after which the yield stress remains constant.[20] There are exceptions

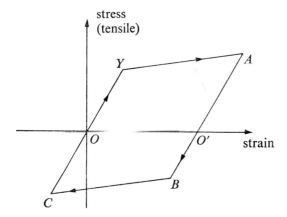

FIGURE 1.21

to this type of behaviour, for example in prestrained materials and some precipitation-hardened materials, which can actually show cyclic softening. During the cycle $OYABCO$ a certain amount of energy is absorbed by the material, proportional in fact to the area of the stress-strain curve. This is called *internal friction*.

The diagram in Figure 1.21 is considerably idealized. Usually the elastic parts COY and $AO'B$ are indeed straight as drawn, but the strain-hardening parts YA and BC are not. Furthermore the yield points Y and B are not usually sharp knees on the curve. It is in fact possible to observe small amounts of plastic flow (called *microplasticity*) at stresses much below yield, Y, and the conventional yield stress is the stress at which this plasticity spreads across the whole specimen. A typical result of microplasticity is the occurrence of internal friction within the 'elastic' range, which has as one of its consequences the decay of free elastic vibrations.

B *Single crystals*

As mentioned earlier, a new and unstrained crystal would normally contain about 10^5 dislocation lines per square centimetre. These dislocations would tend to arrange themselves in a low-energy configuration, rather than being distributed at random. Such rearrangement would be aided by thermal fluctuations, and so would occur readily at the high temperatures encountered during and immediately after crystallization. A section of a configuration of the type that is found is shown in Figure 1.22. It consists of

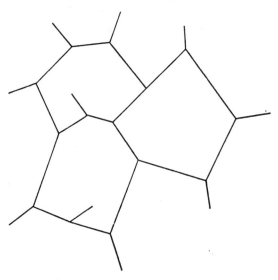

FIGURE 1.22

a three-dimensional network of dislocation segments, intersecting at various nodes of the type seen in Figure 1.15. At each node one has conservation of the Burgers vector. Such configurations are called Frank networks[21] and are relatively stable states, in regard to both their own internal interactions and the effects of external stresses.

One consequence of the presence of a Frank network within a crystal is that some of the segments will be able to act as Frank-Read sources. A second is that the network dislocations will interact with those mobile dislocations which are participating in the plastic process. In this way, the network provides a resistance to the progress of slip – in other words, it increases the yield stress of the material, and the more network dislocations there are, the bigger the increase.

Upon application of stress to a newly formed crystal, one of the first events to occur is that any free dislocations, not involved in a network, begin to move, and either glide out of the crystal or are stopped by the network. This produces only a minute amount of plastic strain, in the microplasticity range. At some higher stress, Frank-Read sources begin to operate, possibly formed by branches of the network, and produce series of concentric loops. These loops expand until they too are stopped by the network. Eventually there will usually be enough arrested loops from each source to allow their combined stress-fields to prevent the source from operating further. Raising the applied stress to a higher level can enable the loops to break through the network, at which time the slip propagates across the crystal, and the original sources begin to operate again.

What happens next depends on the circumstances. If the stressing is such that slip occurs on only one set of parallel slip-planes, the same process continues with very little increase in stress necessary. This produces the type of slip pattern in Figure 1.2, which is associated with very little strain-hardening. It is the general feature of hexagonal metals, in which only one set of slip-planes can be activated easily; when obtained by suitable orientation of the crystal in face-centred cubic metals it is called *easy glide*. The small amount of hardening that occurs during easy glide is thought to be due mainly to the interaction of different gliding dislocations. An example is

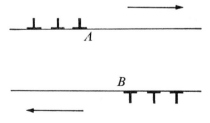

FIGURE 1.23

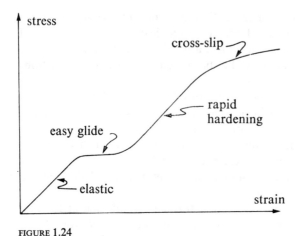

FIGURE 1.24

shown in Figure 1.23, where two edge dislocations A and B of opposite signs glide past one another on neighbouring slip-planes. We shall see later (§3.4) that such pairs of dislocations attract one another, and if the two slip-planes are sufficiently close together, the applied shear stress may not be able to overcome the attraction. The two dislocations thus can adopt some equilibrium position, and so form a barrier for succeeding dislocations on the same or nearby planes. As plastic strain progresses, there is an accumulation of such barriers, and a build-up of more complex ones, leading to a certain amount of hardening.

In face-centred cubic single crystals, a small amount of easy glide will usually change the orientation of the crystal so as to start slip on a second plane. With slip occurring on two planes at once, there is the possibility that much stronger intersections than that of Figure 1.23 will be formed, probably the most important of which is called the Lomer-Cottrell[22,23] barrier. The accumulation of such barriers during plastic deformation produces a strong rate of hardening within the active slip-bands. Eventually, however, the stress becomes sufficiently high that mobile screws are able to cross-slip over the obstacles, and at this stage, the rate of hardening decreases. The resulting stress-strain diagram is shown in Figure 1.24.

c *Polycrystals*

In polycrystals the presence of the grain boundaries produces a fundamental change in the early processes of plastic flow. This arises because these boundaries are strong barriers against dislocation motion, principally due to the fact that a dislocation arriving at the boundary of its own grain does not have the right Burgers vector and slip-plane to glide readily into the next

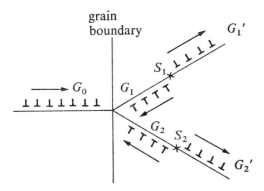

FIGURE 1.25

one. Thus, in polycrystals the grain boundaries play a similar role to the Frank network in single crystals. The first stage of plastic flow in polycrystals is that Frank-Read sources in a few of the grains with the most suitable orientation are made to operate. Each source produces a sequence of concentric loops which pile up against the grain boundaries, unable to penetrate them. Eventually the stresses of the piled-up dislocations stop the source from operating any further. As the applied stress is increased, more and more grains show this behaviour, which results in a small, but often measurable, plastic strain. However, the event which corresponds to macroscopic yield is the propagation of the slip from one grain to another. This occurs normally when the stresses around the piled-up groups at the grain boundaries become large enough to activate Frank-Read sources in the neighbouring grain. In Figure 1.25 the group G_0 of edges produces a stress sufficient to activate sources S_1 and S_2 in the next grain which produce edges as shown (these are actually parts of complete loops). The groups G_1 and G_2 move towards the grain boundary, and to some extent neutralize the stresses of G_0, while the positive edges, $G_1{}'$ and $G_2{}'$, act in such a way as to continue the slip process started by G_0. Once the stresses of G_0 are relaxed, the original source can produce more dislocations to add to G_0, which then cause S_1 and S_2 to continue operating, and so on.

For this process to occur there needs to be in the grain beyond the pile-up a suitable slip-plane on which the stresses of G_0 can activate a source. In hexagonal crystals, where slip occurs readily on only one set of planes, such a suitably oriented slip-plane does not usually exist. Thus hexagonal polycrystals do not show macroscopic plasticity, and deform elastically right up to their fracture stress. (Such materials are called *brittle*.) In cubic crystals, particularly face-centred cubic, propagation of slip from one grain to another is usually relatively easy.

At stresses above yield, strain-hardening occurs, again being especially rapid for face-centred cubic crystals. This follows the same pattern as in single crystals, and is believed to be due in general to the same cause that hardens the individual grains. The grain boundaries play very little role here, in contrast to their importance in the initial yielding.

After some amount of plastic flow, at a point such as A on the idealized stress-strain curve of Figure 1.21, dislocation arrangements such as that in Figure 1.25 remain at the grain boundaries. The actual positions of the dislocations are determined by a set of equilibrium equations according to which the force per unit length on each dislocation due to the external forces $b\sigma$, plus a similar force due to all the other dislocations, which we write as $b\sigma_d$, should exactly equal, but not exceed, the force ($b\sigma_1$, say) which is needed to move the dislocations through the crystal. (Dislocation motion was illustrated in Figure 1.6, and σ_1 must at least equal the shear stress necessary to enable the dislocation to overcome the energy hill between the successive positions shown in that figure. In general, however, σ_1 contains other contributions such as the frictional type of resistance presented to the dislocation by impurity particles.) Thus $b\sigma + b\sigma_d = b\sigma_1$. If now the external stress is reduced to σ' the dislocations do not move back from the equilibrium positions until $b\sigma' + b\sigma_d = -b\sigma_1$. Thus $\sigma - \sigma' = 2\sigma_1$ gives a measure of the stress reduction necessary from A to B in Figure 1.21. Upon further reduction from B some of the dislocations move backwards along the same paths as during the original plastic stage YA and are absorbed by their sources again, by the reverse of the creation mechanism. However, during YA certain dislocation structures will have been formed which are particularly stable, such as Lomer-Cottrell barriers, or non-reversible events such as cross-slip of some screws will have occurred. Thus state C will not be precisely the same, as far as its dislocation structure is concerned, as was state Y. In particular, there will be a much greater density of dislocations, many of which will be locked in stable and unmovable configurations. These form barriers for the motion of free dislocations, so that the yield stress after the cycle is higher than in the virgin specimen. This cyclic hardening increases with further cycles although eventually sufficient purely reversible slip mechanisms develop to cause saturation of hardening. The detail of these mechanisms is not properly understood at present, and certainly depends on the crystal structure involved. One possibility will be discussed later (§6.4) with reference to the initiation of fatigue cracks.

D *Influence of temperature*

Finally, in passing, we mention some of the effects of changes in temperature on the processes described above. The dislocation structures caused by plastic strain are in equilibrium, but since their strain energy is much higher than that of an undislocated crystal, the equilibrium is only metastable.

Given a large enough local fluctuation, they are able to change into a state of lower energy, and eventually to return to the state of very low dislocation density which occurred in the virgin specimen. The occurrence of such local fluctuations of energy increases very rapidly with increase in temperature. Thus, if the temperature is raised sufficiently, a strain-hardened material can be returned to its original softness – a process called *annealing*. The ease with which this can be accomplished depends on the stability of the dislocation structures which are responsible for the hardening. Arrangements such as those in Figure 1.23 which occur in the easy glide region are removed by relatively small increases in temperature, while the barriers, networks, and tangles produced by later stages of rapid hardening in cubic metals are harder to anneal away.

Many of the processes which occur during plastic flow can also occur when promoted by thermal fluctuations. In particular, the motion of a dislocation itself can occur at stresses below that at which AE = BE in Figure 1.6(a) if there is a sufficiently large fluctuation in the positions of the atoms near A. Similarly the operation of a Frank-Read source can proceed at smaller stresses than those indicated in §1.3 if it undergoes a large enough fluctuation. Thus plastic flow can occur at stresses below yield when aided by thermal fluctuations. However, the occurrence of sufficiently large thermal fluctuations is an infrequent event in most metals unless they are raised appreciably above room temperature, so that at ordinary temperatures flow below yield is a slow process. It is called (plastic) *creep*. As expected, the rate of creep strain increases rapidly with temperature.

For a more complete coverage of the topics touched on in this section, the reader is referred, for example, to the books by Cottrell (Chaps. IV, V) and Friedel (Chaps. VII–XI).

1.5
PARTICULAR CRYSTAL STRUCTURES

A *Possible Burgers vectors*

So far we have described dislocations in simple cubic crystals only. However, dislocations can be formed in any crystal structure, and we shall in this section sketch some of their basic properties for the three structures of most practical interest. We shall assume that the reader is conversant with the elementary notations of crystallography.[24]

The line of a dislocation in a crystal marks the boundary of a region of the slip-plane over which slip has occurred. The parts of the crystal on two sides of this region have been displaced relative to one another through one Burgers vector. Since these two sides must still mesh, after slip, in a perfect crystal structure, it follows that the Burgers vector must always be one of the translation vectors of the lattice concerned. That is, it must be a vector between some pair of corresponding lattice points. Conversely, any such translation vector is a possible Burgers vector for a dislocation. In Figure 1.26 we have drawn three possible Burgers vectors for the simple cubic structure which, together with similar vectors of the same types, represent the three shortest Burgers vectors. In order of increasing length, they are [100], [110], and [111].

Two dislocations with Burgers vectors $\mathbf{b}_1$ and $\mathbf{b}_2$ and parallel lines may conceivably join together to make a single dislocation of Burgers vector $\mathbf{b}_3$. Thus, for example, two similar edge dislocations of the type illustrated in Figure 1.10 may combine to produce a composite dislocation with Burgers vector $2\mathbf{b}$. In general, if a combination occurs, it is necessary that $\mathbf{b}_3 = \mathbf{b}_1 + \mathbf{b}_2$, as follows by taking a Burgers circuit as in Figure 1.14 which encloses both $\mathbf{b}_1$ and $\mathbf{b}_2$: the closure failure of such a circuit is unaffected by the combining of the two dislocations into $\mathbf{b}_3$. Conversely, it is possible for the dislocation $\mathbf{b}_3$ to dissociate into a pair of dislocations with Burgers vectors $\mathbf{b}_1$ and $\mathbf{b}_2$. Thus, in the simple cubic case, a [110] dislocation can split into a pair with Burgers vectors [100] and [010]. This is written

$$[110] \rightarrow [100] + [010]. \tag{1.1}$$

Combinations and dissociations of dislocations are termed *dislocation reactions*.

A dislocation reaction will take place automatically if, by doing so, it leads to a reduction in energy. We shall show in Chapter 3 that a dislocation with Burgers vector $\mathbf{b}$ produces a strain energy in the medium equal to $k\mathbf{b}^2$ per unit length of dislocation line. The parameter k depends on the angle between the Burgers vector and the line direction and also on the detailed arrangement of the atoms near the line itself, as expressed through a quantity called the core radius. It is, however, not very sensitive to these factors, and we get a rough idea of dislocation behaviour by taking it to be a constant. Then the two states in the dislocation reaction $\mathbf{b}_3 \rightarrow \mathbf{b}_1 + \mathbf{b}_2$ have

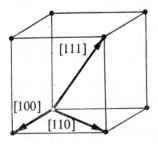

FIGURE 1.26

energies $k\mathbf{b}_3^2 = k(\mathbf{b}_1 + \mathbf{b}_2)^2$ and $k(\mathbf{b}_1^2 + \mathbf{b}_2^2)$. (In the state with two dis-locations, there is also a certain amount of energy of interaction between them, but this is small when they become widely separated.) Thus the re-action is spontaneous if

$$\mathbf{b}_3^2 > \mathbf{b}_1^2 + \mathbf{b}_2^2. \tag{1.2}$$

This is called Frank's rule.[25] If it is satisfied, the dislocation $\mathbf{b}_3$ will be mechanically unstable, and hence will play no role in the slip process.

An elementary illustration of Frank's rule is provided by the double dislocation with Burgers vector [200] in a simple cubic structure, which is unstable against dissociation into two [100] dislocations. The two sides of the reaction (1.1) have exactly the same energy in the present approximation, so that it is impossible to decide whether [110] dislocations should be stable or not in a simple cubic lattice. In the same way, the stability of [111] dis-locations is also indeterminate.

So far we have looked at the conditions necessary for dislocations with any particular Burgers vector to exist in a given crystal. There is the further question as to whether they participate in slip. This is decided by the ease with which the dislocation under consideration can be made to move through the lattice: those dislocations which require the lowest shear stress to make them move can generally be expected to make the largest contribu-tion to the plastic shear strain. In Chapter 4 we shall calculate this critical shear stress, and obtain the result (cf. eqs. (4.25) and (4.75)) that it is a rather sensitive function of the ratio d/b, where b is the magnitude of the Burgers vector and d is the spacing between the glide-plane and the next parallel plane. The shear stress is low for small values of b and large values of d. Thus we expect slip to be caused predominantly by dislocations with small Burgers vectors (which also happen to be the most stable ones) gliding on the planes of closest packing in the crystal (which are, of course, the planes of greatest separation).

With some modification to allow for the existence of partial dislocations in some structures, the above conclusions are in very wide agreement with what is observed in practice. We shall apply them to some particular crystals in the next section.

B *Dislocations in some particular structures*

1. Face-centred cubic
Examples of fcc metals are aluminium, copper, and nickel. The unit cube for an fcc crystal is shown in Figure 1.27, from which it can be seen that the two shortest translation vectors for this structure are $\frac{1}{2}[110]$ and [100] (together with all the possible equivalent vectors). [100] dislocations can dissociate into two of the shorter type according to the reaction

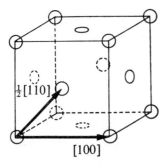

FIGURE 1.27

$[100] \rightarrow \frac{1}{2}[110] + \frac{1}{2}[1\bar{1}0].$ (1.3)

However, this produces equality in Frank's criterion, so that the question of the stability of [100] dislocations must remain open.

The close-packed planes in fcc crystals are those of type {111}, and these planes contain the Burgers vectors of type $\frac{1}{2}\langle110\rangle$. For example $\frac{1}{2}[110]$ lies in both of the planes $(1\bar{1}1)$ and $(1\bar{1}\bar{1})$. Thus it is possible for the dislocations with the shortest Burgers vectors to glide in the planes of closest packing, and this form of slip is highly favoured. The longer $\langle100\rangle$ Burgers vectors do not lie in the {111} planes, but the planes with the maximum packing which do contain them are the {100} planes. Slip in a $\langle100\rangle$ direction on a {100} plane would require a relatively high shear stress compared to the $\langle110\rangle$ slip on {111} planes, and we would expect this latter to be dominant. This prediction is in very firm agreement with experimental observations on fcc metals.

2. Body-centred cubic
Examples of bcc metals are chromium, iron, molybdenum, and tungsten. The shortest Burgers vectors in a bcc crystal (see Fig. 1.28) are $\frac{1}{2}[111]$ and [100], and equivalent vectors. Although the first of these has the lower energy, both are in fact stable. However, a pair of $\langle100\rangle$ dislocations would, if near enough to one another, undergo the more complex reaction

$[100] + [010] \rightarrow \frac{1}{2}[111] + \frac{1}{2}[11\bar{1}],$

which would result in an energy reduction. Thus [100] dislocations tend to eliminate one another to some extent.

The close-packed planes are those of type {110}, which contain both $\frac{1}{2}\langle111\rangle$ and $\langle100\rangle$ Burgers vectors. However, the shorter Burgers vector has a smaller critical shear stress, and hence dominates the slip process. Thus we can expect a dominant slip in $\langle111\rangle$ directions on {110} planes. This is indeed found in the majority of cases, although the situation is not as clear cut as with fcc metals: the direction of slip is correctly predicted, but the plane of slip is often not {110}. To a large extent this is due to separation

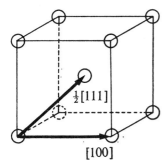

FIGURE 1.28

into partial dislocations (see next section), which favours slip on {112}
planes.

3. Hexagonal close-packed

Examples of hexagonal metals are cobalt, magnesium, and zinc. The basic
hexagonal prism for hcp crystals is drawn in Figure 1.29, showing the two
shortest Burgers vectors, $\frac{1}{3}[2\bar{1}\bar{1}0]$ and [0001], using hexagonal indices. Both
of these give stable dislocations. A third Burgers vector $\frac{1}{3}[2\bar{1}\bar{1}3]$ (not illustrated) can dissociate into one each of the shorter Burgers vectors, and is
predicted to have neutral stability by Frank's rule. The third vector shown
in Figure 1.29, between atoms on neighbouring basal planes, does not correspond to a possible Burgers vector, since the structure is not invariant under
such a translation: the hcp structure has two atoms in each unit cell, and only
atoms on alternate basal planes are equivalent.

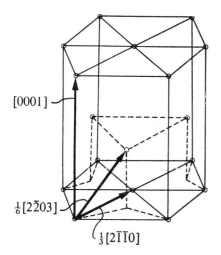

FIGURE 1.29

In a true hcp structure, with an axial ratio $c/a = 1.633$, the closest packing occurs in the basal planes, (0001). Thus we can expect slip to be primarily in $\langle 2\bar{1}\bar{1}0 \rangle$ directions on (0001) planes. This is indeed found to be the case in those hexagonal metals in which c/a approximates to (or exceeds) the ideal value.

c *Partial dislocations*

Instead of the type of atomic arrangement sketched in Figure 1.10(a), it is conceivable that an edge dislocation in a simple cubic crystal might have an arrangement of atoms such as that in Figure 1.30. We can regard this as two dislocations of Burgers vectors $b/2$, one at A and one at B, separated by a region in which there is complete misalignment of atoms across the slip-plane. Such dislocations are called *partial dislocations*, since their Burgers vectors are not translation vectors for the lattice. A dissociation of this kind into two partial dislocations would be compatible with Frank's rule. How-ever, it is necessary to modify this criterion to include the energy contained in the misfitting interatomic bonds in the region AB. For a simple cubic material this energy of misfit is so large that partial dislocations do not occur.

This is not true, however, of many materials with other crystalline struc-tures. In fcc metals dissociations of the type

$$\tfrac{1}{2}[\bar{1}01] \rightarrow \tfrac{1}{6}[\bar{2}11] + \tfrac{1}{6}[\bar{1}\bar{1}2] \tag{1.4}$$

correspond to a relatively low misfit energy between the two partial dis-locations, which in this case are called *Shockley partials*.[26] A classification of such dissociations has been given by Thompson.[27] An analogous dissociation

$$\tfrac{1}{3}[\bar{1}2\bar{1}0] \rightarrow \tfrac{1}{3}[01\bar{1}0] + \tfrac{1}{3}[\bar{1}100] \tag{1.5}$$

occurs in hcp crystals, and we shall discuss these two structures together.

The two close-packed structures, fcc and hcp, can be formed by stacking planes of close-packed spheres on top of one another. Starting with the

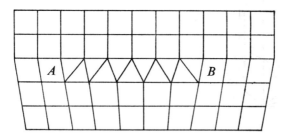

FIGURE 1.30

plane shown in Figure 1.31, in which the spheres are centred at points labelled A, there are two positions for the layer to be stacked next: the spheres can be centred above points such as B or C. Having chosen a position, say B, for the second layer, there are again two positions for the third, namely A and C. Thus we can form two distinct regular stacking arrangements, one being $ABCABC \ldots$, the other being $ABABAB. \ldots$ The first of these corresponds to the fcc lattice with the layers being some set of $\{111\}$ planes; the second corresponds to the stacking of basal planes in the hcp structure.

The shortest Burgers vector of perfect dislocations in each of these two structures is a vector AA between neighbouring corresponding atoms, such as $\mathbf{b}$ in Figure 1.31. As such a dislocation sweeps across a crystal, all layers above the slip-plane are displaced by $\mathbf{b}$. This displacement does not change the type of any layer – that is an A layer becomes again an A layer, and similarly for B and C – and therefore does not alter the stacking arrangement within the crystal. Consider, however, the effect of motion of the partial dislocation $\mathbf{b}_1$ through the crystal. This would change all the layers above the slip-plane according to the transformations $A \to B$, $B \to C$, and $C \to A$ (assuming an appropriate sense of the motion), so that the stacking arrangement is altered. Taking the slip-plane to be between the third and fourth layers, the fcc sequence $ABC|ABCABC \ldots$ becomes $ABC|BCABCA \ldots$, which is still basically fcc but has a subsequence (underlined) with the hcp character. Similarly the hcp sequence $ABA|BABAB \ldots$ becomes $ABA|CBCBC \ldots$, which has an fcc subsequence. Arrangements such as

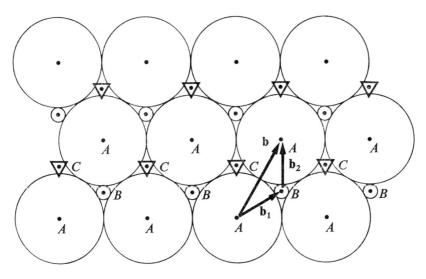

FIGURE 1.31

these are said to have *stacking faults* on the planes on which the partial dislocations have moved.

Stacking faults in close-packed crystals have a low energy because they
do not disturb the close-packed arrangement. Only the geometrical relationships between atoms on second-adjacent planes are changed from those in a
perfect crystal, and these make a smaller contribution to the binding energy
of the crystal than do the nearest-neighbour pairs of atoms. The fault AB in
Figure 1.30, for example, affects the nearest neighbours, and hence has a
high energy.

The dissociations (1.4) and (1.5) into Shockley partials correspond to
$\mathbf{b} \rightarrow \mathbf{b}_1 + \mathbf{b}_2$ in Figure 1.31, provided an appropriate choice of axes is made.
The two partials separate, leaving a strip of stacking fault between them, and
since the stacking-fault energy is low, the extended configuration often has
lower energy than does a concentrated dislocation. In practice, the equilibrium separation turns out to be of the order of ten lattice spacings.

The existence of partial dislocations in the close-packed structures leads
to a number of important effects such as the formation of Lomer-Cottrell
barriers,[22,23] which provide a significant strain-hardening effect in fcc
crystals. Shockley partials also inhibit cross-slip of screw dislocations, since
the two partials cannot both be screws, having non-parallel Burgers vectors,
and at least one of them is restricted to moving in the plane of dissociation.
There is, however, no change in the slip process itself, which still occurs on
the close-packed planes in the close-packed directions: the two partials
simply move across the slip-plane as a unit.

This is not true of bcc crystals however, where the planes $\{112\}$ on which
the stacking-fault energy is least do not coincide with the $\{110\}$ primary slip-
planes. A dislocation with Burgers vector $\frac{1}{2}[111]$ can dissociate into $\frac{1}{6}[111]+$
$\frac{1}{3}[111]$ or perhaps into three dislocations of $\frac{1}{6}[111]$, with the intervening
stacking fault lying on one of the three planes $(11\bar{2})$, $(1\bar{2}1)$, and $(\bar{2}11)$. Such
a dissociation does not prevent cross-slip, since the partial dislocations this
time have parallel Burgers vectors. However, with the presence of extended
dislocations, the easiest slip process becomes the slip of screw dislocations
in the stacking-fault plane, that is in a $\{112\}$ plane, rather than in a $\{110\}$
plane.

D *Twinning*

Figure 1.32 shows a boundary between two crystals of a material having an
oblique lattice in which the lattice on one side of the boundary is the reflection in the boundary of the lattice on the other side. The crystals in such a
case are called *twins* and the boundary a *twin boundary*. The twin configuration can be formed from an initially perfect crystal by shearing the atom
layers above TT through the displacements shown: the nth layer is displaced

by an amount $n\mathbf{t}$. Such a deformation could be accomplished by moving a partial dislocation of Burgers vector $\mathbf{t}$ across every one of the lattice planes above TT.

The process of changing a crystal into its twin offers another mechanism of plastic deformation distinct from dislocation glide. The plastic strains that could be produced by twinning are clearly limited, depending on the particular crystal structure and plane of twinning involved. Nevertheless, twinning under mechanical stress does occur in some cases, and particularly for bcc metals it makes an important contribution to the total plasticity.

There is a close connection between twin boundaries and stacking faults. To illustrate this, consider the following stacking arrangement of {111} planes in an fcc crystal: $AB\overset{\downarrow}{C}BACBA$. The stacking order is inverted in the middle, and the two halves of the crystal on either side of the layer indicated by the arrow are mirror images of one another. This structure is called a *spinel twin*. It can be created from the regular fcc structure by a partial dislocation $\mathbf{b}_1$ moving across every plane to the right of the arrowed twin boundary ($\mathbf{b}_1$ is the same as in Fig. 1.31 and the accompanying discussion). If now we had two opposite spinel twin boundaries on adjacent planes in the stack we would have a stacking arrangement $AB\overset{\uparrow\uparrow}{C}BCABC$. This arrangement is precisely the same as the stacking fault corresponding to motion of a $\mathbf{b}_1$ dislocation. Thus a stacking fault can be viewed as a pair of opposite twin boundaries one layer apart.

In bcc crystals the planes on which twinning is simplest are the {112} planes. As we saw in the last subsection, these are the planes on which stacking faults have the lowest energies. The partial twinning dislocations, analogous to $\mathbf{t}$ in Figure 1.32, have Burgers vectors $\frac{1}{6}\langle 111\rangle$. The problem of how simultaneous slip could occur on all the planes in the twinned region

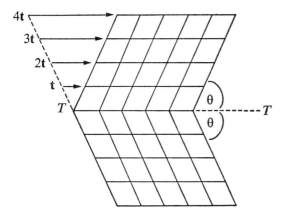

FIGURE 1.32

was resolved for bcc crystals by Cottrell and Bilby.[28] These authors suggested a mechanism similar to the Frank-Read source but with the differences that successive loops do not lie on the same slip-plane but on adjacent planes and that the loops have the Burgers vector of partial twinning dislocations. The operation of such a source produces a thin sheet of twinned crystal. The suggested mechanism is not possible in fcc crystals, which is consistent with the lack of observation of twinning under stress in such materials.

PROBLEMS

1 Find the Burgers vectors of each of the dislocations in Figure 1.1, taking the line direction ξ to be out of the plane of the paper.

2 In the large elastic deformation of Figure 1.4(b), suppose that the stress-strain relation has the form $\sigma_{xy} = \sigma_0 \sin(2\pi u/b)$, where b is the lattice spacing and u is the horizontal displacement of one atom layer relative to the next. Find σ_0 for consistency with Hooke's law at small strains, and hence obtain the maximum shear stress that can be applied.[9]

3 An alternative definition of the Burgers vector can be given by taking a closed circuit in the reference crystal of Figure 1.10(b), mapping this onto the real crystal, and then using the closure failure of the mapped circuit. Establish a sign convention so that $\mathbf{b}$ is as defined in §1.2A.

4 Prove that the Burgers vector is constant along a dislocation line.

5 With the notation of Figure 1.13, show that $\mathbf{b} = \mathbf{u}_+ - \mathbf{u}_-$.

6 If the dislocation loop of Figure 1.13 is formed by removing one layer of atoms over the surface C, it is termed a *vacancy loop*; if by inserting a layer of atoms, an *interstitial loop*. Find the Burgers vectors in these two cases. How can such loops glide? What surface steps are produced when such loops glide out of a free surface?

7 Taking the line tension $T = \mu b^2$, find the shear stress necessary to operate a Frank-Read source of length $10^4 b$.

8 A rectangular dislocation loop lies in a plane $x = $ constant with its sides parallel to the y and z axes. The Burgers vector is in the x-direction. Calculate the force on each side of the loop when an external shear stress σ_{xy} is applied, and discuss the ensuing glide motion.

9 The ionic crystal structure (e.g. NaCl) can be formed from a simple cubic lattice by placing positive and negative ions alternately at each lattice point. Discuss the possible Burgers vectors and slip-planes in this structure (see §4.5).

10 A dislocation moves across its slip-plane when the externally applied stress-field reaches a level at which the shear traction across the slip-plane in the direction of the Burgers vector attains a critical value, σ_c say. If an fcc single-crystal specimen is pulled in uniform simple tension, T, in a direction $[lmn]$ relative to the cube axes, find the value of T at which slip starts on the (111) plane in the $[1\bar{1}0]$ direction. Does the result change if T is compressive?

11 With the same stress as in Problem 10 applied to a bcc crystal, find the values of T at which slip occurs in the [111] direction on the planes $(1\bar{1}0)$ and $(2\bar{3}1)$.

12 In an hcp crystal, a tension T has direction $[lmnp]$. Find the condition for slip on the basal plane in any one of the six directions $\pm a_i$ (see Fig. 1.29). Find also the condition for slip in the direction of the c-axis on any one of the six hexagonal faces, $\{1\bar{1}00\}$. Show that none of these slip processes can occur if T is either in or normal to the basal plane.

13 Apply Frank's rule to the reaction $[110]+[1\bar{1}0] \rightarrow [100]+[100]$ in a simple cubic crystal. Discuss the implications of your result.

14 A vacancy loop (see Prob. 6) is formed on a $\{111\}$ plane in an fcc crystal by removing the atoms on the plane concerned over the area of the loop. Obtain the Burgers vector and the stacking arrangement in the neighbourhood of the loop. Repeat the calculation for an interstitial loop. (These partial dislocations are termed *sessile* by Frank[29,30] owing to their lack of mobility.)

15 Discuss the ability of an hcp crystal to form sessile dislocation loops of the type in Problem 14.

REFERENCES

1 Volterra, V., Ann. Ec. Norm. **24** (1907), 401; Timpe, A., Z. Math. Phys. **52** (1905), 348

2 Somigliana, C., Atti Reale Accad. Lincei, **23** (1914), 463; **24** (1915), 655

3 Love, A.E.H., *A treatise on the mathematical theory of elasticity* (Cambridge, 1927; New York: Dover, 1944), p. 221

4 Taylor, G.I., Proc. Roy. Soc. **A145** (1934), 362

5 Orowan, E., Z. Phys. **89** (1934), 634

6 Polanyi, M., Z. Phys. **89** (1934), 660

7 Burgers, J.M., Proc. Roy. Acad. Sci., Amsterdam **42** (1939), 293 and 378; Proc. Phys. Soc. **52** (1940), 23

8 Schmid, E., and Boas, W., *Plasticity of crystals* (London: Hughes, 1950; Chapman and Hall, 1968)

9 Frenkel, J., Z. Phys. **37** (1926), 572

10 Mackenzie, J.K., Ph.D. Thesis, University of Bristol (1949)

11 Burton, W.K., Cabrera, N., and Frank, F.C., Nature **163** (1949), 398

12 Frank, F.C., Disc. Faraday Soc. **5** (1949)

13 Griffin, L.J., Phil. Mag. **41** (1950), 196

14 Pashley, D.W., Repts. Prog. Phys. **28** (1965), 291

15 Doremus, R.H., Roberts, B.W., and Turnbull, D. (eds.), *Growth and perfection of crystals* (New York: Wiley, 1958)

16 Frank, F.C., *Conference report on strength of solids* (London: Physical Society, 1948), p. 46

17 Koehler, J.S., Phys. Rev. **86** (1952), 52

18 Frank, F.C., and Read, W.T., Phys. Rev. **79** (1950), 722

19 Dash, W.C., J. Appl. Phys. **27** (1956), 1193

20 Wood, W.A., and Segall, R.L., Proc. Roy. Soc. **A242** (1957), 180

21 Frank, F.C., Proc. Phys. Soc. **A62** (1950), 131

22 Lomer, W.M., Phil. Mag. **42** (1951), 1327

23 Cottrell, A.H., Phil. Mag. **43** (1952), 645

24 Schmid, E., and Boas, W., *Plasticity of crystals* (Berlin: Springer, 1935; London: Hughes, 1950; London: Chapman and Hall, 1968), chap. 1; Hull, D., *Introduction to dislocations* (Oxford: Pergamon Press, 1965),

chap. 1; Barrett, C.S., and Massalski, T.B., *Structure of metals* (New York: McGraw-Hill, 1966), chap. 1; Jaswon, M.A., *Introduction to mathematical crystallography* (New York: American Elsevier, 1965), chaps. 4–6

25 Frank, F.C., Physica **15** (1949), 131
26 Heidenreich, R.D., and Shockley, W., *Report on strength of solids* (London: Physical Society, 1948), p. 57
27 Thompson, N., Proc. Phys Soc. **66B** (1953), 481
28 Cottrell, A.H., and Bilby, B.A., Phil. Mag. **42** (1951), 573
29 Frank, F.C., Proc. Phys. Soc. **62A** (1949), 202
30 Frank, F.C., and Nicholas, J.F., Phil. Mag. **44** (1953), 1213

2

THEORY OF ELASTIC CONTINUA

2.1

REVIEW OF CLASSICAL ELASTICITY[1]

The theory of continuum mechanics is concerned with the deformations of macroscopic bodies of material under the action of external forces. Elasticity is a specialization of continuum mechanics to a subclass of materials in which the local internal forces in any element of the body, as measured by the stress tensor, are completely determined by the state of deformation of that element at the instant of time in question. The classical theory of elasticity makes two further restrictions, namely that the stress in any element is determined solely by the strain in the same element, and that the relationship between these two tensors is a linear one.

In the classical theory the displacement vector plays a central role. We consider initially a state of the body called the natural state, in which the stresses are identically zero. The application of external forces to the body then produces some distortion from the natural state, and we denote by $\mathbf{u}(\mathbf{x}) = \{u_i(x_k)\}$ the displacement from the natural to the distorted state of the particle which occupies the position $\mathbf{x}$. The components of $\mathbf{u}$ and $\mathbf{x}$ are taken with respect to some rectangular Cartesian axes. We have been deliberately ambiguous in not stating whether $\mathbf{x}$ refers to positions in the natural or the distorted state since, within the range of application of the linear theory, it does not matter which is intended. A restriction of the theory is that the derivatives $\partial u_i/\partial x_j$ of the displacement components must be small. The components (e_{ij}) of the classical strain tensor are defined in terms of $\mathbf{u}$ by

$$e_{ij} = \tfrac{1}{2}(\partial_i u_j + \partial_j u_i), \qquad (2.1)$$

where we use ∂_i to denote $\partial/\partial x_i$. It is shown in most books on elasticity that (e_{ij}) determines, to first order, the changes in the lengths of elements of material and the changes in the angles between different elements in going from the natural to the deformed state.

The stress tensor (σ_{ij}) is usually defined in terms of the internal forces acting between neighbouring parts of the body. Consider two parts of the body, labelled 1 and 2, separated by a surface whose normal in the sense of pointing from 1 to 2 is in the jth coordinate direction. The particles in 2 exert certain forces on the particles in 1, and it is an assumption of the theory that these forces may be regarded as being localized on the surface separating the two parts. That is, the interaction between the two parts is assumed to consist solely of contact forces acting across their surface of separation. Then σ_{ij} is defined as the ith component of the resultant force produced by 2 on 1, per unit area of interface between them. In terms of (σ_{ij}) the equations of motion take the form

$$\rho\ddot{u}_i = \partial_j\sigma_{ij}+f_i, \qquad \sigma_{ij} = \sigma_{ji} \tag{2.2}$$

(using the usual summation convention for repeated indices). Here ρ is the material density, f_i the components of the body force per unit volume, and dots denote time derivatives. Furthermore, at an external surface of the body, where the unit outward normal is $\mathbf{n}$, we have

$$\sigma_{ij}n_j = t_i, \tag{2.3}$$

with t_i being the applied traction per unit area.

The above conditions are consequences of either kinematical reasoning or general principles of mechanics. Classical elasticity makes one further postulate, which is a constitutive assumption in the sense that it delineates the behaviour of a particular subclass of materials from that of continua as a whole. It is that there exists a linear relationship of the form

$$\sigma_{ij} = c_{ijkl}e_{kl} \tag{2.4}$$

between the stress and strain tensors. This is usually termed (the generalized) Hooke's law. The constants c_{ijkl} are called the *elastic modulus components* of the material in question. Since σ_{ij} and e_{kl} are both symmetric, we must have $c_{ijkl} = c_{jikl}$, and there is no loss of generality if we take $c_{ijkl} = c_{ijlk}$ also.

For materials with certain intrinsic symmetries the modulus components are not all independent of one another. For an isotropic material we have the result that

$$c_{ijkl} = \lambda\delta_{ij}\delta_{kl}+\mu(\delta_{ik}\delta_{jl}+\delta_{il}\delta_{jk}), \tag{2.5}$$

involving just the two Lamé constants, λ and μ. δ_{ij} is the Kronecker delta. For cubic symmetry three moduli appear,[2] and the stress-strain relation (2.4) can be written in the form

$$
\begin{bmatrix} \sigma_{11} \\ \sigma_{22} \\ \sigma_{33} \\ \sigma_{12} \\ \sigma_{13} \\ \sigma_{23} \end{bmatrix}
=
\begin{bmatrix}
c_{11} & c_{12} & c_{12} & \cdot & \cdot & \cdot \\
c_{12} & c_{11} & c_{12} & \cdot & \cdot & \cdot \\
c_{12} & c_{12} & c_{11} & \cdot & \cdot & \cdot \\
\cdot & \cdot & \cdot & 2c_{44} & \cdot & \cdot \\
\cdot & \cdot & \cdot & \cdot & 2c_{44} & \cdot \\
\cdot & \cdot & \cdot & \cdot & \cdot & 2c_{44}
\end{bmatrix}
\begin{bmatrix} e_{11} \\ e_{22} \\ e_{33} \\ e_{12} \\ e_{13} \\ e_{23} \end{bmatrix},
\tag{2.6}
$$

when the cube axes are taken as coordinate axes. Isotropy is a special case of cubic symmetry, for which $c_{11} = \lambda + 2\mu$, $c_{12} = \lambda$, $c_{44} = \mu$, so that the constant $A = 2c_{44}/(c_{11} - c_{12})$ has the value 1. For this reason, A is called the anisotropy constant of a cubic material.

For an isotropic material, the usual elastic constants can be expressed in terms of λ and μ:

$$\left.\begin{array}{l} \text{Young's modulus, } E = \mu(3\lambda + 2\mu)/(\lambda + \mu), \\ \text{Poisson's ratio, } \nu = \lambda/[2(\lambda + \mu)], \\ \text{Shear modulus } = \mu, \\ \text{Bulk modulus, } B = (\text{compressibility})^{-1} = \lambda + \tfrac{2}{3}\mu. \end{array}\right\} \tag{2.7}$$

Then equation (2.4) can be written in the two equivalent forms

$$\sigma_{ij} = \lambda e_{kk}\delta_{ij} + 2\mu e_{ij}, \tag{2.8}$$

$$e_{ij} = \frac{1+\nu}{E} \sigma_{ij} - \frac{\nu}{E} \sigma_{kk}\delta_{ij}. \tag{2.9}$$

The boundary-value problems of elasticity consist of solving the field equations (2.2) within some volume, with the various quantities related by equations (2.1) and (2.4) or (2.8), subject to certain initial and boundary conditions. The initial conditions would consist of prescribing the displacements and velocities throughout the volume at some initial instant. The boundary conditions in their simplest form would involve prescribing either the displacement vector (u_i) or the surface traction $t_i = \sigma_{ij}n_j$ at each point of the boundary of the volume, although more generally some components of displacement and other components of traction could be prescribed at each point.

We have spoken of equations (2.1)–(2.3) as general principles as opposed to the constitutive equation (2.4). This is correct inasmuch as the first three equations are satisfied very well by almost all continuous media, elastic or otherwise, but it is also true that there are certain physical assumptions and approximations underlying these 'general principles'. Thus, for example, matter consists of discrete atoms or, to go further, of nuclei and electrons, rather than being a continuum as the theory supposes. To treat a body as a continuous distribution of matter is certainly a plausible approximation where large-scale deformations are concerned, but it still represents an idealization. Similarly the assumption that the forces between neighbouring parts of the body can be represented by contact forces (i.e. stresses) on the surface between the two parts is an approximation to the true distribution of mutual interactions between all pairs of atoms on the two sides of the surface. The possibility of couple stresses has also been ignored. The validity of these approximations which are made by continuum mechanics rests ultimately on the test of experiment, and here the evidence is overwhelmingly in their favour, at least for macroscopic phenomena.

However, continuum mechanics and particularly elasticity are sometimes used in contexts which are beyond the range of their empirical justification, for example in situations in which the stress varies appreciably on an atomic scale. Examples in this category which will be of major concern to us are the central region of a dislocation and the immediate vicinity of the tip of a sharp crack in a stressed medium. In both of these problems, which we shall discuss later at some length, continuum mechanics on its own does not provide an adequate treatment.

2.2
PLANE AND ANTIPLANE DEFORMATIONS OF ISOTROPIC MATERIALS

A *Plane strain*

A state of plane strain, perpendicular to the x_3-direction, is defined by the conditions that $u_3 = 0$ and u_1 and u_2 are independent of x_3. Then the only non-zero strains are e_{11}, e_{12}, and e_{22}, so that the stress components σ_{13} and σ_{23} are zero. From equation (2.9)

$$e_{33} = [\sigma_{33} - \nu(\sigma_{11} + \sigma_{22})]/E = 0,$$

so that

$$\sigma_{33} = \nu(\sigma_{11} + \sigma_{22}). \tag{2.10}$$

The other stress-strain relations take the form, after eliminating σ_{33},

$$e_{11} = \frac{1+\nu}{E}[(1-\nu)\sigma_{11} - \nu\sigma_{22}], \qquad e_{22} = \frac{1+\nu}{E}[-\nu\sigma_{11} + (1-\nu)\sigma_{22}],$$

$$e_{12} = \frac{1+\nu}{E}\sigma_{12}. \tag{2.11}$$

At equilibrium with a zero body force the equations of motion simplify to

$$\frac{\partial\sigma_{11}}{\partial x_1} + \frac{\partial\sigma_{12}}{\partial x_2} = 0, \qquad \frac{\partial\sigma_{12}}{\partial x_1} + \frac{\partial\sigma_{22}}{\partial x_2} = 0.$$

The first of these implies, by a well-known theorem of vector field theory, that there exists a function $A(x_1, x_2)$ such that $\sigma_{11} = \partial A/\partial x_2$ and $\sigma_{12} = -\partial A/\partial x_1$. In the same way the second equation gives that there exists $B(x_1, x_2)$ such that $\sigma_{12} = -\partial B/\partial x_2$ and $\sigma_{22} = \partial B/\partial x_1$. Thus $\sigma_{12} = -\partial A/\partial x_1 = -\partial B/\partial x_2$ so that, using the theorem a third time, there exists a function $\Psi(x_1, x_2)$ such that $A = \partial\Psi/\partial x_2$ and $B = \partial\Psi/\partial x_1$. In terms of Ψ,

$$\sigma_{11} = \partial^2\Psi/\partial x_2{}^2, \qquad \sigma_{22} = \partial^2\Psi/\partial x_1{}^2, \qquad \sigma_{12} = -\partial^2\Psi/\partial x_1\partial x_2. \tag{2.12}$$

Ψ is called the *Airy stress function*.

In terms of the displacements, the strains are

$$e_{11} = \frac{\partial u_1}{\partial x_1}, \qquad e_{22} = \frac{\partial u_2}{\partial x_2}, \qquad e_{12} = \frac{1}{2}\left(\frac{\partial u_1}{\partial x_2} + \frac{\partial u_2}{\partial x_1}\right). \tag{2.13}$$

Eliminating u_1 and u_2 leads to the *compatibility equation*

$$\frac{\partial^2 e_{11}}{\partial x_2{}^2} + \frac{\partial^2 e_{22}}{\partial x_1{}^2} = 2\,\frac{\partial^2 e_{12}}{\partial x_1 \partial x_2}. \tag{2.14}$$

From its derivation, this equation is a necessary condition on the strains in order that a corresponding displacement field should exist. We shall show in §8.1, in a more general context, that for any simply connected region it is also a sufficient condition (see also Prob. 2.1). Using equations (2.11) and (2.12) in (2.14) gives the compatibility equation as a condition on Ψ, which is, in fact, just the biharmonic equation

$$\left(\frac{\partial^2}{\partial x_1{}^2} + \frac{\partial^2}{\partial x_2{}^2}\right)^2 \Psi \equiv \nabla^4 \Psi = 0. \tag{2.15}$$

In this way we reduce the plane strain problem to a single field equation for Ψ together with certain associated boundary conditions.

B Cylindrical problems

For the plane strain deformation of a circular cylinder it is convenient to use polar coordinates (r, θ) in place of (x_1, x_2). The representations (2.12) of the stress in terms of the Airy stress function can be written in the following way:

$$\sigma^{ij} = \varepsilon^{ik}\varepsilon^{jl}\Psi_{,kl}, \tag{2.16}$$

where ε^{ik} is the permutation tensor in two dimensions. Its Cartesian components are $\varepsilon^{11} = \varepsilon^{22} = 0$, $\varepsilon^{12} = -\varepsilon^{21} = 1$. The comma notation in equation (2.16) denotes covariant differentiation, which for Cartesian coordinates is the same as ordinary differentiation. In this form equation (2.16) is a tensor equation and can be used in any coordinate system. For polars, $\varepsilon^{rr} = \varepsilon^{\theta\theta} = 0$, $\varepsilon^{r\theta} = -\varepsilon^{\theta r} = 1/r$, so that

$$\sigma^{rr} = \frac{1}{r^2}\Psi_{,\theta\theta}, \qquad \sigma^{\theta\theta} = \frac{1}{r^2}\Psi_{,rr}, \qquad \sigma^{r\theta} = -\frac{1}{r^2}\Psi_{,r\theta}. \tag{2.17}$$

The non-zero Christoffel symbols for cylindrical coordinates are

$$\begin{Bmatrix} r \\ \theta\theta \end{Bmatrix} = -r \quad \text{and} \quad \begin{Bmatrix} \theta \\ r\theta \end{Bmatrix} = \frac{1}{r},$$

so that in terms of simple derivatives

$$\Psi_{,rr} = \partial_{rr}\Psi, \qquad \Psi_{,\theta\theta} = \partial_{\theta\theta}\Psi + r\partial_r\Psi, \qquad \Psi_{,r\theta} = \partial_{r\theta}\Psi - (1/r)\partial_\theta\Psi.$$

The tensor components σ^{rr}, $\sigma^{\theta\theta}$, and $\sigma^{r\theta}$ in equation (2.17) do not represent the physical forces per unit area in the medium. These 'physical components' of stress are related to the tensor components by the equations

$$\hat{\sigma}_{rr} = \sigma^{rr}, \qquad \hat{\sigma}_{\theta\theta} = r^2\sigma^{\theta\theta}, \qquad \hat{\sigma}_{r\theta} = r\sigma^{r\theta}.$$

Thus in terms of the stress function the physical components of stress are

$$\hat{\sigma}_{rr} = \frac{1}{r^2}\partial_{\theta\theta}\Psi + \frac{1}{r}\partial_r\Psi,$$

$$\hat{\sigma}_{\theta\theta} = \partial_{rr}\Psi, \tag{2.18}$$

$$\hat{\sigma}_{r\theta} = -\frac{1}{r}\partial_{r\theta}\Psi + \frac{1}{r^2}\partial_\theta\Psi = -\partial_r\left(\frac{1}{r}\partial_\theta\Psi\right).$$

These representations for the physical components of stress may be derived equivalently from the equations of equilibrium in terms of polar variables. It is shown in many books on elasticity that the conditions of equilibrium are

$$\partial_r\hat{\sigma}_{rr} + \frac{1}{r}\partial_\theta\hat{\sigma}_{r\theta} + \frac{1}{r}(\sigma_{rr} - \hat{\sigma}_{\theta\theta}) = 0, \tag{2.19}$$

$$\frac{1}{r}\partial_\theta\hat{\sigma}_{\theta\theta} + \partial_r\hat{\sigma}_{r\theta} + \frac{2}{r}\hat{\sigma}_{r\theta} = 0.$$

It can be shown that these equations imply that there exists a function Ψ such that equations (2.18) hold (see Prob. 2.2).

The biharmonic equation (2.15) is a scalar equation and has the usual form for polar coordinates

$$\frac{1}{r}\left(\partial_r(r\partial_r) + \frac{1}{r^2}\partial_{\theta\theta}\right)^2\Psi = 0. \tag{2.20}$$

c Antiplane strain

A state of antiplane strain is defined by $u_1 = u_2 = 0$ and $u_3 = u_3(x_1, x_2)$. The only non-zero strain components are $e_{13} = \frac{1}{2}\partial_1 u_3$ and $e_{23} = \frac{1}{2}\partial_2 u_3$, so that the non-zero stresses are

$$\sigma_{13} = \mu\partial_1 u_3, \qquad \sigma_{23} = \mu\partial_2 u_3. \tag{2.21}$$

Only one equilibrium equation remains, which leads to

$$\nabla^2 u_3 = 0. \tag{2.22}$$

Thus the antiplane problem reduces to solving Laplace's equation for u_3 together with the relevant boundary conditions.

2.3
ENERGIES AND RECIPROCAL THEOREMS

Consider a body occupying a volume $\mathscr{B}$ with boundary $\partial\mathscr{B}$ which is in a certain equilibrium state of strain with displacements u_i and strains e_{ij} under the action of certain body forces f_i and surface tractions t_i. We wish to calculate the potential energy that is stored in the deformed body. We shall assume that there exists for any state of strain a function $W(e_{ij})$, called the *strain energy density*, depending only on the strain tensor at the point in question, such that the total potential energy U is given by

$$U = \int_{\mathscr{B}} W(e_{ij})dV. \tag{2.23}$$

Furthermore we suppose that U is exactly equal to the work done by the external tractions and body forces on the material in deforming the natural state into the current state of strain.

Consider an additional small displacement du_i leading to an additional strain $de_{ij} = \frac{1}{2}(\partial_i du_j + \partial_j du_i)$. The change in internal energy is

$$dU = \int_{\mathscr{B}} \frac{\partial W}{\partial e_{ij}} de_{ij}dV. \tag{2.24}$$

This must equal, by our hypothesis, the amount of work done by the external forces during this small displacement, so that

$$dU = \int_{\mathscr{B}} f_i du_i dV + \int_{\partial\mathscr{B}} t_i du_i dS. \tag{2.25}$$

In the first term we may replace f_i by $\partial_j \sigma_{ij}$ in view of the equilibrium equations (2.2). In the second we set $t_i = \sigma_{ij}n_j$ and use the divergence theorem to convert the integral to one over the volume of the body, giving

$$dU = \int_{\mathscr{B}} \left[-\partial_j \sigma_{ij} du_i + \partial_j(\sigma_{ij}du_i) \right] dV$$

$$= \int_{\mathscr{B}} \sigma_{ij} \partial_j du_i dV = \int_{\mathscr{B}} \sigma_{ij} de_{ij}dV. \tag{2.26}$$

In the last step we have made use of the symmetry of σ_{ij}.

Comparing the two results (2.24) and (2.26), which must hold for arbitrary small strains de_{ij}, we conclude that

$$\sigma_{ij} = \partial W/\partial e_{ij}. \tag{2.27}$$

(The right-hand side here is required to be symmetrized between i and j.)

From equation (2.4) it follows that $c_{ijkl} = \partial^2 W/\partial e_{ij}\partial e_{kl}$. Therefore it is necessary that the elastic moduli have the symmetry property

$$c_{ijkl} = c_{klij}. \tag{2.28}$$

This symmetry is a consequence of our assumption of the existence of a strain energy density. It is conceivable that there could exist materials for which the symmetry (2.28) did not hold and for which our postulate (2.23) was not applicable. However, no such elastic materials are known, and we may continue to use equations (2.23)–(2.28) without restricting the range of practical applicability of the theory.

Once the moduli satisfy equation (2.28), we may integrate equation (2.27) to give

$$W(e_{ij}) = \tfrac{1}{2}c_{ijkl}e_{ij}e_{kl}. \tag{2.29}$$

The total elastic energy in the body can then be written in the two equivalent forms

$$U = \tfrac{1}{2}\int_{\mathscr{B}}c_{ijkl}e_{ij}e_{kl} = \tfrac{1}{2}\int_{\mathscr{B}}\sigma_{ij}e_{ij}dV. \tag{2.30}$$

For the second of these expressions, using the reverse of the argument which led from equation (2.25) to equation (2.26), we can also write

$$U = \tfrac{1}{2}\int_{\mathscr{B}}f_i u_i dV + \tfrac{1}{2}\int_{\partial\mathscr{B}}t_i u_i dS. \tag{2.31}$$

Now consider two sets of external forces, $(t_i^{(1)}, f_i^{(1)})$ and $(t_i^{(2)}, f_i^{(2)})$, which cause displacement fields $u_i^{(1)}$ and $u_i^{(2)}$ respectively. If both sets of external forces are imposed at the same time, the resulting displacement field will be $u_i^{(1)}+u_i^{(2)}$, since the equations are linear. However, the elastic energy of the combined state, U, is not the sum $U^{(1)}+U^{(2)}$ of the two separate energies, since energy is a quadratic function of the displacement fields. The additional term is called the *interaction energy* between the two deformations, and we write $U = U^{(1)}+U^{(2)}+U_{\text{int}}$. From equation (2.29) we have

$$U = \tfrac{1}{2}\int_{\mathscr{B}}c_{ijkl}(e_{ij}^{(1)}+e_{ij}^{(2)})\,(e_{kl}^{(1)}+e_{kl}^{(2)})dV,$$

so that

$$U_{\text{int}} = \tfrac{1}{2}\int_{\mathscr{B}}c_{ijkl}(e_{ij}^{(1)}e_{kl}^{(2)}+e_{ij}^{(2)}e_{kl}^{(1)})dV$$

$$= \int_{\mathscr{B}}\sigma_{ij}^{(1)}e_{ij}^{(2)}dV = \int_{\mathscr{B}}\sigma_{ij}^{(2)}e_{ij}^{(1)}dV, \tag{2.32}$$

where we have used equations (2.4) and (2.28). Now each of these expressions can be reduced in the same way that was used in obtaining equation (2.31) from equation (2.30), giving

$$U_{\text{int}} = \int_{\partial\mathscr{B}}t_i^{(1)}u_i^{(2)}dS + \int_{\mathscr{B}}f_i^{(1)}u_i^{(2)}dV$$

$$= \int_{\partial\mathscr{B}}t_i^{(2)}u_i^{(1)}dS + \int_{\mathscr{B}}f_i^{(2)}u_i^{(1)}dV. \tag{2.33}$$

In particular cases, either of these two expressions can be the most useful. Their equivalence is known as the *reciprocal theorem* of Betti and Rayleigh.

2.4
ELASTIC GREEN'S FUNCTIONS

A *Equilibrium boundary-value problems*

Substituting from the stress-strain relations (2.4) into the field equation (2.2) we obtain

$$\rho \ddot{u}_i = c_{ijkl} \partial_j e_{kl} + f_i = c_{ijkl} \partial_j \partial_k u_l + f_i, \tag{2.34}$$

after using the symmetry of the elastic moduli. Then the boundary-value problem for a body $\mathcal{B}$ which is in equilibrium under given body forces and prescribed tractions on a part S_1 of its boundary, with the displacement being prescribed over the rest of the boundary, S_2, is contained in the following set of equations:

$$c_{ijkl} \partial_j \partial_k u_l + f_i = 0 \quad \text{in } \mathcal{B},$$

$$c_{ijkl} \partial_k u_l n_j = t_i \quad \text{on } S_1, \tag{2.35}$$

$$u_i = U_i \quad \text{on } S_2.$$

We can construct a solution of this problem by finding an appropriate Green's function. Let $\mathbf{r}'$ be a general point in $\mathcal{B}$, Σ_ε be a small sphere of radius ε centred at $\mathbf{r}'$, and $\mathcal{B}_\varepsilon$ be that part of $\mathcal{B}$ which lies outside Σ_ε. Next suppose the functions $u_{ij}(\mathbf{r}, \mathbf{r}')$ ($i, j = 1, 2, 3$) can be found to satisfy the following conditions:

(a) $c_{ijkl} \partial_j \partial_k u_{lm}(\mathbf{r}, \mathbf{r}') = 0$ for $\mathbf{r}$ in $\mathcal{B}_\varepsilon$,

(b) $c_{ijkl} \partial_k u_{lm}(\mathbf{r}, \mathbf{r}') = 0$ for $\mathbf{r}$ on S_1,

(c) $u_{lm}(\mathbf{r}, \mathbf{r}') = 0$ for $\mathbf{r}$ on S_2, $\qquad$ (2.36)

(d) $\int_{\Sigma_\varepsilon} c_{ijkl} \partial_k u_{lm}(\mathbf{r}, \mathbf{r}') n_j \, dS = \delta_{im}$ (where $\mathbf{n}$ is the inward normal to Σ_ε),

(e) $\varepsilon^2 u_{lm}(\mathbf{r}, \mathbf{r}') \to 0$ as $\varepsilon \to 0$, for $\mathbf{r}$ on Σ_ε.

Then the solution of the boundary-value problem can be expressed simply in terms of these functions. Physically, $u_{lm}(\mathbf{r}, \mathbf{r}')$ represents the l component of the displacement field at $\mathbf{r}$ when there are no body forces except for a unit force in the m-direction localized at $\mathbf{r}'$, and when the tractions on S_1 and the displacements on S_2 are both zero.

Using the Gauss divergence theorem, we get

$$\int_{\partial \mathscr{B}+\Sigma_\varepsilon} [u_i(\mathbf{r})c_{ijkl}\partial_k u_{lm}(\mathbf{r}, \mathbf{r}') - u_{im}(\mathbf{r}, \mathbf{r}')c_{ijkl}\partial_k u_l(\mathbf{r})]n_j dS$$

$$= \int_{\mathscr{B}_\varepsilon} c_{ijkl}\partial_j[u_i(\mathbf{r})\partial_k u_{lm}(\mathbf{r}, \mathbf{r}') - u_{im}(\mathbf{r}, \mathbf{r}')\partial_k u_l(\mathbf{r})]dV$$

$$= \int_{\mathscr{B}_\varepsilon} u_{im}(\mathbf{r}, \mathbf{r}')f_i(\mathbf{r})dV, \tag{2.37}$$

after using the symmetry of the modulus tensor and equations (2.35) and (2.36a). Returning to the surface integral, after using the boundary conditions on u_i and u_{im}, the integral over $\partial \mathscr{B}$ reduces to

$$-\int_{S_1} u_{im}(\mathbf{r}, \mathbf{r}')t_i(\mathbf{r})dS + \int_{S_2} U_i(\mathbf{r})c_{ijkl}\partial_k u_{lm}(\mathbf{r}, \mathbf{r}')n_j dS. \tag{2.38}$$

There remains the surface integral from Σ_ε. In the second term, $\partial_k u_l(\mathbf{r})$ is bounded throughout $\mathscr{B}$ while $u_{im}(\mathbf{r}, \mathbf{r}')dS = \varepsilon^2 u_{im}(\mathbf{r}, \mathbf{r}')d\Omega$ tends to zero as $\varepsilon \to 0$. So the second term can be made arbitrarily small by choosing ε sufficiently small. On the other hand,

$$\int_{\Sigma_\varepsilon} u_i(\mathbf{r})c_{ijkl}\partial_k u_{lm}(\mathbf{r}, \mathbf{r}')n_j dS = [u_i(\mathbf{r}') + o(1)]\int_{\Sigma_\varepsilon} c_{ijkl}\partial_k u_{lm}(\mathbf{r}, \mathbf{r}')n_j dS$$

$$= u_m(\mathbf{r}') + o(1),$$

as $\varepsilon \to 0$, by condition (2.36e).[*] Putting these together gives for equation (2.37) in the limit $\varepsilon \to 0$ the required solution of the original problem:

$$u_m(\mathbf{r}') = \int_{\mathscr{B}} u_{im}(\mathbf{r}, \mathbf{r}')f_i(\mathbf{r})dV + \int_{S_1} u_{im}(\mathbf{r}, \mathbf{r}')t_i(\mathbf{r})dS$$

$$- \int_{S_2} U_i(\mathbf{r})c_{ijkl}\partial_k u_{lm}(\mathbf{r}, \mathbf{r}')n_j dS. \tag{2.39}$$

If the medium extends to infinity, it is necessary to add both to the boundary-value problem (2.35) and to the conditions (2.36) on $u_{im}(\mathbf{r}, \mathbf{r}')$ certain restrictions on the asymptotic behaviour of the two sets of displacements. In the set of equations (2.36) defining the Green's function $u_{im}(\mathbf{r}, \mathbf{r}')$, the integral of the tractions $c_{ijkl}\partial_k u_{lm}(\mathbf{r}, \mathbf{r}')n_j$ over the surface of a large sphere of radius R has to be bounded and, in fact, has to balance the unit force at $\mathbf{r}'$. This will hold if $u_{lm}(\mathbf{r}, \mathbf{r}') \sim \text{const.}/r$ as $r \to \infty$, and we shall suppose that this is an appropriate asymptotic condition to impose. (Note that if the infinite portion of $\mathscr{B}$ subtends a zero solid angle at the origin, this assumption has to be modified.) Then in solving the boundary-value problem (2.36), we use equation (2.37) but applied to that part of $\mathscr{B}_\varepsilon$ which lies within a large sphere Σ_R of radius R and centred at, say, the origin. The left-hand side contains a surface term from Σ_R, and using the assumed behaviour of the Green's function, this term tends to zero as $R \to \infty$ provided that

_____

[*] Here $o(1)$ denotes terms which tend to zero as $\varepsilon \to 0$.

$u_i(\mathbf{r}) \to 0$ and $r\partial_k u_i(\mathbf{r}) \to 0$ as $r \to \infty$. Then on taking the limits $\varepsilon \to 0$ and $R \to \infty$ we return again to equation (2.39).

A further identity involving the Green's function, which will be useful when we consider dislocations, is obtained by applying the above method of deriving equations (2.37) and (2.39) to a subvolume V of the whole body $\mathscr{B}$, with surface ∂V. Equations (2.36a, d, e) still hold for $\mathbf{r}$ and $\mathbf{r}'$ restricted to lie in V, and we obtain, in place of equation (2.39),

$$u_m(\mathbf{r}') = \int_V u_{im}(\mathbf{r}, \mathbf{r}')f_i(\mathbf{r})dV + \int_{\partial V}[u_{im}(\mathbf{r}, \mathbf{r}')\sigma_{ij}(\mathbf{r}) - u_i(\mathbf{r})c_{ijkl}\partial_k u_{lm}(\mathbf{r}, \mathbf{r}')]n_j\,dS.$$

$$(2.40)$$

Here σ_{ij} is the stress-field corresponding to u_i. This result holds for any subset V of $\mathscr{B}$. If ∂V contains some parts of the boundary of the whole body, then the surface terms from this part simplify by virtue of conditions (2.36b, c). Similarly, if for an infinite body V also extends to infinity, then the surface term from infinity vanishes under the same restrictions as in equation (2.39).

B Traction boundary-value problems

When the problem for $u_i(\mathbf{r})$ has boundary conditions of only the traction type (i.e., when S_2 is empty) the above procedure has to be modified, because the requirements (a), (b), and (d) in equations (2.36) become inconsistent when S_1 is the whole of $\partial\mathscr{B}$. Physically the reason for this is clear: there are no longer any boundary tractions in equations (2.36) to balance the point force at $\mathbf{r}'$. But this physical picture also indicates how to overcome the difficulty: in the problem for the Green's function we must impose not only the point force at $\mathbf{r}'$ but an equal and opposite force at some other point, $\mathbf{r}_0$, and a point couple which exactly cancels the torque of the two point forces.

To simplify the considerations, we suppose that we are dealing with generalized functions and combine equations (2.36a, d) in the form

$$c_{ijkl}\partial_j\partial_k u_{lm}(\mathbf{r}, \mathbf{r}') = -\delta_{im}\delta(\mathbf{r}-\mathbf{r}') \quad \text{for } \mathbf{r} \text{ in } \mathscr{B}, \tag{2.41}$$

where $\delta(\mathbf{r}-\mathbf{r}')$ is the three-dimensional delta function. When $S_1 \equiv \partial\mathscr{B}$, we change this by inserting further point forces at or near some point $\mathbf{r}_0$ which are in static equilibrium with the unit force $\mathbf{1}$ at $\mathbf{r}'$ (Fig. 2.1). We take a force $-\mathbf{1}$ at $\mathbf{r}_0$, $\mathbf{F}$ at $\mathbf{r}_0+\boldsymbol{\varepsilon}$ and $-\mathbf{F}$ at $\mathbf{r}_0-\boldsymbol{\varepsilon}$, and $\mathbf{F}'$ at $\mathbf{r}_0+\boldsymbol{\varepsilon}'$ and $-\mathbf{F}'$ at $\mathbf{r}_0-\boldsymbol{\varepsilon}'$, where $\mathbf{F}$ is parallel to $\boldsymbol{\varepsilon}'$ and $\mathbf{F}'$ is parallel to $\boldsymbol{\varepsilon}$. We choose the forces so that the $(\mathbf{F}, -\mathbf{F})$ and $(\mathbf{F}', -\mathbf{F}')$ pairs contribute equal torques, which together just cancel the torque of the $(\mathbf{1}, -\mathbf{1})$ pair, i.e. $\boldsymbol{\varepsilon}\wedge\mathbf{F} = \boldsymbol{\varepsilon}'\wedge\mathbf{F}' = -\frac{1}{4}(\mathbf{r}'-\mathbf{r}_0)\wedge\mathbf{1}$. For this system of forces the right-hand side in the equivalent of equation (2.41) is

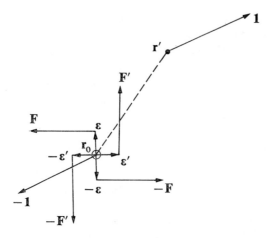

FIGURE 2.1

$$\delta_{im}[-\delta(\mathbf{r}-\mathbf{r}')+\delta(\mathbf{r}-\mathbf{r}_0)]$$

$$-F_i[\delta(\mathbf{r}-\mathbf{r}_0-\boldsymbol{\varepsilon})-\delta(\mathbf{r}-\mathbf{r}_0+\boldsymbol{\varepsilon})]-F_i'[\delta(\mathbf{r}-\mathbf{r}_0-\boldsymbol{\varepsilon}')-\delta(\mathbf{r}-\mathbf{r}_0+\boldsymbol{\varepsilon}')]$$

$$=\delta_{im}[-\delta(\mathbf{r}-\mathbf{r}')+\delta(\mathbf{r}-\mathbf{r}_0)]+2(F_i\varepsilon_p+F_i'\varepsilon_p')\partial_p\delta(\mathbf{r}-\mathbf{r}_0)$$

to order $\boldsymbol{\varepsilon}$. Furthermore, from the construction of the system of forces, it follows that

$$(F_i\varepsilon_p+F_i'\varepsilon_p')=\tfrac{1}{4}[(r_i'-r_{0i})\delta_{pm}-(r_p'-r_{0p})\delta_{im}].$$

Thus equation (2.36a) for the Green's function now reads

$$c_{ijkl}\partial_j\partial_k u_{lm}(\mathbf{r},\mathbf{r}')$$

$$=\delta_{im}[-\delta(\mathbf{r}-\mathbf{r}')+\delta(\mathbf{r}-\mathbf{r}_0)]+\tfrac{1}{2}[(r_i'-r_{0i})\delta_{pm}-(r_p'-r_{0p})\delta_{im}]\partial_p\delta(\mathbf{r}-\mathbf{r}_0).$$
$$(2.42)$$

If we now proceed as in the derivation of equation (2.37) but using the whole of $\mathscr{B}$ as the integration region, we obtain, after using the boundary conditions, on $\partial\mathscr{B}$,

$$\int_{\partial\mathscr{B}}u_{im}(\mathbf{r},\mathbf{r}')t_i(\mathbf{r})dS+\int_{\mathscr{B}}u_{im}(\mathbf{r},\mathbf{r}')f_i(\mathbf{r})dV=-\int_{\mathscr{B}}c_{ijkl}u_i(\mathbf{r})\partial_j\partial_k u_{lm}(\mathbf{r},\mathbf{r})'dV$$

$$=-\int_{\mathscr{B}}u_i(\mathbf{r})\,\{\delta_{im}[-\delta(\mathbf{r}-\mathbf{r}')+\delta(\mathbf{r}-\mathbf{r}_0)]$$

$$+\tfrac{1}{2}[(r_i'-r_{0i})\delta_{pm}-(r_p'-r_{0p})\delta_{im}]\partial_p\delta(\mathbf{r}-\mathbf{r}_0)\}dV$$

$$=u_m(\mathbf{r}')-u_m(\mathbf{r}_0)-(r_p'-r_{0p})\tfrac{1}{2}[\partial_p u_m(\mathbf{r}_0)-\partial_m u_p(\mathbf{r}_0)].\qquad(2.43)$$

The second and third terms on the right-hand side here are a rigid translation and rigid rotation about $\mathbf{r}_0$. For the traction boundary-value problem, the displacement field is unique only up to such a rigid displacement, so that equation (2.43) solves the problem if we regard $u_m(\mathbf{r}_0)$ and $\frac{1}{2}[\partial_p u_m(\mathbf{r}_0) - \partial_m u_p(\mathbf{r}_0)]$ as arbitrary constants which can be specified independently of the rest of the problem.

c Green's function for an infinite region

For an isotropic medium, the modulus tensor is given by equation (2.5), and the solution of equations (2.36a, d) for the case of an infinite medium is

$$u_{lm}(\mathbf{r}, \mathbf{r}') = \frac{1}{8\pi\mu}\left(\delta_{lm}\partial_{kk}R - \frac{\lambda+\mu}{\lambda+2\mu}\partial_{lm}R\right)$$

$$= \frac{1}{8\pi\mu(\lambda+2\mu)}\left((\lambda+3\mu)\frac{\delta_{lm}}{R} + (\lambda+\mu)\frac{(x_l-x_l')(x_m-x_m')}{R^3}\right). \tag{2.44}$$

Here $R = |\mathbf{r}-\mathbf{r}'|$ and $\partial_{lm} = \partial^2/\partial x_l \partial x_m$. This result can be arrived at via Fourier transformation of equation (2.41) and solution of the resulting algebraic equations for the Fourier transform of u_{lm},[3-5] or by using displacement potentials or Papkovich-Neuber functions to solve equations (2.36).[6] We shall content ourselves here by demonstrating that equations (2.36) are satisfied by the above solution.

First of all, we can derive the result

$$c_{ijkl}\partial_k u_{lm}(\mathbf{r}, \mathbf{r}') = \frac{1}{4\pi}\left[-\frac{\mu}{\lambda+2\mu}\left(\delta_{im}\frac{x_j-x_j'}{R^3} + \delta_{jm}\frac{x_i-x_i'}{R^3} - \delta_{ij}\frac{x_m-x_m'}{R^3}\right)\right.$$

$$\left. - \frac{3(\lambda+\mu)}{\lambda+2\mu}\frac{(x_i-x_i')(x_j-x_j')(x_m-x_m')}{R^5}\right]. \tag{2.45}$$

It then follows immediately that equation (2.36a) is satisfied. Equation (2.36d) follows also if we use the results that, with $\mathbf{n}$ the inward normal to Σ_ε,

$$\int_{\Sigma_\varepsilon} \frac{x_j-x_j'}{R^3} n_j dS = -4\pi,$$

$$\int_{\Sigma_\varepsilon} \frac{x_i-x_i'}{R^3} n_j dS = -\frac{4\pi}{3}\delta_{ij},$$

$$\int_{\Sigma_\varepsilon} \frac{(x_i-x_i')(x_j-x_j')(x_m-x_m')}{R^5} n_j dS = -\frac{4\pi}{3}\delta_{im}.$$

In Problem 2.13 we have sketched the derivation of the Green's function of equation (2.44) using the Fourier transform method. The Fourier transform of the partial differential equations (2.41) produces a system of algebraic equations for the transform of the Green's function. For an isotropic medium, the solution of these equations is simple and the transform

c

may be inverted. For media of arbitrary symmetry, however, the algebraic equations give a much more complicated solution for the transform of the Green's function involving a denominator which is a sixth-degree polynomial in the transform variables. In the general case this makes the transform impossible to invert in closed form. A series form for the inverse transform has been obtained by Kröner.[7]

A formal solution for the Green's function in an anisotropic medium was found much earlier by Fredholm.[8] However, this solution contains expressions which involve the roots of a sixth-order equation, and so again, in general, cannot be written in closed form. Using results obtained by Gebbia,[9] which showed that the sixth-order equation can be factorized for hexagonal materials, Kröner[7] was able to obtain the Green's function explicitly in that case.

2.5
KINEMATICS OF FINITE DEFORMATIONS

A *Deformation and change of shape*

The development of the previous four sections is restricted to infinitesimal deformations in the sense that the derivatives $\{\partial_i u_j\}$ of the displacement components are required to be small. In the following sections we shall outline the extension of the theory to finite deformations.[10] Our reason for doing this, apart from the intrinsic interest of the theory itself, is that it will enable us, in Chapters 7 and 9, to discuss the theory of continuous distributions of dislocations from a more rigorous standpoint than would be possible using simply the classical theory.

We shall identify a continuous body with one of the configurations it occupies, or could conceivably occupy, in ordinary space (E^3), and we shall use the letter $\mathscr{B}$ to mean both the body itself and the region of E^3 which it occupies in this one configuration which is selected as a reference. The particles of the body can be specified by their positions X, with components (X_α) $\alpha = 1, 2, 3$, in some Cartesian coordinate system, in the reference configuration, and we shall mean by 'the particle X' the particle which occupies the position X in the reference state. A motion of the body can then be defined as a 1-parameter family of mappings

$$\mathbf{x} = \mathbf{x}(\mathbf{X}, t) \tag{2.46}$$

from $\mathscr{B}$ onto a region of E^3, where the parameter t denotes time. $\mathbf{x}$ is the position occupied by the particle X at time t. In components equation (2.46) takes the form

$$x_i = x_i(\mathbf{X}, t),$$

where we use latin indices for components relative to a second Cartesian basis, which it is advantageous to keep distinct from the (X_α) basis. We shall need to make some smoothness requirements on the mapping (2.46), usually that it is twice continuously differentiable with respect to both X_α and t.

The velocity and acceleration of the particle $\mathbf{X}$ are obtained by partial differentiation of equation (2.46) with $\mathbf{X}$ held fixed:

$$\dot{x}_i = \partial x_i(\mathbf{X}, t)/\partial t, \qquad \ddot{x}_i = \partial^2 x_i(\mathbf{X}, t)/\partial t^2.$$

The *deformation gradients* are defined as the derivatives $\partial_\alpha x_i \equiv \partial x_i/\partial X_\alpha$ and the inverse deformation gradients as $\partial_i X_\alpha \equiv \partial X_\alpha/\partial x_i$. They satisfy the relations

$$\partial_i X_\alpha \partial_\beta x_i = \delta_{\alpha\beta}, \qquad \partial_\alpha x_i \partial_j X_\alpha = \delta_{ij}. \tag{2.47}$$

From these quantities we can construct the two strain tensors appropriate to the current and reference configurations by considering the change in length of a small element of material as it deforms from one configuration to the other. The element $\mathbf{dX}$ of material in the reference state deforms into an element $\mathbf{dx}$ at time t, where

$$dx_i = \partial_\alpha x_i dX_\alpha, \qquad dX_\alpha = \partial_i X_\alpha dx_i.$$

The change in length squared is

$$dx_i dx_i - dX_\alpha dX_\alpha = (\delta_{ik} - \partial_i X_\alpha \partial_k X_\alpha) dx_i dx_k = (\partial_\alpha x_i \partial_\beta x_i - \delta_{\alpha\beta}) dX_\alpha dX_\beta.$$

Defining strain tensors e_{ik} and $E_{\alpha\beta}$ by the requirement that

$$dx_i dx_i - dX_\alpha dX_\alpha = 2e_{ik} dx_i dx_k = 2E_{\alpha\beta} dX_\alpha dX_\beta, \tag{2.48}$$

we obtain

$$e_{ik} = \tfrac{1}{2}(\delta_{ik} - \partial_i X_\alpha \partial_k X_\alpha), \tag{2.49}$$

$$E_{\alpha\beta} = \tfrac{1}{2}(\partial_\alpha x_i \partial_\beta x_i - \delta_{\alpha\beta}). \tag{2.50}$$

An element of volume dV in the reference configuration corresponds to an element dv in the current configuration, where

$$dv = jdV, \tag{2.51}$$

with j being the Jacobian of the mapping (2.46):

$$j = \det(\partial_\alpha x_i). \tag{2.52}$$

If ρ_0 and ρ are the mass densities in the reference and current configuration respectively, conservation of mass requires that $\rho dv = \rho_0 dV$, and hence that

$$j = \rho_0/\rho. \tag{2.53}$$

We shall consider only deformations for which $0 < j < \infty$.

It is also of value to examine the way in which elements of surface are transformed between the two configurations. Let dS be an element of area with unit normal $\mathbf{N} = (N_\alpha)$ on the surface $\partial \mathcal{B}$ of the body in its reference configuration, and suppose that the same particles form a surface element ds with unit normal $\mathbf{n} = (n_i)$ at time t. An application of Stokes' theorem shows that

$$\tfrac{1}{2} \oint \varepsilon_{jil} x_j dx_i = \int n_l ds \approx n_l ds,$$

where the line integral is taken around the boundary of ds. Transforming to an integral around dS and then to a surface integral over dS gives

$$n_l ds = \tfrac{1}{2} \varepsilon_{jil} \oint x_j \partial_\alpha x_i dX_\alpha = \tfrac{1}{2} \varepsilon_{jil} \int \varepsilon_{\alpha\beta\gamma} \partial_\beta (x_j \partial_\gamma x_i) N_\alpha dS$$

$$\approx \tfrac{1}{2} \varepsilon_{jil} \varepsilon_{\alpha\beta\gamma} \partial_\beta x_j \partial_\gamma x_i N_\alpha dS.$$

Using $\varepsilon_{jil} \partial_\beta x_j \partial_\gamma x_i \partial_\alpha x_l = j \varepsilon_{\beta\gamma\alpha}$, which is a well-known property of determinants, gives the required relationship between the two vector elements of area, $n_l ds$ and $N_\alpha dS$:

$$\partial_\alpha x_l n_l ds = j N_\alpha dS. \tag{2.54}$$

B Polar decomposition theorem

A useful result in simplifying constitutive relations is the polar decomposition theorem, which states that, given any non-singular matrix $\mathbf{F}$, there exist an orthogonal matrix $\mathbf{R}$ and two positive-definite symmetric matrices $\mathbf{U}$ and $\mathbf{V}$ such that

$$\mathbf{F} = \mathbf{RU} = \mathbf{VR}. \tag{2.55}$$

This result is proved as follows.

The matrix $\mathbf{F}^T \mathbf{F}$ is symmetric and positive definite, with $\mathbf{F}^T$ denoting the transpose of $\mathbf{F}$ (since for any vector $\mathbf{v} \neq \mathbf{0}$, $\mathbf{v}^T (\mathbf{F}^T \mathbf{F}) \mathbf{v} = (\mathbf{Fv})^2 > 0$). Hence there exists a symmetric positive-definite matrix $\mathbf{U}$ such that $\mathbf{U}^2 = \mathbf{F}^T \mathbf{F}$. $\mathbf{U}$ can be constructed explicitly by taking a coordinate basis in the principal directions of $(\mathbf{F}^T \mathbf{F})$, in which case $\mathbf{U}$ is the diagonal matrix whose elements are the positive square roots of those of $\mathbf{F}^T \mathbf{F}$. Furthermore this construction shows that $\mathbf{U}$ is unique, and we write

$$\mathbf{U} = (\mathbf{F}^T \mathbf{F})^{1/2}. \tag{2.56}$$

Now define

$$\mathbf{R} = \mathbf{FU}^{-1}, \tag{2.57}$$

so that we need to show that $\mathbf{R}$ is orthogonal. But

$$\mathbf{R}^T \mathbf{R} = (\mathbf{U}^{-1})^T \mathbf{F}^T \mathbf{F} \mathbf{U}^{-1} = \mathbf{U}^{-1} \mathbf{U}^2 \mathbf{U}^{-1} = \mathbf{1}.$$

It follows similarly that $\mathbf{V}$ can be uniquely defined as

$$\mathbf{V} = (\mathbf{FF}^T)^{1/2}$$

and that $\mathbf{R}' = \mathbf{V}^{-1}\mathbf{F}$ is orthogonal. It remains to show that $\mathbf{R}$ and $\mathbf{R}'$ are identical. From the first decomposition,

$$\mathbf{F} = \mathbf{RU} = (\mathbf{RUR}^{-1})\mathbf{R} = \mathbf{VR}'.$$

But the second decomposition, into the product $\mathbf{VR}'$, is unique, and since $\mathbf{RUR}^{-1}$ is also symmetric, it is necessary that $\mathbf{RUR}^{-1} = \mathbf{V}$ and $\mathbf{R} = \mathbf{R}'$.

The polar decomposition has a significant physical interpretation when $F_{i\alpha} = \partial_\alpha x_i$ is a deformation gradient. If dX_α is an infinitesimal material element in the reference configuration, $\mathbf{F}d\mathbf{X} = (F_{i\alpha}dX_\alpha)$ gives the element into which it deforms in the current configuration. This mapping $d\mathbf{X} \to d\mathbf{x} = \mathbf{F}d\mathbf{X}$ can be decomposed into two mappings: $d\mathbf{X} \to d\mathbf{X}' = \mathbf{U}d\mathbf{X}$ and $d\mathbf{X}' \to d\mathbf{x} = \mathbf{R}d\mathbf{X}'$. The second of these is just a rigid rotation of the element. For the first, if we take as basis for $d\mathbf{X}$ and $d\mathbf{X}'$ the principal vectors of $\mathbf{U}$, which form an orthonormal set, the mapping becomes simply $dX_\alpha' = \lambda_\alpha dX_\alpha$ where $\{\lambda_\alpha\}$ are the three eigenvalues of $\mathbf{U}$. This mapping is just a stretching of the material elements by the factors λ_α along the coordinate directions. Thus the theorem states that any deformation of material elements can be obtained by first stretching along some three mutually perpendicular directions, then rotating.

The second form $\mathbf{F} = \mathbf{VR}$ shows that the deformation can be obtained by rotating the elements first and then applying the stretches. Since $\mathbf{V} = \mathbf{RUR}^{-1}$, $\mathbf{V}$ has the same eigenvalues $\{\lambda_\alpha\}$ as $\mathbf{U}$, so that the magnitudes of the stretches are the same in the two cases. The sets of principal directions of $\mathbf{U}$ and $\mathbf{V}$ differ by the rotation $\mathbf{R}$.

2.6
EQUATIONS OF MOTION

A *The action principle*

The usual derivation of the equations of motion for a continuum proceeds via the construction of a stress tensor in terms of the forces between neighbouring parts of the body and thence to an application of Newton's laws to an arbitrary material volume. This approach's drawback is that it is not at all clear to what extent it is valid to replace the non-local interatomic forces of a real body by a perfectly localized stress vector which depends only on the normal direction of the surface element across which it acts. For this reason we shall obtain the equations of motion from an action principle. This has the advantage that the assumptions being made can be displayed more explicitly, and can also to a large extent be reduced by deeper investigation.

Let $\{x_i(\mathbf{X}, t): \mathbf{X} \in \mathcal{B}, t \in \mathcal{T}\}$ be a motion of the body $\mathcal{B}$ during a time interval $\mathcal{T}$ under the action of an external body force and external surface tractions. We suppose that the former have magnitude $F_i dV$ on each volume dV of $\mathcal{B}$ and the latter $T_i dS$ on each element of $\partial\mathcal{B}$, where components are taken in the x_i coordinate system. Then it is assumed that there is an action $A\{\mathcal{B}, \mathcal{T}\}$ associated with the motion such that if $\{x_i(\mathbf{X}, t) + \delta x_i(\mathbf{X}, t)\}$ is a slightly different motion with $\delta x_i = 0$ at the initial and final times, then the difference in the action to first order in the δx_i is

$$\delta A = -\int_{\mathcal{B}} \int_{\mathcal{T}} F_i \delta x_i dV dt - \int_{\partial\mathcal{B}} \int_{\mathcal{T}} T_i \delta x_i dS dt. \tag{2.58}$$

The postulate of an action principle is in itself perhaps not very drastic – if the atoms of the body were classical particles obeying Newtonian mechanics, for instance, no assumption would be involved. It is in the following assumptions about the explicit form of A that we restrict attention to conventional elastic media. The first assumption is that the action is local, i.e. there exists a function $L(\mathbf{X}, t)$, called the *Lagrangian density*, depending only on dynamical variables at $\mathbf{X}$ and t such that

$$A\{\mathcal{B}, \mathcal{T}\} = \int_{\mathcal{B}} \int_{\mathcal{T}} L(\mathbf{X}, t)\, dV dt. \tag{2.59}$$

Secondly, we assume that the material is of *first grade* (or *simple*), i.e. that L depends only on the variables

$$L = L(\dot{x}_i, \partial_\alpha x_i, X_\alpha, x_i, t) \tag{2.60}$$

and not on any derivatives of $x_i(\mathbf{X}, t)$ higher than the first, or on any other kinematic quantities. And, finally, we shall at a later stage assume that L can be split into separate kinetic and potential parts,

$$L = \tfrac{1}{2}\rho_0 \dot{x}_i^2 - W(\partial_\alpha x_i, X_\alpha), \tag{2.61}$$

where W is the *strain energy density*. This last case corresponds to what is often called non-linear *hyperelasticity*.

B *Derivation of the equations*

With equations (2.59) and (2.60) we have to first order

$$\delta A = \int_{\mathcal{B}} \int_{\mathcal{T}} \left(\frac{\partial L}{\partial \dot{x}_i} \delta \dot{x}_i + \frac{\partial L}{\partial(\partial_\alpha x_i)} \delta(\partial_\alpha x_i) + \frac{\partial L}{\partial x_i} \delta x_i \right) dV dt.$$

If we define a *momentum density* P_i and a *stress tensor* $T_{i\alpha}$ by the equations

$$P_i = \partial L/\partial \dot{x}_i, \qquad T_{i\alpha} = -\partial L/\partial(\partial_\alpha x_i), \tag{2.62}$$

we have

$$\delta A = \int_{\mathscr{B}} \int_{\mathscr{T}} \left(\frac{\partial}{\partial t} (P_i \delta x_i) - \dot{P}_i \delta x_i - \partial_\alpha (T_{i\alpha} \delta x_i) + \partial_\alpha T_{i\alpha} \delta x_i + \frac{\partial L}{\partial x_i} \delta x_i \right) dV dt.$$

The first term vanishes since δx_i is restricted to be zero at the ends of the time interval, and the third term can be transformed by Gauss's theorem to an integral over $\partial \mathscr{B}$. The action principle (2.58) then takes the form

$$\int \int \left(-\dot{P}_i + \partial_\alpha T_{i\alpha} + \frac{\partial L}{\partial x_i} + F_i \right) \delta x_i dV dt + \int_{\partial \mathscr{B}} \int_{\mathscr{T}} (T_i - T_{i\alpha} N_\alpha) \delta x_i dS dt = 0.$$

This has to hold for arbitrary small variations δx_i, so that

$$-\dot{P}_i + \partial_\alpha T_{i\alpha} + \frac{\partial L}{\partial x_i} + F_i = 0 \quad \text{throughout } \mathscr{B}, \tag{2.63}$$

$$T_{i\alpha} N_\alpha - T_i = 0 \quad \text{on } \partial \mathscr{B}. \tag{2.64}$$

If we consider the hyperelastic special case (2.61), we have $P_i = \rho_0 \dot{x}_i$ so that equation (2.63) becomes

$$\rho_0 \ddot{x}_i = \partial_\alpha T_{i\alpha} + F_i, \tag{2.65}$$

where

$$T_{i\alpha} = \partial W / \partial (\partial_\alpha x_i). \tag{2.66}$$

c *Spatial variables*

The equations of motion (2.63) and (2.64) are inconvenient inasmuch as they contain quantities which refer to both the current and the reference configuration. The stress tensor $T_{i\alpha}$ is an obvious case in point, with one suffix in each camp. This problem can be remedied by using quantities which refer entirely to one or the other configuration. Choosing the spatial one, which is the most useful, we define new quantities by

$$F_i = j f_i, \tag{2.67}$$

$$P_i = j p_i, \tag{2.68}$$

$$\sigma_{ij} = j^{-1} T_{i\alpha} \partial_\alpha x_j, \tag{2.69}$$

$$t_i = T_i dS / ds, \tag{2.70}$$

where dS and ds are related elements of surface area. f_i is the body force per unit volume in the current configuration and p_i, similarly, is the momentum per unit current volume, which equals $\rho \dot{x}_i$ in the hyperelastic case (2.61). σ_{ij} is the usual Cauchy stress tensor and t_i is the external traction per unit current surface area.

Substituting these into equation (2.64) and using equation (2.54) shows that the boundary condition takes the form

$$\sigma_{ij} n_j = t_i. \tag{2.71}$$

Thus σ_{ij} represents the components of forces acting across surface elements in the body, as it must to agree with the usual formulation of elasticity theory.

To obtain the spatial field equations, we use the fact that, from equation (2.47), the cofactor of $\partial_\beta x_k$ in the determinant j equals $j \partial_k X_\beta$, and hence

$$\partial_\alpha j = j \partial_k X_\beta \partial_\alpha \partial_\beta x_k. \tag{2.72}$$

It then follows that

$$\partial_j \sigma_{ij} = j^{-1} \partial_\alpha T_{i\alpha},$$

so that equation (2.63) becomes, for a hyperelastic material,

$$-\rho \ddot{x}_i + \partial_j \sigma_{ij} + j^{-1} \frac{\partial L}{\partial x_i} + f_i = 0. \tag{2.73}$$

Except for the third term, these equations are the usual Cauchy equations of motion, phrased entirely in terms of spatial variables and coordinates. In almost all cases, L would be independent of x_i (see next section) and the third term would therefore not appear. In this connection we observe that for a conservative body force, L can be made to include the body force. For, suppose that $f_i = -\partial V / \partial x_i$, where $V(x_i)$ is the potential function for the external force, and define $L' = L - V$. The momentum and stress from L' are the same as those derived from L, and the last two terms in equation (2.73) are equal to $j^{-1} \partial L' / \partial x_i$. However, we shall here choose to separate out explicitly all the external forces in f_i.

2.7
SYMMETRIES, INVARIANCES, AND CONSERVATION LAWS[11]

A *Spatial invariance*

If we suppose that the Lagrangian density is independent of x_i, then it may be shown that the motion of the body obeys the conservation law of linear momentum. The total momentum of the body $\mathcal{B}$ is defined as

$$P_i(\mathcal{B}, t) = \int_{\mathcal{B}} P_i \, dV. \tag{2.74}$$

Then

$$\frac{\partial}{\partial t} P_i(\mathcal{B}, t) = \int_{\mathcal{B}} \left(\partial_\alpha T_{i\alpha} + F_i + \frac{\partial L}{\partial x_i} \right) dV$$

$$= \int_{\partial \mathcal{B}} T_i \, dS + \int_{\mathcal{B}} \left(F_i + \frac{\partial L}{\partial x_i} \right) dV, \tag{2.75}$$

after using equations (2.63) and (2.64) and the divergence theorem. Hence, if $\partial L/\partial x_i = 0$, the rate of change of total momentum equals the total resultant of the external tractions T_i on the boundary and the external body force F_i acting throughout the volume. This is the required conservation law.

B *Temporal invariance*

If L is independent of t, we can derive a conservation law for energy. The total energy is defined in the general case as

$$U(\mathcal{B}, t) = \int_{\mathcal{B}} (\dot{x}_i P_i - L)\, dV, \tag{2.76}$$

which, in the event equation (2.61) holds, assumes the form

$$U(\mathcal{B}, t) = \int_{\mathcal{B}} (\tfrac{1}{2}\rho_0 \dot{x}_i^2 + W)\, dV. \tag{2.77}$$

Then

$$
\begin{aligned}
\frac{\partial}{\partial t} U(\mathcal{B}, t) &= \int_{\mathcal{B}} \left[\ddot{x}_i P_i + \dot{x}_i \left(\partial_\alpha T_{i\alpha} + \frac{\partial L}{\partial x_i} + F_i \right) \right. \\
&\qquad \left. - \left(\frac{\partial L}{\partial \dot{x}_i} \ddot{x}_i + \frac{\partial L}{\partial(\partial_\alpha x_i)} \partial_\alpha \dot{x}_i + \frac{\partial L}{\partial x_i} \dot{x}_i + \frac{\partial L}{\partial t} \right) \right] dV \\
&= \int_{\partial \mathcal{B}} T_i \dot{x}_i \, dS + \int_{\mathcal{B}} \left(F_i \dot{x}_i - \frac{\partial L}{\partial t} \right) dV
\end{aligned} \tag{2.78}
$$

with the use of equations (2.62)–(2.64). Then, if $\partial L/\partial t = 0$, the rate of change of energy equals the rate of working of the external forces, as required.

C *Rotational invariance*

If L is invariant under rotation of the x_i coordinate system, it follows that angular momentum is conserved. The total angular momentum of the body is defined as

$$H_i(\mathcal{B}, t) = \int_{\mathcal{B}} \varepsilon_{ijk} x_j P_k\, dV. \tag{2.79}$$

Then

$$
\begin{aligned}
\frac{\partial}{\partial t} H_i(\mathcal{B}, t) &= \int_{\mathcal{B}} \varepsilon_{ijk} \left[\dot{x}_j P_k + x_j \left(\partial_\alpha T_{k\alpha} + \frac{\partial L}{\partial x_k} + F_k \right) \right] dV \\
&= \int_{\mathcal{B}} \varepsilon_{ijk} x_j F_k\, dV + \int_{\partial \mathcal{B}} \varepsilon_{ijk} x_j T_k\, dS + \int_{\mathcal{B}} \varepsilon_{ijk} M_{jk}\, dV,
\end{aligned} \tag{2.80}
$$

where

$$M_{jk} = \dot{x}_j P_k - \partial_\alpha x_j T_{k\alpha} + x_j \frac{\partial L}{\partial x_k} = \dot{x}_j \frac{\partial L}{\partial \dot{x}_k} + \partial_\alpha x_j \frac{\partial L}{\partial(\partial_\alpha x_k)} + x_j \frac{\partial L}{\partial x_k}. \quad (2.81)$$

The first two terms in equation (2.80) represent the torques of the external body force and surface traction. The last term vanishes provided M_{jk} is symmetric. If this condition is met, we have the required result that the rate of change of angular momentum equals the total torque on the body.

Let x_i' be a new set of spatial coordinates obtained from x_i by a rigid rotation. Then $x_i' = R_{ij} x_j$ with (R_{ij}) an orthogonal matrix. The rotational invariance principle now states that L is the same whether expressed as a function of the x_i or the x_i', i.e. that

$$L(\dot{x}_i', \partial_\alpha x_i', x_i', X_\alpha, t) = L(\dot{x}_i, \partial_\alpha x_i, x_i, X_\alpha, t).$$

This must hold for an arbitrary choice of coordinate system, so that an equivalent statement is that

$$L(R_{ij} \dot{x}_j, R_{ij} \partial_\alpha x_j, R_{ij} x_j, X_\alpha, t)$$

is independent of (R_{ij}) for all orthogonal matrices $\mathbf{R}$. In determining the implications of this we cannot directly set $\partial L/\partial R_{ij}$ equal to zero, since the elements R_{ij} are not independent, but satisfy the constraints

$$\phi_{lm} \equiv R_{ln} R_{mn} - \delta_{lm} = 0.$$

Introducing Lagrange multipliers λ^{lm} gives the required conditions:

$$\frac{\partial L}{\partial R_{ij}} + \lambda^{lm} \frac{\partial \phi_{lm}}{\partial R_{ij}} = 0,$$

i.e.

$$\frac{\partial L}{\partial R_{ij}} = -(\lambda^{im} + \lambda^{mi}) R_{mj},$$

so that

$$\frac{\partial L}{\partial R_{ij}} R_{kj} = -(\lambda^{ik} + \lambda^{ki}).$$

Thus the left-hand side here is symmetric between i and k, giving the three conditions associated with the three rotational degrees of freedom. It follows then that

$$\frac{\partial L}{\partial R_{ij}} R_{kj} = \left(\frac{\partial L}{\partial \dot{x}_i'} \dot{x}_j + \frac{\partial L}{\partial(\partial_\alpha x_i')} \partial_\alpha x_j + \frac{\partial L}{\partial x_i'} x_j \right) R_{kj} = M_{ki}',$$

where M_{ki}' corresponds to definition (2.81) in terms of the primed coordinates. Hence M_{ki}' (and therefore M_{jk}, which is a particular case of

M_{jk}' when $\mathbf{R} = \mathbf{1}$) is indeed symmetric. Thus we have shown that the invariance of L under rotations of the x_i coordinate system guarantees the usual conservation law of angular momentum.

D *Consequences of invariances*

From now on we shall assume that our system obeys the above three invariance principles. The consequences of the first two are trivial, namely that L is not explicitly dependent on the variables x_i and t. The restrictions on L that follow from rotational invariance, however, are less immediate. If we suppose that L has the form of equation (2.61), then the invariance implies that W is a function not of the nine deformation gradient components $(\partial_\alpha x_i)$ but only of the six strain components, $(E_{\alpha\beta})$.

We can see this geometrically as follows. For fixed α, the three quantities $(\partial_\alpha x_i)$ are components of a vector $\mathbf{v}_\alpha$ with respect to the x_i coordinates. The vector $\mathbf{v}_\alpha$ represents in fact the deformation from the reference configuration of a unit vector along the X_α coordinate direction if this vector is treated as an infinitesimal material element. The three vectors $\{\mathbf{v}_\alpha: \alpha = 1, 2, 3\}$ are independent and form a right-handed basis if we assume that $0 < j < \infty$. Then rotational invariance states that W, which is a function of the components of these three vectors, is independent of the coordinate frame in which these components are measured. In other words, W is a function of the rigid framework formed by the $\{\mathbf{v}_\alpha\}$ and is independent of the orientation of this framework with respect to the coordinate directions. Now a rigid framework of three vectors in three dimensions is fixed, apart from a rigid rotation and a reflection, by giving the lengths of the three vectors and their three scalar products when taken in pairs. These six quantities are $\mathbf{v}_\alpha \cdot \mathbf{v}_\beta = \partial_\alpha x_i \partial_\beta x_i = \delta_{\alpha\beta} + 2E_{\alpha\beta}$, so that W must be expressible as a function of the $(E_{\alpha\beta})$. The freedom of a reflection mentioned above cannot occur in fact since we know that $\{\mathbf{v}_\alpha\}$ form a right-handed triad. If negative values of j were allowed, corresponding to a material inversion, we would have to allow for this reflection possibility, and in this case W would depend on the sign of j as well as on the strain tensor.

The same result can be obtained purely algebraically by using the polar decomposition theorem (eq. (2.55)) applied to the deformation gradients: $\partial_\alpha x_i = R_{ij} U_{j\alpha}$, where $\mathbf{R}$ is a certain proper orthogonal matrix and $\mathbf{U}$ is a positive-definite symmetric matrix. Rotational invariance requires that

$$W(\partial_\alpha x_i) = W(R_{ij}' \partial_\alpha x_j) \tag{2.82}$$

for all orthogonal matrices $\mathbf{R}'$. If we now choose $\mathbf{R}' = \mathbf{R}^T$, then $R_{ij}' \partial_\alpha x_j = R_{ij}' R_{jk} U_{k\alpha} = U_{i\alpha}$ so that $W(\partial_\alpha x_i) = W(U_{i\alpha})$, which shows that W is a function of $\mathbf{U}$ only. But since $\mathbf{U}$ is the positive-definite square root of the matrix

$\mathbf{F}^T\mathbf{F} = \mathbf{1} + 2\mathbf{E}$, it follows that W can equally well be regarded as a function of the strain tensor $\mathbf{E}$.

Equations (2.66) and (2.69) for the stresses can now be rewritten in a form more useful when W is given as a function of $(E_{\alpha\beta})$:

$$T_{i\alpha} = \frac{\partial W}{\partial E_{\beta\gamma}} \frac{\partial E_{\beta\gamma}}{\partial(\partial_\alpha x_i)} = \frac{\partial W}{\partial E_{\beta\gamma}} \tfrac{1}{2}(\partial_\alpha x_i \delta_{\alpha\gamma} + \partial_\gamma x_i \delta_{\alpha\beta}), \tag{2.83}$$

$$\sigma_{ij} = j^{-1} \partial_\alpha x_i \partial_\beta x_j \frac{\partial W}{\partial E_{(\alpha\beta)}}, \tag{2.84}$$

where we write

$$\frac{\partial W}{\partial E_{(\alpha\beta)}} = \frac{1}{2}\left(\frac{\partial W}{\partial E_{\alpha\beta}} + \frac{\partial W}{\partial E_{\beta\alpha}}\right), \tag{2.85}$$

The significance of equation (2.84) is that it expresses the stresses in terms of the finite deformation gradient and strain tensors once the strain energy function is specified. It is therefore the non-linear analogue of Hooke's law.

E *Material homogeneity*

Most conventional situations concern materials that are uniform in the sense that their physical composition and structure are, on a macroscopic scale, constant throughout the body. This does not necessarily mean, however, that the strain energy density $W(E_{\alpha\beta}, \mathbf{X})$ is independent of $\mathbf{X}$, since the reference configuration might be non-uniformly strained. We shall refer to a body as being homogeneous if there exists a reference configuration such that W is in fact independent of $\mathbf{X}$ when the strains are measured with respect to this reference. Such a reference configuration is called a (*global*) *uniform reference*.

It is important to recognize that a body made of uniform material is not necessarily homogeneous. Consider, for example, a cylinder with a Volterra edge dislocation (Fig. 1.1(b)) when the undislocated cylinder (Fig. 1.1(a)) is a homogeneous body in a uniform reference state. Then if we are concerned with deformations of the dislocated cylinder, the strain energy density depends on the particle when strains are measured from the state in Figure 1.1(b), because of the presence of the initial stress-field of the dislocation. Furthermore, there is no way in which the dislocated cylinder can be deformed continuously into a uniform reference state. The only way this can be achieved is by making a certain cut in the cylinder and applying a deformation which is discontinuous across the cut. In this way we see a connection between dislocations and certain types of inhomogeneity – that is, certain ways in which W can depend on $\mathbf{X}$. This connection forms the basis for Noll's approach to dislocation theory,[12,13] to which we shall return in Chapter 7. For the moment we shall restrict our attention to homogeneous materials.

If one uniform reference exists, then in fact many exist. For let (X_α) be a uniform reference and let (Y_β) be another state obtained from it by a homogeneous deformation:

$$Y_\beta = L_{\beta\alpha}X_\alpha + C_\beta, \tag{2.86}$$

where $L_{\beta\alpha}$ and C_β are constants. The deformation gradients and strains with respect to these two configurations are related by

$$\frac{\partial x_i}{\partial X_\alpha} = \frac{\partial x_i}{\partial Y_\beta}L_{\beta\alpha}, \tag{2.87}$$

$$E_{\alpha\beta}^{(x)} = E_{\gamma\delta}^{(y)}L_{\gamma\alpha}L_{\delta\beta} + \tfrac{1}{2}(L_{\gamma\alpha}L_{\gamma\beta} - \delta_{\alpha\beta}). \tag{2.88}$$

Then if W is a function of $\mathbf{E}^{(x)}$, independent of the particle $\mathbf{X}$, it is also a similar function of $\mathbf{E}^{(y)}$, independent of $\mathbf{X}$.

The two strain energy functions W_x and W_y with respect to the $\mathbf{X}$ and $\mathbf{Y}$ configurations are related by

$$W_x(E_{\alpha\beta}^{(x)}) = W_y(E_{\alpha\beta}^{(y)}) = W_x[E_{\gamma\delta}^{(y)}L_{\gamma\alpha}L_{\delta\beta} + \tfrac{1}{2}(L_{\gamma\alpha}L_{\gamma\beta} - \delta_{\alpha\beta})]. \tag{2.89}$$

By varying $\mathbf{L}$ we obtain access to a variety of reference states $\mathbf{Y}$; in fact there are nine degrees of freedom in $\mathbf{L}$, three of rotation and six of strain, and it seems reasonable to expect that we can find a reference configuration with some simplifying properties. We propose to show that in any physically reasonable situation there will exist a reference configuration for which the stresses vanish whenever the strains vanish. This configuration is called the *natural state*.

If $\mathbf{L}$ is symmetric, the deformation $\mathbf{X} \rightarrow \mathbf{Y}$ represents a pure strain. The six elements of a symmetric $\mathbf{L}$ are allowed to vary in the region $0 < \det \mathbf{L} < \infty$, bounded by the hypersurface $\det \mathbf{L} = 0$ and extending to infinity elsewhere. Now as the boundary $\det \mathbf{L} = 0$ is approached, the configuration $\mathbf{Y}$ becomes highly compressed, and for a fixed value of $\mathbf{E}^{(y)}$, $W_y(E_{\alpha\beta}^{(y)})$ must become very large. On the other hand, if any element of $\mathbf{L}$ becomes large, the configuration $\mathbf{Y}$ becomes distended, and again $W_y(E_{\alpha\beta}^{(y)})$ must be large. Thus, varying $\mathbf{L}$ with $\mathbf{E}^{(y)}$ fixed leads to a strain energy which diverges as any point on the boundary of the allowed region of $\mathbf{L}$ is approached, and hence there must exist at least one particular value of $\mathbf{L}$ within the region at which W_y is a minimum. This value of $\mathbf{L}$ depends, of course, on the strain $\mathbf{E}^{(y)}$; we choose $\mathbf{E}^{(y)}$ to be zero, so that the state of minimum energy is determined by

$$\partial W_y/\partial L_{(\lambda\mu)}|_{\mathbf{E}^{(y)}=0} = 0.$$

Now

$$\frac{\partial W_y}{\partial L_{(\lambda\mu)}}\bigg|_{E^{(y)}=0} = \frac{\partial W_x}{\partial E_{\alpha\beta}{}^{(x)}}\frac{\partial E_{\alpha\beta}{}^{(x)}}{\partial L_{(\lambda\mu)}}\bigg|_{E^{(y)}=0}$$

$$= \frac{\partial W_x}{\partial E_{\alpha\beta}{}^{(x)}}\tfrac{1}{4}(\delta_{\alpha\mu}L_{\lambda\beta}+\delta_{\beta\mu}L_{\lambda\alpha}+\delta_{\alpha\lambda}L_{\mu\beta}+\delta_{\beta\lambda}L_{\mu\alpha})$$

$$= \frac{\partial W_x}{\partial E_{(\alpha\beta)}{}^{(x)}}\tfrac{1}{2}(\delta_{\alpha\mu}L_{\lambda\beta}+\delta_{\alpha\lambda}L_{\mu\beta}).$$

Thus for the particular state that concerns us, $\partial W_x/\partial E_{(\alpha\beta)}{}^{(x)} = 0$. It now follows that

$$\frac{\partial W_y}{\partial E_{(\lambda\mu)}{}^{(y)}}\bigg|_{E^{(y)}=0} = \frac{\partial W_x}{\partial E_{\alpha\beta}{}^{(x)}}\frac{\partial E_{\alpha\beta}{}^{(x)}}{\partial E_{(\lambda\mu)}{}^{(y)}} = \frac{\partial W_x}{\partial E_{(\alpha\beta)}{}^{(x)}}L_{\lambda\alpha}L_{\mu\beta} = 0. \qquad (2.90)$$

Hence, if we use this configuration (Y_α) as the reference state, equation (2.84) shows that the stresses vanish when $\mathbf{E}^{(y)} = \mathbf{0}$.

2.8
LINEARIZATION

In this final section we shall show how the equations of classical elasticity are obtained as an approximation to the non-linear theory when the deformation is small. Here we shall take the two coordinate systems for x_i and X_α to be identical, and make no distinction between greek and latin indices. Then we define the displacement vector $\mathbf{u}$ as having components

$$u_i = x_i - X_i, \qquad (2.91)$$

the difference between the deformed and undeformed positions of the particle $\mathbf{X}$. Regarding u_i as a function of $\mathbf{X}$ and t, the particle velocity and acceleration are $\partial u_i/\partial t$ and $\partial^2 u_i/\partial t^2$. The classical theory of elasticity is concerned with the case of small displacement gradients:

$$\partial_\alpha u_i \ll 1,$$

which condition does in fact hold in the majority of practical cases. Then, with this assumption, we can proceed to linearize the theory, keeping only terms of the lowest order in the displacement gradients.

Using equation (2.47), we get, to first order,

$$\partial_\alpha x_i = \delta_{i\alpha} + \partial_\alpha u_i, \qquad \partial_j X_\alpha = \delta_{j\alpha} - \partial_j u_\alpha.$$

(In equations of this type it is unnecessary to distinguish between $\partial u_\alpha/\partial x_j$ and $\partial u_\alpha/\partial X_j$, which are equal to first order.) Then from equations (2.49) and (2.50) the two strain tensors are equal to first order and are given by the usual expressions

$$e_{ik} = E_{ik} = \tfrac{1}{2}(\partial_i u_k + \partial_k u_i).$$

The dilatation, from equation (2.9), is to first order

$$j = \rho_0/\rho = 1 + \partial_k u_k = 1 + e_{kk}.$$

The boundary conditions and equations of motion, from equations (2.71) and (2.73), are

$$\sigma_{ij} n_j = t_i, \qquad -\rho \ddot{u}_i + \partial_j \sigma_{ij} + f_i = 0.$$

Approximating equation (2.84) to first order we obtain the stress-strain relation

$$\sigma_{ij} = \partial W / \partial e_{(ij)}.$$

For small strains we suppose that W can be expanded in a Taylor series about $\mathbf{e} = \mathbf{0}$. Keeping only the first terms we have

$$W = a + b_{ij} e_{ij} + \tfrac{1}{2} c_{ijkl} e_{ij} e_{kl} + \ldots . \tag{2.92}$$

For the symmetry of the strain tensor, it is meaningful to use only co-efficients which satisfy the relations

$$b_{ij} = b_{ji}, \qquad c_{ijkl} = c_{jikl} = c_{ijlk} = c_{klij}.$$

If now the undisplaced configuration (X) is the natural state,

$$\partial W / \partial e_{(ij)} \big|_{e=0} = b_{ij} = 0,$$

and this leaves the stresses in the simple form

$$\sigma_{ij} = c_{ijkl} e_{kl}.$$

Thus we have retrieved all the equations of linear elasticity.

PROBLEMS

1 Show that equation (2.14) implies the existence of a function ω such that $\partial_2 e_{11} - \partial_1 e_{12} = \partial_1 \omega$ and $\partial_1 e_{22} - \partial_2 e_{12} = -\partial_2 \omega$. Hence show that there also exist functions u_1 and u_2 such that equations (2.13) hold. Express ω in terms of u_1 and u_2 and give its physical interpretation.

2 From the second of equations (2.19) deduce that there exists a function $A(r, \theta)$ such that $\hat{\sigma}_{r\theta} = -\partial_\theta (A/r^2)$ and $r\hat{\sigma}_{\theta\theta} = \partial_r A$. Using this result in the first of equations (2.19) show that there exists a function $C(r, \theta)$ such that $r\hat{\sigma}_{rr} - B = \partial_\theta C$ and $\hat{\sigma}_{r\theta} = -\partial_r C$. Here $B(r, \theta)$ is defined by the equation $r\partial_r B = \partial_r A$. Hence deduce the representations (2.18).

3 Write down the equilibrium equation for antiplane strain and show that there exists a function $\psi(x_1, x_2)$ such that $\sigma_{13} = \partial_2 \psi$ and $\sigma_{23} = \partial_1 \psi$. Show that ψ is harmonic ($\nabla^2 \psi = 0$). What is the relationship between ψ and μu_3? Find ψ for the stress-field of a screw dislocation which is given in equation (3.2).

4 Show that a traction boundary-value problem in antiplane strain can be reduced either to a Neumann problem for u_3 or to a Dirichlet problem for ψ (with ψ as in Prob. 3). In the second case, are the boundary values of ψ single-valued? Consider separately the case of multiply connected regions.

5 Show that for a state of plane strain there exist functions $\phi(x_1, x_2)$ and $\psi(x_1, x_2)$ (called *displacement potentials*) such that $u_1 = \partial_1\phi + \partial_2\psi$ and $u_2 = \partial_2\phi - \partial_1\psi$. Express the stress in terms of ϕ and ψ. Hence show that the equilibrium equations reduce to the pair of equations

$$\partial_1\Phi + \partial_2\Psi = 0, \qquad \partial_2\Phi - \partial_1\Psi = 0,$$

where $\Phi = (\lambda + 2\mu)\nabla^2\phi$ and $\Psi = \mu\nabla^2\psi$.

6 Deduce equation (2.31) from equation (2.30).

7 A point force $\mathbf{F}$ acts at the origin in an infinite isotropic medium and a second point force $-\mathbf{F}$ acts at the point $\mathbf{h}$. If $\mathbf{h}$ is small, show that the displacement at $\mathbf{r}$ has an l component equal to $-F_m h_k \partial_k u_{lm}(\mathbf{r}, \mathbf{0})$. Evaluate the displacement fields explicitly, using equation (2.44), when (a) $\mathbf{F}$ and $\mathbf{h}$ are along the x-axis and (b) $\mathbf{h}$ is along the x-axis and $\mathbf{F}$ along the y-axis. (Such applied loads are called *double forces*.)

8 In Problem 7 find the solution which is obtained by superimposing three double forces without moment (i.e. case (a)), one in each of the coordinate directions. The displacement field should be

$$u_l = \frac{Fh}{4\pi(\lambda + 2\mu)} \frac{x_l}{r^3}.$$

(This applied load is called a *centre of dilatation*.)

9 In Problem 7, superimpose two double forces of type (b) with equal torques about the z-axis, with $\mathbf{h}$ along the x- and y-axes and $\mathbf{F}$ along the y- and $(-x)$-axes respectively. The displacement field should be

$$u_l = \frac{Fh(2\lambda + 5\mu)}{8\pi\mu(\lambda + 2\mu)} \frac{x_1\delta_{12} - x_2\delta_{11}}{R^3}.$$

10 Look for a solution of the plane strain equations in which $u_1 = x_1 f(r)$ and $u_2 = x_2 f(r)$. Show that $f'' + 3f'/r = 0$ and hence find $f(r)$. What boundary-value problems can be solved using this solution?

11 Find the plane strain solutions for which $u_r = 0$ and $u_\theta = g(r)$ depends only on r. A hollow circular cylinder is deformed in plane strain by means of shear tractions uniformly distributed over its inner and outer surfaces. Find the resulting displacement field (see Fig. 2.2).

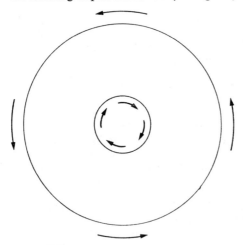

FIGURE 2.2

12 An infinite isotropic elastic solid has a spherical cavity centered at the origin. The cavity is filled with gas at pressure p. Find the resulting displacement field in the solid. How is the solution related to Problem 8?

13 Let $\bar{u}_{lm}(\mathbf{k})$ denote the three-dimensional Fourier transform of $u_{lm}(\mathbf{r}, \mathbf{r}')$ with respect to $\mathbf{r} - \mathbf{r}'$, in an infinite medium. Show from equation (2.41) that $c_{ijkl}k_jk_k\bar{u}_{lm} = \delta_{im}$. Write down these equations explicitly for an isotropic medium, and solve them for $\bar{u}_{lm}(\mathbf{k})$. Invert the transform to obtain u_{lm} as given in equation (2.41).

14 Establish the integral formulae at the end of §2.4c.

15 Calculate the strain tensors $E_{\alpha\beta}$ and e_{ik} when the deformation (2.46) has the following forms:

   (a) $x_i = \lambda X_i$;   (b) $x_1 = X_1 + KX_2,\ x_2 = X_2,\ x_3 = X_3.$

   Give physical interpretations of these two deformations.

16 Find $\mathbf{U}$, $\mathbf{V}$, and $\mathbf{R}$ in equation (2.55) for the two deformations in the previous problem (using $\mathbf{F}$ as the deformation gradient matrix, $F_{i\alpha} = \partial_\alpha x_i$).

17 Use equation (2.72) to show that $\partial j/\partial t = j\partial\dot{x}_i/\partial x_i$, where the time derivative is evaluated with the reference coordinate held fixed. Hence show that the mass conservation equation (2.53) is equivalent to the equation

$$\frac{\partial\rho}{\partial t} + \rho\frac{\partial\dot{x}_i}{\partial x_i} = 0.$$

18 Express j in terms of each of the two strain tensors (e_{ik}) and $(E_{\alpha\beta})$.

19 Let $d\mathbf{X}^{(1)}$ and $d\mathbf{X}^{(2)}$ be two small material elements in the reference configuration. Under the deformation $\mathbf{x} = \mathbf{x}(\mathbf{X}, t)$, the angle between these elements changes. Express this change in angle in terms of each of the strain tensors.

REFERENCES

1 General references on elasticity: Sokolnikoff, I.S., *Mathematical theory of elasticity* (New York: McGraw-Hill, 1956); Murnaghan, F.D., *Finite deformations of an elastic solid* (New York: Wiley, 1951); Green, A.E., and Zerna, W., *Theoretical elasticity* (Oxford, 1954); Love, A.E.H., *A treatise on the mathematical theory of elasticity* (Cambridge, 1927; New York: Dover, 1944); Landau, L.D., and Lifschitz, E.M., *Theory of elasticity* (Oxford: Pergamon, 1959); Fung, Y.C., *Foundations of solid mechanics* (Englewood Cliffs, N.J.: Prentice-Hall, 1965); Luré, A.I., *Three dimensional problems in the theory of elasticity* (New York: Interscience Publishers, 1964); Sneddon, I.N., and Hill, R. (eds.), *Progress in solid mechanics*, vols. 1–4 (Amsterdam: North Holland, 1960–3); Muskhelishvili, N.I., *Some basic problems of the mathematical theory of elasticity*, 4th ed. (Amsterdam: Noordhoff, 1963)

2 Sokolnikoff, I.S., ref. 1, pp. 62–3; Nye, J.F., *Physical properties of crystals* (Oxford, 1957); Wooster, W.A., *Crystal physics* (Cambridge, 1938)

3 Zeilon, N., Ark. Mat. **6**, no. 23 (1911)

4 Leibfried, G., Z. Phys. **135** (1953), 23

5 de Wit, R., Solid State Phys. **10** (1960), 247

6 Luré, A.I., ref. 1, chap. 2

7 Kröner, E., Z. Phys. **136** (1953), 402

8 Fredholm, I., Acta Math. Stockholm **23** (1900), 1

9 Gebbia, M., Ann. di Math. **10** (1904), 157

10 General references on continuum mechanics: Truesdell, C., *The elements of continuum mechanics* (Berlin: Springer-Verlag, 1966); Eringen, A.C., *Mechanics of continua* (New York: Wiley, 1967); *Non-linear theory of continuous media* (New York: McGraw-Hill, 1962); Leigh, D.C., *Nonlinear continuum mechanics* (New York: McGraw-Hill, 1968); Jaunzemis, W., *Continuum mechanics* (London: Macmillan, 1967); Truesdell, C., and Toupin, R., *Handbuch der Physik* III/1 (Berlin: Springer-Verlag, 1960); Truesdell, C., and Noll, W., *Handbuch der Physik* III/3 (Berlin: Springer-Verlag, 1965)
11 Toupin, R., Arch. Rat. Mech. Anal. **17** (1964), 85
12 Truesdell, C., and Noll, W., ref. 10, §34
13 Noll, W., Arch Rat. Mech. Anal. **27** (1967), 1

3

BASIC THEORY OF
SINGLE DISLOCATIONS

3.1
STRAIGHT SCREW DISLOCATION

A *Stress-field*

We consider a screw dislocation in the Volterra sense[1] in a hollow circular cylinder of inner and outer radii r_0 and R respectively, made of isotropic elastic material. This is illustrated in Figure 3.1 for a right-handed screw where $\boldsymbol{\xi}$ and $\mathbf{b}$ are taken as pointing in the positive z-direction. For definiteness we suppose that, in making the dislocation, the cylinder is cut on the half-plane $y = 0$, $x > 0$, the top face of the cut ($y = 0+$) being held fixed and the bottom face displaced by b in the z-direction, although as far as the resulting stress-field is concerned, the particular plane on which the cut is made is irrelevant.

If we use (r, θ, z) as cylindrical polar coordinates, the displacement component u_z increases from 0 to b as θ increases from 0 to 2π. It is natural to suppose that this increase occurs linearly with θ, so that $u_z = b\theta/2\pi$. We also assume, for the moment, that the deformation is antiplane, i.e. $u_x = u_y = 0$.

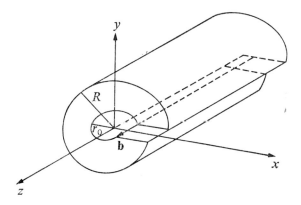

FIGURE 3.1

Corresponding to this displacement field, the only non-zero strain components (cf. eq. (2.1)) are

$$e_{zx} = \frac{1}{2}\frac{\partial u_z}{\partial x} = \frac{b}{4\pi}\frac{-y}{x^2+y^2},$$

$$e_{zy} = \frac{1}{2}\frac{\partial u_z}{\partial y} = \frac{b}{4\pi}\frac{x}{x^2+y^2}. \qquad (3.1)$$

From equation (2.8), the only non-zero stress components are then

$$\sigma_{zx} = -\frac{\mu b}{2\pi}\frac{y}{x^2+y^2} = -\frac{\mu b}{2\pi}\frac{\sin\theta}{r},$$

$$\sigma_{zy} = \frac{\mu b}{2\pi}\frac{x}{x^2+y^2} = \frac{\mu b}{2\pi}\frac{\cos\theta}{r}. \qquad (3.2)$$

These stresses take a transparent form if written in cylindrical components. Their physical components are

$$\hat{\sigma}_{z\theta} = \mu b/2\pi r, \qquad \hat{\sigma}_{zr} = \hat{\sigma}_{zz} = \hat{\sigma}_{rr} = \hat{\sigma}_{r\theta} = \hat{\sigma}_{\theta\theta} = 0. \qquad (3.3)$$

We need to verify that the above displacements and stresses satisfy the equilibrium equations and the boundary conditions appropriate to the screw dislocation of Figure 3.1. For antiplane strain, the equilibrium equations reduce simply to the requirement that u_z be harmonic (eq. (2.22)), which $b\theta/2\pi$ is. u_z has the correct discontinuity across the cut, and the second condition on the cut, that the tractions on the two faces be equal and opposite, is satisfied since the stress-field is single-valued. Finally we need the surfaces $r = r_0$ and $r = R$ to be traction-free, or in other words $\hat{\sigma}_{rr}$, $\hat{\sigma}_{r\theta}$, and $\hat{\sigma}_{rz}$ should all vanish there; these stress components are, in fact, zero everywhere. Thus all the requirements of the problem are indeed met.

B *Energy and the dislocation core*

The total elastic energy in the cylinder per unit length in the z-direction may be calculated most simply using equation (2.31). We take the boundary $\partial\mathcal{B}$ to consist of the two cylindrical surfaces plus the two faces of the cut: $y = 0\pm$, $x > 0$. On the former, the traction $\mathbf{t}$ vanishes. On $y = 0+$, $t_z = -\sigma_{yz}(x,0)$, and on $y = 0-$, $t_z = +\sigma_{yz}(x,0)$. Thus the energy per unit length is given by

$$\frac{U^d}{L} = \frac{1}{2}\int_{r_0}^{R}\sigma_{yz}(x,0)\,[u_z(x,0-)-u_z(x,0+)]dx = \frac{1}{2}\int_{r_0}^{R}\sigma_{yz}(x,0)b\,dx$$

$$= \frac{\mu b^2}{4\pi}\ln\frac{R}{r_0}. \qquad (3.4)$$

The divergence of this expression as $R \to \infty$ or $r_0 \to 0$ prevents these cases from being discussed as limiting cases of the present results, even though the stress and deformation fields are independent of r_0 and R. As we shall see later, the divergence with large R disappears with more complicated systems of dislocations which have no net Burgers vector (for example, pairs of opposite dislocations or systems of closed loops). In applying these results to a practical case of a dislocation in a crystal, formula (3.4) would be used with R being twice the distance of the screw from the nearest part of the boundary (cf. eq. (3.49)). For a crystal containing many dislocations, R would be taken as the average spacing between them (cf. eq. (3.46)).

In actual application to dislocations in crystals, the problem of an appropriate choice for the *core radius* r_0 raises the whole question of the validity of using the Volterra hollow-core model for crystal dislocations. Equation (3.4) shows that the energy of a Volterra screw decreases as r_0 increases. However, as r_0 increases, so do the areas of the two cylindrical surfaces, hence so also do their surface energies. The total of strain energy plus surface energy is a minimum at some equilibrium value of r_0 (see Prob. 3.2).[2] For a typical material with lattice spacing a, the equilibrium value of r_0 is about $a/9$ for a simple dislocation ($b = a$) and increases to a when $b = 3a$. We conclude from this, since the equilibrium value of r_0 is so small, that it is highly unlikely that the hollow-core model is applicable to the simple dislocation. But it becomes increasingly plausible to use it for dislocations of large Burgers vector ($b \geq 3a$) when the equilibrium core radius exceeds one lattice spacing. The above model, however, not only assumes a hollow core but also uses linear elasticity, and we need to make the additional requirement that the strains are sufficiently small – say less than 0.05 – before the assumptions are properly legitimate. The maximum strain is attained at $r = r_0$, and is equal to $b/4\pi r_0$, so we must require in addition that $r_0 > 5b/\pi$. Using the equilibrium radius from Problem 3.2 implies approximately that $b \geq 12a$. Hence the above Volterra model can only be applied with confidence to composite dislocations with about twelve times the smallest possible Burgers vector.

For the single dislocation ($b = a$), the strain of equation (3.1) reaches the value 0.05 when r is of the order of $5b/3$. Assuming this to be a valid limit to the applicability of linear elasticity, we conclude that the material outside a cylinder of radius $\sim 5b/3$ can be treated in the way we have done earlier, but that within this cylinder an atomistic calculation should be used. (The cylinder $r_0 = 5b/3$ is then no longer a free surface.) Estimates that have been made[3] by such means indicate that the total energy is still given by equation (3.4), with r_0 of the order $b/2$. The semi-continuum model of Peierls and Nabarro[4] which will be described in Chapter 4 makes a similar prediction.

Within the theory of continuum mechanics, the most satisfactory way of dealing with single dislocations is to spread the distribution of Burgers vector throughout some neighbourhood of the dislocation line, so that the line of singularities is removed and replaced by a tube of weaker singularities. The Peierls-Nabarro model accomplishes this in one particular way, and in §3.6 and Chapter 8 we shall discuss such tubes of dislocations in a more general context, which is not tied to one particular model of the structure of the core tube. We shall there obtain expressions such as equation (3.4) and its corresponding result for edge dislocations for the energy of a dislocation tube, and r_0 will be a certain average radius of the core distribution.

c Correction for finite cylinder length

The stress-field (3.2) corresponds to non-zero tractions across planes $z = $ constant. Hence, if the dislocation occurs in a cylinder of finite length $(0 < z < L)$, end tractions are needed to support the antiplane state. The traction on $z = L$, say, is in the θ-direction, of magnitude $t_\theta = \hat{\sigma}_{\theta z} = \mu b / 2\pi r$. This has zero resultant force, but a non-zero resultant torque about the cylinder axis, of magnitude

$$M_z = \int_{r_0}^{R} r t_\theta 2\pi r dr = \frac{1}{2}\mu b(R^2 - r_0^2). \tag{3.5}$$

If this torque is not applied, which would usually be the case in practice, the cylinder will twist. The exact deformation during the twist can be obtained by solving the boundary-value problem with tractions $-t_\theta$ on the ends of the cylinder. However, if $L \gg R$, we can invoke St Venant's principle to claim that at distances from the ends of the cylinder which are large compared to the radius R the stress and strain fields produced by the torque M_z are independent of the way in which the tractions producing the torque are distributed over the end surfaces. Consequently, at such distances from the ends, the twist deformation produced by tractions $-t_\theta$ is almost identical with the well-known torsion deformation for a torque $-M_z$. This latter corresponds to an angular displacement $\hat{u}_\theta = \alpha rz$, where α is the twist per unit length, and to a total torque of $\frac{1}{2}\pi\mu\alpha(R^4 - r_0^4)$. Setting this equal to $-M_z$ in equation (3.5) gives

$$\alpha = -b/\pi(R^2 + r_0^2).$$

The only non-zero stress component for the torsion displacement is $\hat{\sigma}_{\theta z} = \alpha\mu r$. Adding this to the stress-field of equation (3.3) gives the total stress

$$\hat{\sigma}_{\theta z} = \frac{\mu b}{2\pi}\left(\frac{1}{r} - \frac{2r}{R^2 + r_0^2}\right). \tag{3.6}$$

The Cartesian components are

$$\sigma_{zx} = -\frac{\mu b}{2\pi}\left(\frac{y}{x^2+y^2} - \frac{2y}{R^2+r_0^2}\right),$$

$$\sigma_{zy} = \frac{\mu b}{2\pi}\left(\frac{x}{x^2+y^2} - \frac{2x}{R^2+r_0^2}\right). \tag{3.7}$$

It is of interest to calculate the effect of this change on the elastic energy. We obtain

$$\frac{U^d}{L} = \frac{1}{2}\int_{r_0}^{R}\sigma_{yz}(x, 0)b\,dx = \frac{\mu b^2}{4\pi}\left(\ln\frac{R}{r_0} - \frac{R^2-r_0^2}{R^2+r_0^2}\right)$$

$$\approx \frac{\mu b^2}{4\pi}\left(\ln\frac{R}{r_0} - 1\right) \tag{3.8}$$

for $R \gg r_0$.

3.2
STRAIGHT EDGE DISLOCATION

A *Stress-field*

Consider the Volterra edge dislocation of Figure 3.2 in a hollow circular cylinder $r_0 < r < R$ parallel to the z-axis. If we take $\boldsymbol{\xi} = (0, 0, 1)$, then the Burgers vector is $\mathbf{b} = (b, 0, 0)$ for the edge shown. For a cylinder of infinite length, the deformation will be plane strain.

Taking (r, θ) as polar coordinates in the xy-plane, the Airy stress function $\Psi(r, \theta)$ satisfies the equation $\nabla^4\Psi = 0$ (eq. (2.20)). Thus $\nabla^2\Psi$ is harmonic, and so has an expansion of the form

$$\nabla^2\Psi = \alpha_0+\beta_0\ln r+ \sum_{n=1}^{\infty} [(\alpha_n r^n+\beta_n r^{-n})\sin n\theta+(\gamma_n r^n+\delta_n r^{-n})\cos n\theta]. \tag{3.9}$$

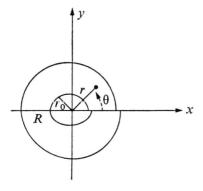

FIGURE 3.2

(Note that $\nabla^2\Psi = \hat{\sigma}_{rr}+\hat{\sigma}_{\theta\theta}$; hence $\nabla^2\Psi$ is a single-valued function of θ.) Integrating this then gives

$$\Psi = \frac{\alpha_0}{4}r^2 + \frac{\beta_0}{4}r^2(\ln r-1) + \left(\frac{\beta_1}{2}r\ln r + \frac{\alpha_1}{8}r^3\right)\sin\theta$$

$$+\left(\frac{\delta_1}{2}r\ln r + \frac{\gamma_1}{8}r^3\right)\cos\theta + \sum_{n=2}^{\infty}\left[\left(\frac{\alpha_n}{4(n+1)}r^{n+2} - \frac{\beta_n}{4(n-1)}r^{-n+2}\right)\sin n\theta\right.$$

$$+\left.\left(\frac{\gamma_n}{4(n+1)}r^{n+2} - \frac{\delta_n}{4(n-1)}r^{-n+2}\right)\cos n\theta\right] + \Psi_1(r, \theta), \tag{3.10}$$

where Ψ_1 is a harmonic function, and so also has an expansion of the type of equation (3.9). For the dislocation problem we expect the stress $\hat{\sigma}_{\theta\theta}$ to be large and negative on $\theta = \pi/2$ and large and positive on $\theta = -\pi/2$ while $\hat{\sigma}_{r\theta}$ is large on the planes $\theta = 0, \pi$ and zero on $\theta = \pm\pi/2$. Thus we might expect $\hat{\sigma}_{\theta\theta}$ to have a $\sin\theta$ dependence and $\hat{\sigma}_{r\theta}$ a $\cos\theta$ one, which are both accomplished if we take $\Psi \propto \sin\theta$. Collecting all of the terms in equation (3.10) which are proportional to $\sin\theta$ we have

$$\Psi = \left(\frac{\beta_1}{2}r\ln r + \frac{\alpha_1}{8}r^3+\alpha_1'r+\beta_1'r^{-1}\right)\sin\theta, \tag{3.11}$$

with the last two terms coming from Ψ_1. Since the third term can be re-written as $\alpha_1'y$, it makes no contribution to the stresses which are second derivatives of Ψ, and hence can be dropped.

Using equation (2.18) with this stress function gives

$$\hat{\sigma}_{rr} = \left(\frac{\beta_1}{2}\frac{1}{r} + \frac{\alpha_1}{4}r - \frac{2\beta_1'}{r^3}\right)\sin\theta,$$

$$\hat{\sigma}_{\theta\theta} = \left(\frac{\beta_1}{2}\frac{1}{r} + \frac{3\alpha_1}{4}r + \frac{2\beta_1'}{r^3}\right)\sin\theta, \tag{3.12}$$

$$\hat{\sigma}_{r\theta} = -\left(\frac{\beta_1}{2}\frac{1}{r} + \frac{\alpha_1}{4}r - \frac{2\beta_1'}{r^3}\right)\cos\theta.$$

The free-surface conditions are that $\hat{\sigma}_{rr} = \hat{\sigma}_{r\theta} = 0$ on both $r = r_0$ and $r = R$, and these are satisfied if $\beta_1/2r+\alpha_1r/4-2\beta_1'/r^3 = 0$ at these two values of r. Solving for α_1 and β_1' in terms of β_1 we obtain

$$\alpha_1 = -\frac{2\beta_1}{R^2+r_0^2}, \qquad \beta_1' = \frac{\beta_1}{4}\frac{r_0^2R^2}{r_0^2+R^2}.$$

In most cases of interest, $R \gg r_0$, and we can approximate these as $\alpha_1 \approx -2\beta_1/R^2$ and $\beta_1' \approx \beta_1r_0^2/4$. The stress function is then

$$\Psi = \frac{\beta_1}{2}\left(r\ln r - \frac{r^3}{2R^2} + \frac{r_0{}^2}{2r}\right)\sin\theta \tag{3.13}$$

and the stress becomes

$$\hat{\sigma}_{rr} = \frac{\beta_1}{2}\left(\frac{1}{r} - \frac{r}{R^2} - \frac{r_0{}^2}{r^3}\right)\sin\theta,$$

$$\hat{\sigma}_{\theta\theta} = \frac{\beta_1}{2}\left(\frac{1}{r} - \frac{3r}{R^2} + \frac{r_0{}^2}{r^3}\right)\sin\theta, \tag{3.14}$$

$$\hat{\sigma}_{r\theta} = -\frac{\beta_1}{2}\left(\frac{1}{r} - \frac{r}{R^2} - \frac{r_0{}^2}{r^3}\right)\cos\theta.$$

The remaining constant β_1 is determined from the condition that the change in the displacement field on encircling the z-axis must equal the Burgers vector. To avoid some algebraic complications, we suppose that we can find a value of r such that $r_0 \ll r \ll R$, so that the last two terms in the stresses are small, and we can approximate them as

$$\hat{\sigma}_{rr} \approx \frac{\beta_1}{2}\frac{1}{r}\sin\theta, \qquad \hat{\sigma}_{\theta\theta} \approx \frac{\beta_1}{2}\frac{1}{r}\sin\theta, \qquad \hat{\sigma}_{r\theta} \approx -\frac{\beta_1}{2}\frac{1}{r}\cos\theta. \tag{3.15}$$

(In fact the two terms we have neglected lead to single-valued displacements, so that the value of β_1 we shall find is exactly correct.) The Cartesian components of stress, obtained either by transforming equation (3.15) or directly from the stress function via equation (2.12), are for $r_0 \ll r \ll R$

$$\sigma_{xx} \approx \frac{\beta_1}{2}\frac{y(3x^2+y^2)}{(x^2+y^2)^2}, \qquad \sigma_{yy} \approx -\frac{\beta_1}{2}\frac{y(x^2-y^2)}{(x^2+y^2)^2},$$

$$\sigma_{xy} \approx -\frac{\beta_1}{2}\frac{x(x^2-y^2)}{(x^2+y^2)^2}, \qquad \sigma_{zz} \approx \nu\beta_1\frac{y}{x^2+y^2}. \tag{3.16}$$

From equation (2.11) the strains turn out to be

$$e_{xx} = \frac{\beta_1}{4\mu}\left(\frac{y(3x^2+y^2)}{(x^2+y^2)^2} - 2\nu\frac{y}{x^2+y^2}\right),$$

$$e_{yy} = -\frac{\beta_1}{4\mu}\left(\frac{y(x^2-y^2)}{(x^2+y^2)^2} + 2\nu\frac{y}{x^2+y^2}\right), \tag{3.17}$$

$$e_{xy} = -\frac{\beta_1}{4\mu}\frac{x(x^2-y^2)}{(x^2+y^2)^2}.$$

Finally integrating equations (2.13) gives the displacement components

$$u_x = \frac{\beta_1}{4\mu}\left(-2(1-\nu)\tan^{-1}\frac{y}{x} - \frac{xy}{x^2+y^2}\right) + Ay + B,$$

$$u_y = \frac{\beta_1}{4\mu}\left(\frac{1-2\nu}{2}\ln(x^2+y^2) + \frac{x^2}{x^2+y^2}\right) - Ax + C.$$

(3.18)

Here A, B, and C are constants of integration, and correspond to a rigid body displacement. Now as θ increases from 0 to 2π, u_x in equation (3.18) increases by $-\pi\beta_1(1-\nu)/\mu$, which equals b, so that

$$\beta_1 = -\mu b/[\pi(1-\nu)].$$

(3.19)

When the cylinder has finite length, the above plane strain deformation requires tractions on the ends of the cylinder in order to support it, just as for a screw dislocation. For the edge, the end tractions are normal to the ends, of magnitude $\pm\sigma_{zz}$. From equation (2.10), for plane strain

$$\sigma_{zz} = \nu(\hat{\sigma}_{rr}+\hat{\sigma}_{\theta\theta}) = \nu\beta_1\left(\frac{1}{r} - \frac{2r}{R^2+r_0{}^2}\right)\sin\theta$$

after using equation (3.12). It is easily seen that this traction has both zero resultant force and zero resultant torque. Consequently, by St Venant's principle, the effect of leaving the cylinder ends free will produce a negligible change in the stress-field at distances away from the ends which are large compared to R. For the edge dislocation, there is no long-range correction for a finite cylinder such as the torsion correction which appears for a screw.

B Energy

The energy of the edge dislocation is found, as in equation (3.4) for the screw case, to be

$$\frac{U^d}{L} = \frac{1}{2}\int_{r_0}^{R}\sigma_{xy}(x,0)b\,dx = \frac{b}{2}\int_{r_0}^{R}\hat{\sigma}_{r\theta}\bigg|_{\theta=0}dr$$

$$= \frac{\mu b^2}{4\pi(1-\nu)}\left(\ln\frac{R}{r_0} - 1\right)$$

(3.20)

after using equation (3.14). Frequently in the literature the energy is derived from the approximate stresses in equation (3.15), when the second term in equation (3.20) does not appear. For typical values of R/r_0 in a crystal (ranging from $\sim 10^5$ in an annealed specimen to $\sim 10^2$ in a heavily worked one) this last term represents a 7–20 per cent correction.

A mixed dislocation in a cylinder along the z-axis would have Burgers vector $\mathbf{b} = (b\sin\psi, 0, b\cos\psi)$, where ψ is the angle between the Burgers vector and the dislocation line. Owing to the linearity of the theory, the stress-field is obtained as a superposition of those of a screw dislocation of strength $b \cos \psi$ and an edge of strength $b \sin \psi$. The energy per unit length is

$$\frac{U^d}{L} = \frac{1}{2} \int_{r_0}^{R} [\sigma_{xy}(x, 0)b \sin \psi + \sigma_{xz}(x, 0)b \cos \psi] \, dx$$

$$= \frac{\mu b^2}{4\pi} \left(\frac{\sin^2 \psi}{1-\nu} + \cos^2 \psi \right) \left(\ln \frac{R}{r_0} - 1 \right) \tag{3.21}$$

for $r_0 \ll R$.

3.3
ENERGIES AND FORCES

A *Dislocations and cracks in external stress-fields*

It is a matter of considerable interest to calculate the elastic strain energy involved in dislocated configurations, and in particular to calculate the change in energy produced by a change in the position of the dislocation lines. The latter will lead us to the idea of a force on a dislocation line, and later on (for example in §§5.1 and 8.4) will enable us to investigate the equilibrium positions of dislocation lines in certain particular problems. In the present section we shall develop some general formulae for the energies connected with arbitrary dislocation configurations, and in §3.4 we shall apply these results to the special case of straight dislocations.

The energy calculations for dislocations and for bodies with cracks are closely related and we shall encompass both situations in the present discussion. We shall not, however, apply the results to crack problems until Chapter 5.

The boundary-value problem in which we are ultimately interested concerns a body $\mathscr{B}$ whose external boundary consists of two parts S_1 and S_2, on one of which the displacement is prescribed while on the other the traction is prescribed. The body also contains one or more cuts in its interior, whose two surfaces we label C_+ and C_-, on which certain displacement or traction conditions are imposed. Denoting the displacement and stress-fields by u_i and σ_{ij}, we have the conditions

$$u_i = U_i \text{ on } S_1, \qquad \sigma_{ij}n_j - t_i \text{ on } S_2,$$

$$u_{i+} - u_{i-} = \Delta u_i, \qquad \sigma_{ij}n_j\big|_{C_+} = -\sigma_{ij}n_j\big|_{C_-} = t_i^c \text{ on } C_{\pm}. \tag{3.22}$$

Here Δu_i defines the displacement discontinuity across the cut and t_i^c the tractions applied to the cut. These tractions will always be supposed equal and opposite at corresponding points on the two faces.

There are two cases of interest.

1. *Dislocation case*   If Δu_i is given a prescribed value corresponding to the Burgers vector of the dislocation, we obtain a dislocation whose line is the edge of the cuts $C_\pm$. The quantities t_i^c are then equal to the tractions necessary on the cut in order to create the dislocation. A general dislocation of this kind was illustrated in Figure 1.13. More simply, an edge dislocation could be formed by taking the cut on the $y = 0$ plane with $x > 0$ (Fig. 3.3), The tractions necessary to make the relative displacement, $\Delta u = (-b, 0, 0)$, are given in equations (3.16) and (3.19) as $t^c = (-\sigma_{xy}, -\sigma_{yy}, -\sigma_{zy}) = (-\mu b/2\pi(1-v)x, 0, 0)$ for an infinite isotropic medium with no applied stresses at infinity.

The discussion of the present section will allow us to consider more general dislocations in which Δu_i is not constant on $C_\pm$. Furthermore, the problem described by conditions (3.22) corresponds in general to a dislocation in a medium which is also subjected to certain external loads.

2. *Crack case*   We can also consider a body containing one or more cracks whose surfaces are $C_\pm$. Usually in such a case the traction t_i^c on the faces of the crack would be prescribed, while the relative displacement Δu_i of the faces would not be known a priori. In the most common situations the crack surfaces would in fact be free, so that $t_i^c = 0$. Thus conditions (3.22) can be used to describe a body, externally loaded, and containing a crack which may be free surface or may have prescribed internal loads on it.

It is convenient to consider a second state of strain in the same body which would be obtained by imposing the same boundary conditions on S_1 and S_2, but also requiring that the relative displacement of C_+ and C_- should be zero. Denoting this displacement and stress-field by u_i^0 and σ_{ij}^0, we have the conditions

$$u_i^0 = U_i \quad \text{on } S_1, \qquad \sigma_{ij}^0 n_j = t_i \quad \text{on } S_2,$$
$$u_{i+}^0 - u_{i-}^0 = 0, \qquad \sigma_{ij}^0 n_j|_{c+} = -\sigma_{ij}^0 n_j|_{c-}. \tag{3.23}$$

In this deformation, certain tractions must, of course, be applied to the faces of the cut in order to prevent their relative displacement, and we set

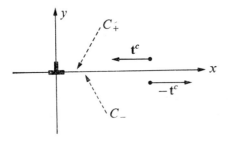

FIGURE 3.3

$t_i{}^0 = \sigma_{ij}{}^0 n_j$ on C_+. The displacement field $u_i{}^0$ is the one which would be obtained by subjecting the same body, but without a cut or crack in it, to the given conditions on S_1 and S_2.

Finally we introduce the displacement field $u_i{}^d = u_i - u_i{}^0$ and its corresponding stress-field $\sigma_{ij}{}^d$. By combining equations (3.22) and (3.23), we obtain the boundary conditions on these fields:

$$u_i{}^d = 0 \quad \text{on } S_1, \qquad \sigma_{ij}{}^d n_j = 0 \quad \text{on } S_2,$$

$$u_{i_+}{}^d - u_{i_-}{}^d = \Delta u_i, \qquad \sigma_{ij}{}^d n_j\big|_{C+} = -\sigma_{ij}{}^d n_j\big|_{C-} = t_i{}^d \equiv t_i{}^c - t_i{}^0. \tag{3.24}$$

In the dislocation case this deformation corresponds to the dislocation alone, with homogeneous conditions on the external boundaries. In the crack case it corresponds to the problem of an internally loaded crack, again with zero displacements or traction at the external boundary.

Using equation (2.31) the elastic energies in the last two states of deformation are respectively

$$U^0 = \tfrac{1}{2} \int_S \mathbf{t}^0 \cdot \mathbf{u}^0 dS, \tag{3.25}$$

$$U^d = \tfrac{1}{2} \int_C \mathbf{t}^d \cdot \mathbf{u}^d dS = \tfrac{1}{2} \int_{C+} \mathbf{t}^d \cdot \Delta \mathbf{u} \, dS \tag{3.26}$$

We have used S to denote $S_1 + S_2$ and C to denote $C_+ + C_-$ in these expressions. This last result often provides a useful means of calculating the energies of dislocations and, in fact, we have already made use of it for the straight screw and edge dislocations in the last two sections.

The interaction energy, U_{int}, between these two states has two equivalent forms in equation (2.33), which are

$$U_{\text{int}} = \int_{\partial \mathcal{B}} \mathbf{t}^d \cdot \mathbf{u}^0 dS = \int_{\partial \mathcal{B}} \mathbf{t}^0 \cdot \mathbf{u}^d dS.$$

Here $\partial \mathcal{B}$ denotes the total boundary $S_1 + S_2 + C_+ + C_-$. Considering the first of these forms, $\mathbf{t}^d$ is zero on S_2, and on C $\mathbf{u}^0$ is continuous between the two faces while $\mathbf{t}^d$ is equal and opposite. Thus all the integrals vanish except that from S_1. In the second form the integral over S_1 is zero, so that after using the conditions (3.23) and (3.24) we obtain

$$U_{\text{int}} = \int_{S_1} \mathbf{t}^d \cdot \mathbf{U} \, dS = \int_{S_2} \mathbf{t} \cdot \mathbf{u}^d dS + \int_{C+} \mathbf{t}^0 \cdot \Delta \mathbf{u} \, dS. \tag{3.27}$$

An important special case occurs when the external boundary conditions are entirely of traction type. Then S_1 is empty and $U_{\text{int}} = 0$. There is zero interaction energy between a dislocation and a state of strain produced by prescribed external tractions. The interaction energy is non-zero, however, if some of the external boundary conditions are of displacement type.

The energy U of the state of strain described by conditions (3.22) is equal to $U^0 + U^d + U_{\text{int}}$. We are interested in actual fact in the difference

$U-U^0$, which we denote by ΔU. ΔU represents the increase in the internal energy of the body if the dislocation (or crack) is formed after the external loads on S_1 and S_2 have already been applied. Using equations (3.26) and (3.27), we have

$$\Delta U = \int_{S_2} \mathbf{t} \cdot \mathbf{u}^d dS + \tfrac{1}{2} \int_{C_+} (\mathbf{t}^0 + \mathbf{t}^c) \cdot \Delta \mathbf{u} \, dS, \tag{3.28}$$

after noting that $\mathbf{t}^d = \mathbf{t}^c - \mathbf{t}^0$ on C_+.

Later on we shall be concerned with the question as to whether, under given external loads, such processes as the motion of a dislocation or the formation or growth of a crack can occur. Thus we shall need to calculate the change in energy of the body plus its surroundings during the process under investigation. For the body itself, equation (3.28) provides the required energy. However, there are also certain external agents which are providing tractions on S_1 and S_2, and they also will suffer an energy change upon formation of the dislocation or crack. This change in energy between the states $\mathbf{u}^0$ and $\mathbf{u}$ of deformation is

$$\Delta U_{\text{ext}} = - \int_{S_2} \mathbf{t} \cdot (\mathbf{u} - \mathbf{u}^0) \, dS, \tag{3.29}$$

since the applied traction is maintained at the constant value $\mathbf{t}$ on S_2 and on S_1 there is no motion of the boundary to cause any work to be done. The total energy change upon formation of the dislocation is the sum $\Delta U + \Delta U_{\text{ext}}$, giving

$$\Delta U_{\text{tot}} = \tfrac{1}{2} \int_{C_+} (\mathbf{t}^0 + \mathbf{t}^c) \cdot \Delta \mathbf{u} \, dS. \tag{3.30}$$

An equivalent formula, obtained by replacing $\mathbf{t}^c$, is

$$\Delta U_{\text{tot}} = U^d + \int_{C_+} \mathbf{t}^0 \cdot \Delta \mathbf{u} \, dS. \tag{3.31}$$

Most of the above results were first obtained by Colonnetti,[5] later being rederived and extended by Steketee[6] and Eshelby.[7]

In applying equation (3.31) to the dislocation case, we would calculate U^d by solving the boundary-value problem (3.24) for the dislocation alone and then using equation (3.26). The second term requires $\Delta \mathbf{u}$, which is prescribed, and $\mathbf{t}^0$, which is found by solving the boundary-value problem (3.23) for the uncut medium. For the crack case, equation (3.30) is more transparent. Again $\mathbf{t}^0$ is needed, and is found in the same way. The traction $\mathbf{t}^c$ is prescribed (in fact it is usually zero) and we need to find $\Delta \mathbf{u}$, the displacement discontinuity across the crack, by solving the boundary-value problem (3.22).

It is instructive to examine the two contributions to ΔU_{tot} in two special situations for a body with a crack.

1. *Fixed load case*   Here we take the crack to be unloaded ($\mathbf{t}^c = \mathbf{0}$) and

the boundary conditions to be entirely of traction type, so that S_1 is empty. Then the right-hand side of equation (3.27) is zero, and we can use the resulting relation to eliminate the integrals over S_1 in equations (3.28) and (3.29). Comparing the expressions so obtained with equation (3.30), it follows that

$$\Delta U = -\Delta U_{\text{tot}} \quad \text{and} \quad \Delta U_{\text{ext}} = 2\Delta U_{\text{tot}}.$$

Since ΔU_{tot} is negative, we see that the formation of a crack in a body under fixed loads leads to an increase in the strain energy of the material but a decrease of double the magnitude in the external energy.

2. *Fixed grips case* In this case we take the boundary to be either stress-free or to have prescribed displacements. Thus either S_2 is empty or, if not, $\mathbf{t} = \mathbf{0}$ at each point of S_2. From equations (3.29) and (3.30), we obtain

$$\Delta U = \Delta U_{\text{tot}}, \qquad \Delta U_{\text{ext}} = 0.$$

There is in the fixed grips case no change in the external energy, the entire change in total energy coming from a decrease in strain energy in the body upon formation of the crack. This result holds whether or not there are internal loads on the crack.

These two cases can be illustrated by the simple tension experiment. Here the body is a cylinder with generators parallel to the z-axis, the sides are stress-free while the ends are subject to constant normal tractions which with no crack give rise to a state of stress with constant σ_{zz}. If now a crack is formed while maintaining the end tractions at their original constant value, we have an example of case 1: the elastic energy in the cylinder increases, but the cylinder ends move a little and reduce the external energy. On the other hand, if the crack is formed while holding the cylinder ends fixed in their deformed positions, we have a fixed grips case and the elastic energy in the cylinder decreases upon forming the crack.

B *Force on a dislocation*[7-9]

We now consider the case when $\Delta \mathbf{u} = \mathbf{b}$ is constant over the cut C. C is bounded by the curve Γ of the dislocation line, and we calculate the change in U_{tot} when Γ moves through an amount $\delta \mathbf{r}$. From equation (3.31)

$$\delta U_{\text{tot}} - \delta U^d \mid \delta \left[\mathbf{b} \cdot \int_{C_+} \mathbf{t}^0 \, dS \right] = \delta U^d + \mathbf{b} \cdot \int_{\delta C_+} \mathbf{t}^0 \, dS,$$

where δC_+ is the new cut swept out as Γ moves through $\delta \mathbf{r}$ (see Fig. 3.4). If $\mathbf{n}$ is the unit normal to δC_+ and $\boldsymbol{\sigma}^0$ the external stress tensor, then on δC_+, $\mathbf{t}^0 = \boldsymbol{\sigma}^0 \cdot \mathbf{n}$. Here we use dyadic notation in which $\boldsymbol{\sigma}^0 \cdot \mathbf{n}$ denotes the vector with components $\sigma_{ij}^0 n_j$. Also $\mathbf{n} \, dS = -\boldsymbol{\xi} \wedge \delta \mathbf{r} \, dl$, where dl is an element of arc length along Γ and $\boldsymbol{\xi}$ is a unit vector along Γ whose sense has

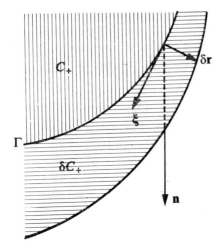

FIGURE 3.4

been chosen so that $\mathbf{\Delta u} = \mathbf{b}$ is the dislocation Burgers vector as defined in Chapter 1. Thus

$$\delta U_{\text{tot}} = \delta U^d - \int_{\Gamma} \mathbf{b} \cdot \mathbf{\sigma}^0 \cdot (\mathbf{\xi} \wedge \delta\mathbf{r}) dl = \delta U^d - \int_{\Gamma} (\mathbf{b} \cdot \mathbf{\sigma}^0 \wedge \mathbf{\xi}) \cdot \delta\mathbf{r} \, dl. \qquad (3.32)$$

Here $\mathbf{b} \cdot \mathbf{\sigma}^0 \wedge \mathbf{\xi}$ denotes the cross-product of the vectors $(b_i \sigma_{ij}^0)$ and $\mathbf{\xi}$.

It is customary, by analogy with general procedures in mechanics, to introduce the notion of a force per unit length, $\mathbf{F}/L$, on a dislocation line, which is minus the rate of change of energy with respect to the dislocation position. In other words we write

$$\delta U_{\text{tot}} = - \int (\mathbf{F}/L) \cdot \delta\mathbf{r} \, dl. \qquad (3.33)$$

Then the contribution to the force provided by the external tractions $\mathbf{t}^0$ at any point on the dislocation line is

$$\mathbf{F}^0/L = \mathbf{b} \cdot \mathbf{\sigma}^0 \wedge \mathbf{\xi}, \qquad (3.34)$$

where $\mathbf{\sigma}^0$ is the stress tensor corresponding to these tractions evaluated a the position of the dislocation line.

For a straight dislocation in a uniform infinite medium, the dislocation energy U^d is independent of the dislocation position, and equation (3.34) gives the total force from equation (3.32). In a finite medium, however, the energy does depend on the position of the dislocation relative to the external surfaces and δU^d is non-zero. This leads to what are often called *image forces*, which we shall investigate in §3.5. Another case in which δU^d provides a contribution to the force is an infinite medium which is non-uniform, and the most important example of this from the dislocation viewpoint is the crystalline solid. Here along the crystal axes there is a periodic variation in material constitution, which leads to a periodic variation of U^d and therefore a periodic force. This force is called the Peierls-Nabarro force,[4] and in Chapter 4 we shall investigate it more fully.

As an example of the application of equation (3.34), consider a screw dislocation lying along the x_3-axis, so that $\boldsymbol{\xi} = (0, 0, 1)$, $\mathbf{b} = (0, 0, b)$. Then

$$\mathbf{F}^0/L = b(\sigma_{zx}{}^0, \sigma_{zy}{}^0, \sigma_{zz}{}^0) \wedge (0, 0, 1) = (b\sigma_{yz}{}^0, -b\sigma_{xz}{}^0, 0). \tag{3.35}$$

Thus, for example, the force tending to make the screw glide in the x-direction is $b\sigma_{yz}{}^0$.

For the edge dislocation with $\boldsymbol{\xi} = (0, 0, 1)$, $\mathbf{b} = (b, 0, 0)$,

$$\mathbf{F}^0/L = (b\sigma_{xy}{}^0, -b\sigma_{xx}{}^0, 0). \tag{3.36}$$

The motion of such an edge in the x-direction is illustrated in the sequence in Figure 1.7, and it is clear from this why the force driving it is $b\sigma_{xy}{}^0$: when the dislocation moves across the crystal, say a distance D, the top face moves distance b in the x-direction, giving a reduction in external potential energy of $(\sigma_{xy}{}^0 LD)b$ for a distance L in the z-direction. Motion of the edge in the y-direction, driven by the force $-b\sigma_{xx}{}^0$, is called *climb*: the compressive stress $-\sigma_{xx}{}^0$ squeezes out the extra half-plane of atoms like toothpaste from a tube. However, it is a rather difficult motion for a dislocation in the sense that there are large intrinsic forces resisting it, in contrast to glide, where the resistances, such as the Peierls-Nabarro force and even the blocking effects of impurities and grain boundaries, are comparatively small. Climb can usually occur only at high temperatures.

c Interactions between dislocations

The above considerations apply to the effect of external tractions on a dislocation and say nothing about the effects of two dislocations on one another. We suppose that the two dislocations are formed by making cuts $C_{+}^{(1)}$ and $C_{+}^{(2)}$ and providing certain discontinuities of displacement: on $C^{(1)}$, $\mathbf{u}_{+}^{(1)} - \mathbf{u}_{-}^{(1)} = \Delta\mathbf{u}^{(1)}$, and $\mathbf{u}_{+}^{(2)} - \mathbf{u}_{-}^{(2)} = 0$, and on $C^{(2)}$, $\mathbf{u}_{+}^{(1)} - \mathbf{u}_{-}^{(1)} = 0$, and $\mathbf{u}_{+}^{(2)} - \mathbf{u}_{-}^{(2)} = \Delta\mathbf{u}^{(2)}$. For the tractions $\mathbf{t}_{+}^{(1)} = -\mathbf{t}_{-}^{(1)}$ and $\mathbf{t}_{+}^{(2)} = -\mathbf{t}_{-}^{(2)}$ on both $C^{(1)}$ and $C^{(2)}$. Then using equation (2.33) with $\partial\mathscr{B} = S + C^{(1)} + C^{(2)}$ gives

$$U_{\text{int}} = \int_{C_{+}{}^{(1)}} \mathbf{t}^{(2)} \cdot \Delta\mathbf{u}^{(1)} \, dS = \int_{C_{+}{}^{(2)}} \mathbf{t}^{(1)} \cdot \Delta\mathbf{u}^{(2)} \, dS. \tag{3.37}$$

These expressions resemble the second term in equation (3.31). They show that if, say, dislocation 2 is fixed, its effect on dislocation 1 is the same as if it were an external stress. Thus, for example, the work of section B goes through if everywhere $\mathbf{u}^0$ and $\boldsymbol{\sigma}^0$ are replaced by $\mathbf{u}^{(2)}$ and $\boldsymbol{\sigma}^{(2)}$, the fields

D

corresponding to another dislocation. In the expression (3.34) for the force on a dislocation we can include in $\boldsymbol{\sigma}^0$ the stress-fields due to both external tractions and any other dislocations.

3.4
FORCES BETWEEN STRAIGHT DISLOCATIONS

A *Parallel screws*

Consider the situation in Figure 3.5 with two screws parallel to the z-axis with Burgers vectors b_1 and b_2 and each with line direction $\boldsymbol{\xi} = (0, 0, 1)$. There will be a force on, say, dislocation 2 due to the stress-field $\boldsymbol{\sigma}^{(1)}$ of dislocation 1. According to the arguments in §3.3C, this force is given by the same formula as it would be if $\boldsymbol{\sigma}^{(1)}$ were an external stress-field, so that, from equation (3.35), it has components

$$F_x/L = b_2 \sigma_{yz}^{(1)}, \qquad F_y/L = -b_2 \sigma_{xz}^{(1)}.$$

If we now use the appropriate stress for dislocation 1, from equation (3.2), these become

$$\frac{F_x}{L} = \frac{\mu b_1 b_2}{2\pi} \frac{x}{x^2+y^2}, \qquad \frac{F_y}{L} = \frac{\mu b_1 b_2}{2\pi} \frac{y}{x^2+y^2}.$$

The resultant of these two is a purely radial force

$$F_r/L = \mu b_1 b_2/2\pi r. \tag{3.38}$$

This is a central repulsion if b_1 and b_2 are of the same sign, and an attraction if they are of opposite signs.

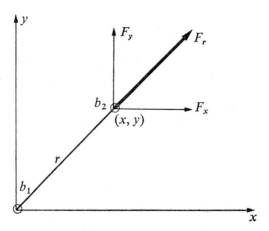

FIGURE 3.5

The force was defined as the gradient of the interaction energy between the two dislocations, so that the latter can be calculated by integrating the force

$$\frac{U_{\text{int}}}{L} = -\int_R^r \frac{F_r}{L} \, dr = \frac{\mu b_1 b_2}{2\pi} \ln \frac{R}{r}. \tag{3.39}$$

Adding in the self-energies of the two dislocations (eq. (3.4)) gives the total energy

$$U_{\text{tot}} = \frac{\mu}{4\pi} (b_1{}^2 + b_2{}^2) \ln \frac{R}{r_0} + \frac{\mu b_1 b_2}{2\pi} \ln \frac{R}{r}. \tag{3.40}$$

For two equal and opposite dislocations ($b_2 = -b_1$) this becomes

$$U_{\text{tot}} = \frac{\mu b_1{}^2}{2\pi} \ln \frac{r}{r_0}. \tag{3.41}$$

In this particular case we can quite validly allow $R \to \infty$ and still keep the energy finite. With two opposite dislocations, the dominant terms in the two long-range stress-fields cancel one another, leading to a convergent energy integral even in an infinite medium.

In all of these cases, the free surface conditions on the two core cylinders and on the outer cylinder are not exactly met by the stress-fields of equation (3.2) for the two dislocations. In order to keep the violations small, we must assume that $r_0 \ll r \ll R$.

It is of interest to compare the magnitude of forces between dislocations with the force $b\sigma$ provided by an external shear stress σ. When $r = 150b$, the force in equation (3.36) is equivalent to an external shear stress of $\mu/1000$, indicating the great significance of interaction effects.

B *Parallel edges with parallel Burgers vectors*

A similar situation, but with edge dislocations, is shown in Figure 3.6, where the x-axis has been chosen as parallel to the two Burgers vectors. Again $\xi = (0, 0, 1)$ for each dislocation. The force per unit length on dislocation 2 is, from equations (3.16) and (3.36),

$$\frac{F_x}{L} = b_2 \sigma_{xy}{}^{(1)} = \frac{\mu b_1 b_2}{2\pi(1-\nu)} \frac{x(x^2 - y^2)}{(x^2 + y^2)^2},$$

$$\frac{F_y}{L} = -b_2 \sigma_{xx}{}^{(1)} = \frac{\mu b_1 b_2}{2\pi(1-\nu)} \frac{y(3x^2 + y^2)}{(x^2 + y^2)^2}. \tag{3.42}$$

The components of this force in the radial and transverse polar directions are

$$F_r = \frac{\mu b_1 b_2}{2\pi(1-\nu)} \frac{1}{r}, \qquad F_\theta = \frac{\mu b_1 b_2}{2\pi(1-\nu)} \frac{\sin 2\theta}{r}. \tag{3.43}$$

In this case the force of interaction is not purely central.

Under normal circumstances it is difficult for an edge to execute a climb motion, and it is therefore only of interest to consider the force in the glide direction. In this case, the glide force is F_x. Considering b_1 and b_2 to be of the same sign, a sketch of F_x/L is given in Figure 3.7 for a range of positions of dislocation 2 on the glide-plane $y = $ const. There are three positions of

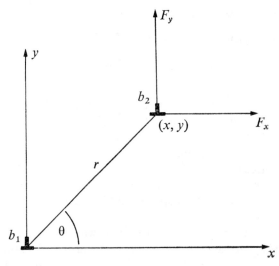

FIGURE 3.6

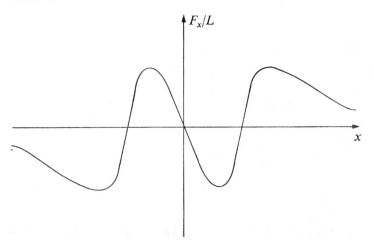

FIGURE 3.7

equilibrium, at $x = 0$ and $\pm y$, but it is clear from the signs of F_x in the neighbourhoods of these points that only the $x = 0$ position is stable. Thus the two dislocations, if free to glide, will tend to align themselves one directly above the other.

The situation is exactly reversed for edges of opposite signs. The position $x = 0$ is then unstable and $x = \pm y$ are stable, so that the dislocations will tend to arrange themselves in a 45° configuration.

The interaction energy is

$$\frac{U_{int}}{L} = -\int_R^x \frac{F_x}{L} dx = -\int_R^x b_2 \sigma_{xy}^{(1)} dx = \int_R^x b_2 \frac{\partial^2 \Psi}{\partial x \partial y} dx = b_2 \left[\frac{\partial \Psi}{\partial y} \right]_R^x,$$

where Ψ is the Airy stress function for the stress-field of dislocation 1. From equation (3.13), for $r \gg r_0$, we have

$$\Psi(x, y) = -\frac{\mu b_1}{4\pi(1-\nu)} \left(y \ln (x^2 + y^2) - \frac{y(x^2 + y^2)}{R^2} \right). \tag{3.44}$$

Thus, for $r \ll R$,

$$\frac{U_{int}}{L} = \frac{\mu b_1 b_2}{2\pi(1-\nu)} \left(\ln \frac{R}{r} - \sin^2 \theta - \frac{1}{2} \right). \tag{3.45}$$

Again the total energy is obtained by adding in the two self-energies from equation (3.20). For equal and opposite Burgers vectors, it is[10]

$$\frac{U_{tot}}{L} = \frac{\mu b^2}{2\pi(1-\nu)} \left(\ln \frac{r}{r_0} - \frac{1}{2} \cos 2\theta \right) \tag{3.46}$$

c *Parallel edges with general Burgers vectors*

For an angle α between the two Burgers vectors (Fig. 3.8), the force on dislocation 2 has components $F_\xi/L = b_2 \sigma_{\xi\eta}^{(1)}$, $F_\eta/L = -b_2 \sigma_{\xi\xi}^{(1)}$ (ξ, η are axes as shown). The tensor transformation law gives $\sigma_{\xi\eta} = (\sigma_{yy} - \sigma_{xx}) \sin \alpha \cos \alpha + \sigma_{xy} (\cos^2\alpha - \sin^2\alpha)$. Therefore the glide force, F_ξ, after using the stresses of equation (3.16), is

$$\frac{F_\xi}{L} = \frac{\mu b_1 b_2}{2\pi(1-\nu)} \frac{x}{r^2} \cos 2(\theta - \alpha).$$

Again on the glide-plane of dislocation 2 there are three positions of equilibrium: at $x = 0$ and $\theta - \alpha = \pi/4 \pm n\pi/2$ (θ covers either the range $\alpha < \theta < \alpha + \pi$ or $-\pi + \alpha < \theta < \alpha$, so that there are always two points of the latter type). The stability of these positions can be established by examining the behaviour of the long-range force: if the force F_ξ is repulsive

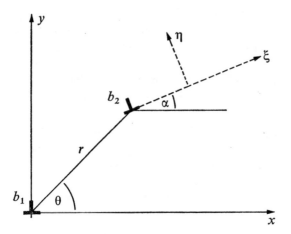

FIGURE 3.8

at large distances, the central position is stable, and if it is attractive at large distances the two outer positions are stable. We take b_1 and b_2 to be always positive, the relative sign of the two dislocations being determined by α ($\mathbf{b}_1 \cdot \mathbf{b}_2 = b_1 b_2 \cos \alpha$). Then at the two limits of the glide-plane F_ξ has the same sign as x. Thus the long-range force is repulsive if $-\pi/2 < \alpha < \pi/2$ (i.e. $\mathbf{b}_1 \cdot \mathbf{b}_2 > 0$) and is attractive if $\pi/2 < \alpha < 3\pi/2$ (i.e. $\mathbf{b}_1 \cdot \mathbf{b}_2 < 0$).

The cases $\alpha = \pm \pi/2$ are special in that one of the three equilibrium positions moves off to infinity, and of the two remaining equilibria (at the 45° positions), one is stable and the other unstable. The stable position is always, as one might expect, the one in which the extra half-plane of one dislocation crosses the region of the missing half-plane of the other. A sequence of equilibrium positions for different values of α is shown in Figure 3.9, with the stable positions labelled S.

The interaction energy in the general case of Figure 3.8 is derived as was equation (3.43):

$$\frac{U_{int}}{L} = -\int_R^r b_2 \sigma_{\xi\eta}^{(1)} \, d\xi = b_2 \left[\frac{\partial \Psi}{\partial \eta} \right]_R^r = b_2 \left[-\frac{\partial \Psi}{\partial x} \sin \alpha + \frac{\partial \Psi}{\partial y} \cos \alpha \right]_R^r,$$

where the integration is carried out on the glide-plane $\eta = $ const. Using the stress function in equation (3.42) then gives[10]

$$\frac{U_{int}}{L} = \frac{\mu b_1 b_2}{2\pi(1-\nu)} \left[\left(\ln \frac{R}{r} - \frac{1}{2} \right) \cos \alpha - \sin \theta \sin (\theta - \alpha) \right]$$

$$= \frac{\mu}{2\pi(1-\nu)} \left[\mathbf{b}_1 \cdot \mathbf{b}_2 \left(\ln \frac{R}{r} - \frac{3}{2} \right) + \frac{(\mathbf{b}_1 \cdot \mathbf{r})(\mathbf{b}_2 \cdot \mathbf{r})}{r^2} \right]. \qquad (3.47)$$

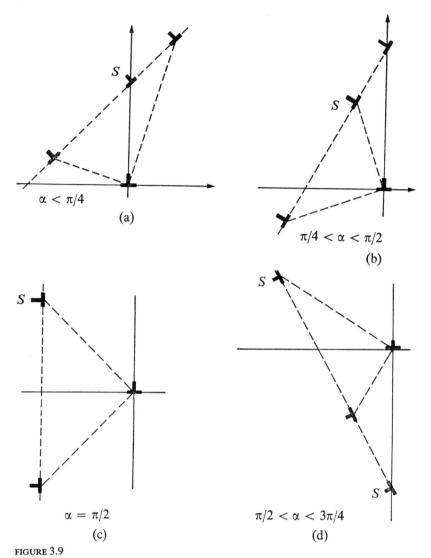

FIGURE 3.9

3.5
IMAGE FORCES

A *Screw dislocation in a half-space*

The change in the total energy due to a small motion of a dislocation, which is given in equation (3.32), contains a contribution δU^d from the change in internal strain energy of the dislocation stress-field. In an infinite homogeneous medium this contribution vanishes. If the body has finite boundaries however, any motion of the dislocation will in general change its position relative to the boundaries and hence will change the stress-field. Consequently there exists a force on a dislocation in such bodies, even in the absence of stress-fields other than that of the dislocation itself. Forces of this kind are called *image forces* owing to the fact that they can often be found using the method of images.

In this subsection we shall consider the case of a straight screw dislocation situated in the half-space $x < 0$ with its line parallel to the boundary $x = 0$. We take the z-axis to lie along the dislocation line, which we suppose intersects the xy-plane at the point $(-l, 0)$ (S in Fig. 3.10). In this section the boundary is supposed to be traction-free; the calculation of image forces with other boundary conditions will be set as problems at the end of the chapter.

The stress-field of equation (3.2) with x replaced by $x+l$, which would be appropriate if the boundary were absent, does not satisfy the conditions that $\sigma_{xk} = 0$ on $x = 0$, since $\sigma_{xz}(0, y) = -(\mu b/2\pi)y/(l^2+y^2)$. However, the combined field of S together with a left-hand screw S' at the image point $(l, 0)$ does satisfy the boundary conditions, and therefore has to solve the problem. The stress components at a general point (x, y) in the region $x < 0$ are then

$$\sigma_{xz}(x, y) = -\frac{\mu b}{2\pi}\left(\frac{y}{(l+x)^2+y^2} - \frac{y}{(l-x)^2+y^2}\right),$$

$$\sigma_{yz}(x, y) = \frac{\mu b}{2\pi}\left(\frac{l+x}{(l+x)^2+y^2} + \frac{l-x}{(l-x)^2+y^2}\right).$$

(3.48)

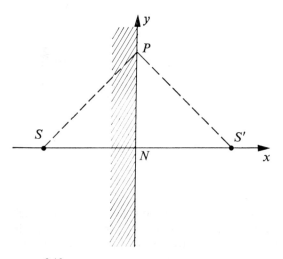

FIGURE 3.10

This stress-field tends to zero at large distances faster than that of equation (3.2) because of a cancellation of the leading asymptotic stresses from the original screw and its image. As a consequence, the energy integral converges for an infinite medium, and we do not need to require that there be a free surface at $r = R$. However, we do still need to make the hollow-core assumption in order to have a convergent energy.

In this respect, it must be pointed out that the stresses of equation (3.48) do not produce zero tractions on a cylinder of radius r_0 centered on the dislocation line at S. Thus the image construction does not give the correct core conditions for a Volterra dislocation. For $l \gg r_0$ the tractions on the cylinder are small, so that equation (3.48) does give approximately the correct stress-field provided that the dislocation does not approach within a few core radii of the boundary. However, it is possible, with a small modification, to obtain the exact stress-field from equation (3.48), which will allow us to draw conclusions no matter how close the dislocation is to the boundary.

Consider a cylinder of radius r_0, parallel to the z-axis, whose axis passes through the point $(-p, 0)$ in the xy-plane (Fig. 3.11). The general point on the cylinder is $(-p+r_0 \cos \theta, r_0 \sin \theta)$ and the traction acting across the cylinder surface is $t_z = \sigma_{xz} \cos \theta + \sigma_{yz} \sin \theta$. It is not hard to show, using the stresses of equation (3.48), that t_z is identically zero on the cylinder, provided that $p^2 = l^2 + r_0^2$. Now let us turn this result around. Suppose that we are given a Volterra dislocation of core radius r_0 and centered at $(-p, 0)$. Then the stresses of equation (3.48) satisfy the free boundary conditions on both the core cylinder and the plane $x = 0$, provided that we take $l = (p^2 - r_0^2)^{1/2}$. Thus the effect of the free surface at $x = 0$ is to shift the effective centre of the dislocation from $(-p, 0)$ to $(-(p^2 - r_0^2)^{1/2}, 0)$ as well as to cause the image stresses.

The configuration of Figure 3.11 can be established by making a cut from the origin to the cylinder surface at $(-p+r_0, 0)$ and displacing the faces

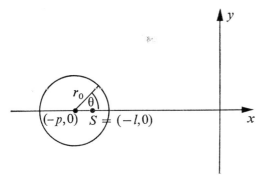

FIGURE 3.11

relative to one another by the Burgers vector. From equation (3.26) we obtain the energy of the dislocated state, per unit length:

$$
\frac{U^d}{L} = \frac{1}{2} \int_{-p+r_0}^{0} \sigma_{yz}(x, 0) b \, dx
$$

$$
= \frac{\mu b^2}{4\pi} \ln \left(\frac{(p^2 - r_0^2)^{1/2} + p - r_0}{(p^2 - r_0^2)^{1/2} - p + r_0} \right) \approx \frac{\mu b^2}{4\pi} \ln \frac{2l}{r_0}, \tag{3.49}
$$

when $l \gg r_0$.

As p (or l) decreases, the dislocation energy also decreases, so that there is a tendency for the dislocation to be attracted towards the free surface. If we define the force on the dislocation line as minus the rate of change of energy with dislocation position, as in equation (3.33), we get a force per unit length in the x-direction equal to

$$
\frac{F}{L} = \frac{\partial}{\partial p} \left(\frac{U^d}{L} \right) = \frac{\mu b^2}{4\pi l}. \tag{3.50}
$$

This result is exact, no matter how close the dislocation is to the surface, although it must be remembered that $l = (p^2 - r_0^2)^{1/2}$ in terms of the true position of the dislocation centre. The force in equation (3.50) is called the *image force* on S.

The image force is in fact the same force that would be felt by S if the two dislocations S and S' were real dislocations in an infinite medium. The stress-field σ' of S' would then be given by the second terms in equations (3.48), and would produce a force on S of amount $b\sigma_{yz}'(-l, 0) = \mu b^2/4\pi l$ according to the discussion in §3.4A. This result can also be seen when $l \gg r_0$ in the following way. If S and S' are both real dislocations in an infinite medium, the change in total energy when $l \to l + \delta l$ is $(F_x/L)(2\delta l)$, where from §3.4A $F_x/L = b\sigma_{yz}'(-l, 0) = \mu b^2/4\pi l$ (since the distance between S and S' changes by $2\delta l$). The stress-field is symmetrical about $x = 0$, so that the change in the energy of the half-space $x < 0$ is just half the total change, namely $(F_x/L)\delta l = (\mu b^2/4\pi l)\delta l$, agreeing with equation (3.50).

The effect of the attraction between a dislocation and a free surface leads in many practical cases to a depletion in the number of dislocations in the surface layers of crystals. It should be stated, however, that the form of the image force depends very sensitively on the condition of the relevant surface. If the surface is not free but rigidly fixed, the dislocation is repelled from the surface, with a force equal in magnitude to that in equation (3.50). Many metals form an oxide layer on their surfaces when they are exposed to the atmosphere, and this layer, having greater elastic moduli than the parent metal, has a similar effect to that of the rigid boundary condition and repels the dislocations when they get close to it.[11] (See Problems 3.3–3.6.)

B *Edge dislocation in a half-space*[11,12]

It is more complicated to calculate the effect of free surfaces on an edge dislocation than for a screw because the image construction does not give the complete solution. It is necessary to solve a further auxiliary boundary-value problem. We shall consider the geometry in Figure 3.12 in which a medium occupies the half-space $y > 0$, with an edge dislocation E at a distance l from the boundary. The Burgers vector is supposed to be inclined at an angle α to the y-axis, so that the slip-plane is EG.

As a first attempt, we put an equal and opposite edge at the mirror image point, E'. Unfortunately the two dislocations together do not give stresses which satisfy free boundary conditions on $y = 0$. The boundary tractions can easily be calculated using equations (3.16) with appropriate changes of coordinates and, in fact, they turn out to be

$$\sigma_{yy}^{(E+E')}(x, 0) = -\frac{\mu b \sin \alpha\, l}{\pi(1-\nu)}\frac{x^2-l^2}{(x^2+l^2)^2},$$

$$\sigma_{xy}^{(E+E')}(x, 0) = \frac{\mu b \cos \alpha\, l}{\pi(1-\nu)}\frac{x^2-l^2}{(x^2+l^2)^2}.$$

(3.51)

Consequently, in order to complete the solution we must find an auxiliary stress-field, say $\sigma_{ij}'(x, y)$ for the region $y > 0$, which satisfies the boundary conditions

$$\sigma_{yy}'(x, 0) = -\sigma_{yy}^{(E+E')}(x, 0), \qquad \sigma_{xy}'(x, 0) = -\sigma_{xy}^{(E+E')}(x, 0). \qquad (3.52)$$

The superposition of the two stress-fields, $\sigma_{ij} = \sigma_{ij}^{(E+E')} + \sigma_{ij}'$, then will satisfy the required boundary conditions on $y = 0$.

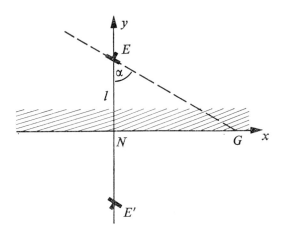

FIGURE 3.12

The problem of finding σ_{ij}' is a regular elastic boundary-value problem for a half-space. In §5.5 we shall discuss such problems at some length, and shall, by using dislocation methods, develop general formulae for their solution. The resulting stress-field solutions are obtained by superposing the expressions in equations (5.104) and (5.110), where $s(x) = \sigma_{xy}'(x, 0)$ and $p(x) = \sigma_{yy}'(x, 0)$. Combining these with equations (3.51) and (3.52), we then have the solutions to our present problem

$$\sigma_{xx}'(x, y) = \frac{2\mu b l}{\pi^2(1-\nu)}(-I_3 \cos \alpha + I_2 \sin \alpha),$$ (3.53)

$$\sigma_{xy}'(x, y) = \frac{2\mu b l}{\pi^2(1-\nu)}(-I_2 \cos \alpha + I_1 \sin \alpha),$$ (3.54)

$$\sigma_{yy}'(x, y) = \frac{2\mu b l}{\pi^2(1-\nu)}(-I_1 \cos \alpha + I_0 \sin \alpha),$$ (3.55)

where

$$I_p = \int_{-\infty}^{\infty} \frac{(x-x')^p y^{3-p}}{[(x-x')^2+y^2]^2} \frac{x'^2-l^2}{(x'^2+l^2)^2} dx'.$$

These integrations can readily be performed, by contour methods for example. However, in the following we shall make use only of σ_{xx}' and σ_{xy}' on the line $x = 0$, which necessitates an evaluation of just I_2, since I_1 and I_3 vanish identically on this line. We obtain

$$I_2(x = 0, y > 0) = \pi(y-l)/2(y+l)^3.$$

Then from equations (3.53) and (3.54) $\sigma_{xx}'(x, 0)$ and $\sigma_{xy}'(x, 0)$ can be obtained directly. These stresses are now combined with the contributions from E and E' to give the total stresses on $x = 0$ as

$$\sigma_{xx}(0, y) = \frac{\mu b \sin \alpha}{2\pi(1-\nu)}\left(\frac{1}{l-y} + \frac{1}{l+y} - \frac{2l(l-y)}{(l+y)^3}\right),$$ (3.56)

$$\sigma_{xy}(0, y) = -\frac{\mu b \cos \alpha}{2\pi(1-\nu)}\left(\frac{1}{l-y} + \frac{1}{l+y} - \frac{2l(l-y)}{(l+y)^3}\right).$$ (3.57)

To calculate the energy contained in the dislocated half-space, we can suppose that the Volterra cut is made from E to N and a relative displacement $\Delta \mathbf{u} = (-b \sin \alpha, b \cos \alpha)$ is given to the two faces. From equation (3.26) the energy is then

$$\frac{U^d}{L} = \frac{1}{2}\int_0^{l-r_0} [\sigma_{xx}(0, y)b \sin \alpha - \sigma_{xy}(0, y)b \cos \alpha] dy.$$ (3.58)

We have again supposed that a tube of material is removed from about the dislocation line in writing the upper limit in this integral. Substituting from equations (3.56) and (3.57) gives finally

$$\frac{U^d}{L} = \frac{\mu b^2}{4\pi(1-\nu)}\left(\ln\frac{2l}{r_0} - \frac{1}{2}\right). \tag{3.59}$$

It can be seen from this that, just as for the screw dislocation, there is a force of attraction on an edge dislocation towards a free surface. The magnitude of the force is

$$\frac{F}{L} = \frac{\partial}{\partial l}\frac{U^d}{L} = \frac{\mu b^2}{4\pi(1-\nu)l}, \tag{3.60}$$

and is independent of the orientation of the Burgers vector relative to the surface. However, when $\alpha \neq 0$, part of the force will be balanced by the very large intrinsic resistance to dislocation climb. The part of the force which can drive the dislocation along its glide-plane is $(F/L)\cos\alpha$, and this would in most cases be the most important contribution.

In the above calculation we have ignored the boundary conditions on the core cylinder of E, which are not satisfied by the stress-field of equation (3.16), which we have used in the present section. In order to satisfy them we need the third terms in equation (3.14). Now, if $r_0 \ll l$, these terms produce tractions on the free surface at $x = 0$ which are smaller than the tractions we have dealt with by a factor of order $(r_0/l)^2$. Hence the corresponding image stresses will be smaller by the same factor, as will their contribution to the energy in equation (3.59). However, the core terms themselves give a sizable contribution to the energy, just as they do for a dislocation in a cylinder, and including them gives a total energy per unit length of

$$\frac{U^d}{L} = \frac{\mu b^2}{4\pi(1-\nu)}\left(\ln\frac{2l}{r_0} - \frac{1}{2} - \frac{1}{2}\cos 2\alpha\right) \tag{3.59'}$$

instead of the result (3.59).

3.6
DISLOCATION LOOPS

A Burgers formula for the displacement

The simplest way of constructing the stress and displacement fields for an arbitrary dislocation loop is to use the method of the Green's function[6,13] described in §2.4. We suppose that we are concerned with a body $\mathscr{B}$, which may be finite or infinite, with boundary $\partial\mathscr{B}$. The dislocation is formed by choosing a closed curve Γ and a surface C spanning it, making a cut on the surface C, and displacing the two faces of the cut by an amount $\mathbf{b}$ relative to one another. More precisely, calling the faces $C_{\pm}$, $\Delta\mathbf{u} \equiv \mathbf{u}_+ - \mathbf{u}_- = \mathbf{b}$, and the line direction $\boldsymbol{\xi}$ on Γ is right-handed with respect to unit vectors $\mathbf{n}$ pointing from C_+ to C_- (cf. Fig. 1.13). As always with Volterra dislocations, the tractions on the two faces at corresponding points are equal and opposite.

The external boundary conditions are taken to be homogeneous. That is (comparing with eqs. (2.35)) $\sigma_{ij} n_j = 0$ on a part, S_1, of the boundary while $u_i = 0$ on the remainder, S_2.

Equation (2.40) holds for any subvolume V of $\mathscr{B}$, with $u_i(\mathbf{r})$ any displacement field with corresponding stresses σ_{ij} and body force f_i. $u_{im}(\mathbf{r}, \mathbf{r}')$ is the Green's function corresponding to the boundary-value problem in $\mathscr{B}$ for u_i. We want to apply this result to the displacement field $u_i(\mathbf{r})$ produced by a dislocation loop, and we choose for V the whole of $\mathscr{B}$ except for the cut C. Then $\partial V = \partial \mathscr{B} + C_+ + C_-$. In equation (2.40), the body force vanishes by assumption, and the surface term from $\partial \mathscr{B}$ is zero, since the dislocation deformation and the Green's function satisfy the same homogeneous boundary conditions on S_1 and S_2. (Some care is necessary here if $\mathscr{B}$ is a finite body with S_2 empty. The Green's function must then satisfy equation (2.42), and this produces rigid motion terms in equation (2.40), just as in equation (2.43).) For the contributions from $C_\pm$, we note that $\sigma_{ij} n_j|_{C_+} = -\sigma_{ij} n_j|_{C_-}$ and $u_{i+} - u_{i-} = b_i$, so that

$$u_m(\mathbf{r}') = -b_i \int_{C_+} c_{ijkl} \partial_k u_{lm}(\mathbf{r}, \mathbf{r}') n_j dS. \tag{3.61}$$

This result holds for any type of material symmetry and for any body $\mathscr{B}$, except that for a bounded body extra rigid terms may be necessary. If we now specialize to the case of a body made of isotropic material and occupying the whole of space, the Green's function is given in equation (2.44), and it is possible to simplify the expression for the displacement field of a dislocation loop. From equations (2.45) and (3.61),

$$u_m(\mathbf{r}') = -\frac{b_i}{8\pi} \int_{C_+} (-\delta_{ij} \partial_{mpp} R + \delta_{jm} \partial_{ipp} R + \delta_{im} \partial_{jpp} R) n_j dS$$

$$-\frac{b_i}{4\pi} \frac{\lambda+\mu}{\lambda+2\mu} \int_{C_+} (\partial_{mpp} R \delta_{ij} - \partial_{ijm} R) n_j dS.$$

Now by Stokes' theorem we can transform to integrals around Γ using the following identities:

$$\int_{C_+} (\partial_{mpp} R \delta_{ij} - \partial_{ijm} R) n_j dS = \int_{C_+} \varepsilon_{klj} \partial_l (\varepsilon_{kpi} \partial_{mp} R) n_j dS$$

$$= -\oint_\Gamma \varepsilon_{kpi} \partial_{mp} R \, dl_k, \tag{3.62}$$

$$\int_{C_+} (\delta_{jm} \partial_{ipp} R - \delta_{ij} \partial_{mpp} R) n_j dS = \int_{C_+} \varepsilon_{klj} \partial_l (\varepsilon_{kim} \partial_{pp} R) n_j dS$$

$$= -\oint_\Gamma \varepsilon_{kim} \partial_{pp} R \, dl_k, \tag{3.63}$$

$$\int_{C_+} \partial_{jpp} R n_j dS = -2 \int_{C_+} \frac{(x_j - x_j')}{R^3} n_j dS = 2\Omega(\mathbf{r}'), \tag{3.64}$$

where (dl_k) is an element of arc on Γ and $\Omega(\mathbf{r}')$ is the solid angle subtended by Γ at $\mathbf{r}'$. Thus we obtain

$$u_m(\mathbf{r}') = -\frac{b_m \Omega(\mathbf{r}')}{4\pi} + \frac{b_i}{8\pi} \oint_\Gamma \varepsilon_{kim} \partial_{pp} R \, dl_k + \frac{b_i}{4\pi} \frac{\lambda+\mu}{\lambda+2\mu} \oint_\Gamma \varepsilon_{kpi} \partial_{mp} R \, dl_k$$

$$= -\frac{b_m \Omega(\mathbf{r}')}{4\pi} + \frac{1}{4\pi} \oint_\Gamma \varepsilon_{kim} \frac{b_i dl_k}{R} - \frac{1}{4\pi} \frac{\lambda+\mu}{\lambda+2\mu} \partial_m' \oint_\Gamma \varepsilon_{kpi} \frac{b_i(x_p - x_p')}{R} dl_k. \tag{3.65}$$

Here we use ∂_m' to denote $\partial/\partial x_m'$. This result is known as Burgers formula.[14] It explicitly demonstrates the fact that the displacement (and stress) field is dependent only on the dislocation line Γ and not on the cut C. If C is moved from one position to another, $u_m(\mathbf{r}')$ is unaltered, unless of course during the movement C should cross through $\mathbf{r}'$, in which case the first term in Burgers formula suffers a discrete jump.

B Peach and Koehler's formula for the stress

Once we have found the displacement field, it is a straightforward matter to derive the stress and strain fields. For the derivative of the solid angle,[13]

$$\partial_n' \Omega(\mathbf{r}') = \partial_n' \int_{C_+} \partial_j \left(\frac{1}{R}\right) n_j dS = -\int_{C_+} \partial_{nj} \left(\frac{1}{R}\right) n_j dS$$

$$= -\oint_\Gamma \varepsilon_{nkl} \partial_l \left(\frac{1}{R}\right) dl_k = -\frac{1}{2} \oint_\Gamma \varepsilon_{nkl} \partial_{lpp} R \, dl_k$$

after using Stokes' theorem and the fact that $\nabla^2(1/R) = 0$, provided $\mathbf{r}'$ is not on C_+. It then follows that

$$\partial_n' u_m(\mathbf{r}') = \frac{1}{8\pi} \oint_\Gamma \left(\varepsilon_{nkl} b_m \partial_{lpp} R - \varepsilon_{kim} b_i \partial_{npp} R - \frac{2(\lambda+\mu)}{\lambda+2\mu} \varepsilon_{kpi} b_i \partial_{mpn} R\right) dl_k,$$

and after contraction that

$$\partial_n' u_n(\mathbf{r}') = \frac{1}{8\pi} \frac{1-2\nu}{1-\nu} \oint_\Gamma \varepsilon_{kmn} b_n \partial_{mpp} R \, dl_k,$$

where ν is Poisson's ratio (cf. eq. (2.7)). Then, substituting in the stress-strain relations (2.8) and after some manipulation, we arrive at the result

$$\sigma_{ij}(\mathbf{r}') = -\frac{\mu b_n}{4\pi} \oint_\Gamma \left(\frac{1}{2} \partial_{mpp} R(\varepsilon_{jmn} dl_i + \varepsilon_{imn} dl_j) \right.$$

$$\left. + \frac{1}{1-\nu} \varepsilon_{kmn} (\partial_{mij} R - \delta_{ij} \partial_{mpp} R) dl_k \right). \tag{3.66}$$

This was first derived by Peach and Koehler.[9]

c Energies of dislocation loops

Consider two dislocation loops whose lines are $\Gamma^{(A)}$ and $\Gamma^{(B)}$ with corresponding cuts $C_{\pm}^{(A)}$ and $C_{\pm}^{(B)}$ and Burgers vectors $\mathbf{b}^{(A)}$ and $\mathbf{b}^{(B)}$. The interaction energy between the two corresponding states of strain is obtained from equation (3.37) as

$$U_{\text{int}}^{(AB)} = \int_{C_{+}^{(B)}} b_i^{(B)} \sigma_{ij}^{(A)} (\mathbf{r}^{(B)}) \, n_j^{(B)} \, dS^{(B)}. \tag{3.67}$$

Here $\sigma_{ij}^{(A)}(\mathbf{r}^{(B)})$ denotes the stress-field of the loop A evaluated at the point $\mathbf{r}^{(B)}$ on the cut spanning $\Gamma^{(B)}$. Since we have now found an expression for this stress-field, namely the Peach-Koehler formula of equation (3.66), we can substitute to obtain the energy as a double integral, with one integral running over the loop $\Gamma^{(A)}$ and the other over the surface $C_{+}^{(B)}$.

This surface integral may be reduced to a line integral around $\Gamma^{(B)}$ as follows. The term in U_{int} arising from the last bracket in equation (3.66) can be immediately reduced using equation (3.62). The first term in equation (3.66) gives rise to a surface integral which can be expressed directly as a line integral by means of Stokes' theorem. The second term in equation (3.66) produces a contribution to U_{int} equal to

$$-\varepsilon_{imn} \frac{\mu b_n^{(A)} b_i^{(B)}}{8\pi} \oint_{\Gamma^{(A)}} dl_j^{(A)} \int_{C_{+}^{(B)}} \partial_{mpp} R \, n_j^{(B)} \, dS^{(B)},$$

which may be transformed using equation (3.63). This gives rise to an additional surface integral over $C_{+}^{(B)}$, which, however, vanishes when integrated around $\Gamma^{(A)}$, since it involves a derivative with respect to $x_j^{(A)}$. This follows from the general result that

$$\oint_{\Gamma} \partial_j f \, dl_j = 0 \tag{3.68}$$

for any single-valued function f integrated around a closed loop.

After these transformations we obtain

$$U_{\text{int}}^{(AB)} = b_n^{(A)} b_i^{(B)} \frac{\mu}{8\pi} \oint_{\Gamma^{(A)}} \oint_{\Gamma^{(B)}} \left(\partial_{pp} R (2 dl_i^{(A)} dl_n^{(B)} - dl_n^{(A)} dl_i^{(B)}) \right.$$
$$\left. + \frac{2}{1-\nu} \varepsilon_{kmn} \varepsilon_{jpi} \partial_{mp} R \, dl_k^{(A)} \, dl_j^{(B)} \right). \tag{3.69}$$

Here $R = |\mathbf{r}^{(A)} - \mathbf{r}^{(B)}|$ is the distance between the running points on the two loops, and the derivatives ∂_p, etc. are with respect to the components of $\mathbf{r}^{(A)}$.

A number of formulae equivalent to equation (3.69) have been given in the literature, the two most well-known being those of Kröner[15] and Blin.[16] These will be derived from a somewhat different standpoint in Chapter 8, and are given in equations (8.33)–(8.35). The equivalence of equation (3.69) and, for example, equation (8.34) may be established by subtracting one from the other and expanding the product of the two permutation symbols by means of the formula quoted in Problem 2 of Chapter 8. The terms which do not cancel out immediately can be shown to vanish by using the result (3.68).

The similarity between the interaction energy in equation (3.37) and the expression (3.26) for the strain energy of a single loop might lead one to apply the results of the present section to such self-energies. Thus, apart from an extra factor of $\frac{1}{2}$, the self-energy $U^{(AA)}$ of a single loop $\Gamma^{(A)}$ might be taken to be given by either equation (3.69) or equation (8.34), with both line integrals running over $\Gamma^{(A)}$. Unfortunately, this double integral diverges. This is essentially the same problem of divergent energy which led us to use the hollow-core model for straight dislocations.

The use of hollow cores for general curved dislocation lines presents an intractable problem, since it would require a knowledge of the Green's function for the region exterior to the core. An alternative approach, which is in fact physically somewhat more realistic than the hollow-core model, is to treat the dislocation line as a tube of dislocation, with the Burgers vector being distributed throughout the tube rather than concentrated on a single line. The configuration is illustrated in Figure 8.1. We let $\mathbf{l}$ denote a general point on the dislocation line Γ, which forms the axis of the dislocation tube, and $\mathbf{q}$ denote a vector from the axis to the general point in each cross-section of the tube. The Burgers vector passing through an element d^2q of the cross-sectional area is assumed to be $\mathbf{b}\gamma(\mathbf{q})d^2\mathbf{q}$. The function $\gamma(\mathbf{q})$ specifies the distribution of Burgers vector throughout the tube and is referred to as the core-structure function. Now for fixed $\mathbf{q}$ and $\mathbf{q}'$ the core elements with Burgers vector $\mathbf{b}\gamma(\mathbf{q})d^2\mathbf{q}$ and $\mathbf{b}\gamma(\mathbf{q}')d^2\mathbf{q}'$ follow curves with position vectors $\mathbf{r} = \mathbf{l}+\mathbf{q}$ and $\mathbf{r}' = \mathbf{l}'+\mathbf{q}'$ as $\mathbf{l}$ and $\mathbf{l}'$ move around Γ. The interaction energy between these two core elements may be written using equation (3.69) or equation (8.34) as long as $\mathbf{q} \neq \mathbf{q}'$, and the integrals remain bounded. By integrating the resulting energy over $\mathbf{q}$ and $\mathbf{q}'$, we then obtain the results expressed in equations (8.39) and (8.40) for the self-energy of the tube of dislocation.

A discussion of the appropriate forms for the core-structure function is given at the end of §8.2. A number of applications of the integral formulae for energies which we have just derived are given in §§8.3 and 8.4.

D *A note on anisotropic media*

The result (3.61) provides the dislocation displacement field in a medium with any material symmetry. Except for the simplest cases, however, it has only formal significance since the Green's functions cannot be found explicitly, even for an infinite medium. It is possible to find the solutions for straight dislocations in an isotropic medium. These were first given by Eshelby, Read, and Shockley[17] and later in a more general context by Stroh.[18] In the particular case of hexagonal symmetry, the situation is a little simpler. For such materials, Chou and Eshelby[19] have found the energy of a circular loop in the basal plane, and Chou[20] has discussed the interaction between pairs of parallel straight dislocations.

PROBLEMS

1 Derive equations (3.8), (3.20), and (3.49) using the form (2.30) for the energy and the appropriate stress-fields for the three cases.
2 A screw dislocation in a cylinder $r_0 < r < R$ of infinite length has energy per unit length

$$\frac{U}{L} = \frac{\mu b^2}{4\pi} \ln \frac{R}{r_0} + 2\pi\gamma(r_0 + R),$$

where γ is the surface energy per unit area. Supposing that R and r_0 can change subject to the volume $\pi(R^2 - r_0^2)$ remaining constant, show that the energy is a minimum when

$$\frac{\mu b^2}{4\pi r_0}\left(1 - \frac{r_0}{R}\right) = 2\pi\gamma.$$

For most metals $\gamma \approx \mu a/10$, where a is the interatomic spacing. Then for $r_0 \ll R$ the equilibrium value of r_0 is $r_0 \approx 5b^2/4\pi^2 a$.[2]
3 In Figure 3.10, suppose that the surface $x = 0$ is held rigid (i.e. $\mathbf{u} = \mathbf{0}$ on $x = 0$) and find a suitable image S' to solve the boundary-value problem in $x < 0$. If the outer boundary of the material is the cylinder $x^2 + y^2 = R^2$, show that if $R \gg l$ traction-free boundary conditions on this surface are approximately satisfied. Show that the elastic energy is, for $r_0 \ll l \ll R$,

$$\frac{U^d}{L} \approx \frac{\mu b^2}{4\pi} \ln \frac{R^2}{2lr_0},$$

and hence find the force on S.
4 In Figure 3.10, suppose that the regions $x < 0$ and $x > 0$ are filled with materials of different shear moduli (μ and μ' respectively) which are glued together at $x = 0$. The solution in $x < 0$ can be constructed from the dislocation of strength b at S together with an image of strength αb at S' while in $x > 0$ it is equivalent to a dislocation of strength βb at S. Use the boundary conditions on $x = 0$ to determine α and β and show that the energy of the system, for $R \gg l \gg r_0$, is

$$\frac{U^d}{L} = \frac{\mu b^2}{4\pi}\left(\ln \frac{R}{r_0} - \frac{\mu - \mu'}{\mu + \mu'}\ln\frac{R}{2l}\right).$$

Obtain the results in equation (3.49) and Problem 3 as special cases of this. The dislocation is attracted towards the interface for $\mu > \mu'$ and is repelled for $\mu < \mu'$.[11]
5 Consider a right-hand screw dislocation lying at S in a circular cylinder of radius R whose surface is free (Fig. 3.13).[21] Show that the boundary

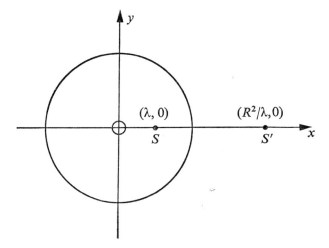

FIGURE 3.13

conditions are satisfied by the stress-fields of S and an image S' of equal and opposite strength at the inverse point (such that $OS.OS' = R^2$). Use equation (2.86) to find the energy and show that there is a force per unit length attracting S towards the cylinder surface equal to

$(\mu b^2/2\pi)\,[\lambda/(R^2 - \lambda^2)]$.

6 In Problem 5 for a cylinder of finite length, the stress-fields of S and S' do not vanish on the cylinder ends. Show that they lead to non-zero torque $M_z = \mu b(R^2 - \lambda^2)/2$. Follow the procedure in §3.1c for constructing the torsion correction, and show that it leads to an additional term in the energy equal to $-(\mu b^2/4\pi)\,[(R^2 - \lambda^2)/R^2]^2$. With this correction, show that there are three positions of equilibrium of the dislocation and discuss their stability.[21]

7 Use an image construction to find the stress-field of a screw dislocation (a) in the quadrant $x > 0$, $y > 0$, and (b) in an infinite wedge of angle π/n. In both cases the plane boundaries are free surfaces and the screw is parallel to these planes.

8 In Problem 7, for each case find the energy of the dislocation stress-field and the force on the dislocation. The method of calculating these quantities can be based either on equation (3.26) or on the method in the paragraph after equation (3.50).

9 Two edge dislocations of Burgers vectors $\mathbf{b}_1$ and $\mathbf{b}_2$ and with lines parallel to the z-axis intersect the xy-plane at (x_1, y_1) and (x_2, y_2) in the half-space $y > 0$. The boundary $y = 0$ is traction-free. Calculate the force on each dislocation for the special cases (a) $y_1 = y_2$ with $\mathbf{b}_1$ and $\mathbf{b}_2$ either both parallel or both perpendicular to the boundary, and (b) $x_1 = x_2$, again with the two special orientations of the Burgers vectors.

10 Use the method of §3.5B to find the force on the edge dislocation in Figure 3.12 when the boundary $y = 0$ is held rigid.

11 In Problem 5 of Chapter 2 we defined the displacement potentials for a state of plane strain. Show that the potentials

$$\phi = -\frac{b(1-2\nu)}{8\pi(1-\nu)}\, y \ln(x^2 + y^2), \qquad \psi = \frac{b}{2\pi}\left(x + y \tan^{-1}\frac{y}{x}\right)$$

correspond to the edge dislocation of Figure 3.2 in the region $r_0 \ll r \ll R$.

12 Obtain the displacement field corresponding to the stresses in equation
   (3.14). Show that the terms additional to those in equation (3.18) are single-
   valued.

13 Use equation (3.65) to obtain the displacement fields of straight edge and
   screw dislocations.

14 Use equation (3.69) to find the interaction energy between two straight
   parallel dislocations of screw or edge types of unequal Burgers vectors, each
   of length $2R$. Why does the result differ from equations (3.39) and (3.45)?
   (See §§8.3A and B.)

15 Two screw dislocations lie in the same glide-plane, but are non-parallel.
   Calculate the force at any point on one of them due to the stress-field of the
   other.

16 Repeat Problem 15 for edge dislocations.

17 Repeat Problem 15 for two dislocations which are non-parallel and which
   do not lie in a common place. The two Burgers vectors may be arbitrary.
   [It is perhaps simplest to take axes such that one of the dislocations lies along
   the z-axis and one of the other two axes is the mutual perpendicular of the
   two lines.]

18 Follow through the derivation of equation (3.69) from equation (3.67) and
   the proof of the equivalence of equations (3.69) and (8.34).

REFERENCES

 1 Volterra, V., Ann. Ec. Norm. **24** (1907), 401
 2 Frank, F.C., Acta Cryst. **4** (1952), 497
 3 Huntingdon, H.B., Dickey, J.E., and Thomson, R., Phys. Rev. **100** (1955),
   1117
 4 Peierls, R.E., Proc. Phys. Soc. **52** (1940), 23; Nabarro, F.R.N., Proc. Phys.
   Soc. **59** (1947), 256
 5 Colonnetti, G., Atti Reale Accad. Lincei **24** (1915), 404
 6 Steketee, J.A., Can. J. Phys. **36** (1958), 192, 1168
 7 Eshelby, J.D., Solid State Phys. **3** (1956), 79
 8 Mott, N.F., and Nabarro, F.R.N., *Report of conference on strength of solids*
   (London: Physical Society, 1948), p. 1
 9 Peach M., and Koehler, J.S., Phys. Rev. **80** (1950), 436
10 Nabarro, F.R.N., Adv. Phys. **1** (1952), 271
11 Head, A.K., Phil. Mag. **44** (1953), 92
12 Yoffe, E., Phil. Mag. **6** (1961), 1147
13 de Wit, R., Solid State Phys. **10** (1960), 247
14 Burgers, J.M., Proc. Kon. Ned. Akad. Wet. **42** (1939), 293
15 Kröner, E., *Kontinuumstheorie der Versetzungen und Eigenspannungen*,
   Ergeb. angew. Math. **5** (1958).
16 Blin, J., Acta Metall. **3** (1955), 199
17 Eshelby, J.D., Read, W.T., and Shockley, W., Acta Metall. **1** (1953), 251
18 Stroh, A.N., Phil. Mag. **3** (1958), 625
19 Chou, Y.T., and Eshelby, J.D., J. Mech. Phys. Solids **10** (1962), 27
20 Chou, Y.T., J. Appl. Phys. **33** (1962), 2747
21 Eshelby, J.D., J. Appl. Phys. **24** (1953), 176

4
DISLOCATIONS IN A PERIODIC LATTICE

4.1
FRENKEL-KONTOROVA DISLOCATIONS

A *The basic model*

A dislocation in a continuous medium has a singularity in stress at the position of the dislocation line, a difficulty which was avoided in Chapter 3 by supposing the dislocation to possess a hollow core in the form of a circular cylinder centred on the dislocation line. Such a hollow-core model is indeed physically meaningful, as we saw in §3.1B, for dislocations with large Burgers vectors, when the strain energy density near the centre is great enough to provide the surface energy for a hollow core of radius greater than the interatomic spacing. However, for dislocations with single Burgers vectors, neither the hollow-core model nor the continuum model with its singularities provides an adequate theory of dislocations in crystalline structures. It is possible (cf. §§3.6 and 8.2) to modify the theory of dislocations in a continuum in such a way as to remove the core singularity, essentially by smearing out the displacement discontinuity over some neighbourhood of the dislocation line. But in this chapter we shall examine a different approach to resolving these difficulties in which an attempt is made to include the discrete atomic nature of the material, at least in the neighbourhood of the slip-plane.

The problem of treating correctly the atomic arrangements near the core of a dislocation is a difficult one, because of both the complexity of the large number of equilibrium equations for all of the atoms involved and the complicated nature of the interatomic forces. Thus the solutions which have been given are inevitably highly idealized. They are instructive, however, in indicating certain qualitative features of the situation which can be expected to be true of the complete problem.

Historically the earliest model of dislocations in a crystal was that of Frenkel and Kontorova,[1] which was later amplified by Indenbom.[2] For the edge dislocation in Figure 1.10(a), this model regards the atoms below the slip-plane (that is those on and below the plane OC) as providing a certain periodic potential field for the atoms of the plane ND directly above the slip-plane. The period of the potential is, of course, the lattice spacing, b, and the potential minima occur, for a cubic crystal, exactly opposite the atoms on OC. If there were no dislocation present, the atoms on ND would simply sit at the bottoms of their respective potential wells. But, with the dislocation, there is one more atom on ND than there are wells, so that a configuration results such as Figure 4.1 in which are drawn the atoms above the slip-plane and the potential background provided for them by the part of the crystal below the slip-plane. We label the atoms from left to right with an index k running from $-\infty$ to $+\infty$ and we let ψ_k denote the displacement of the kth atom from the kth well, defined in such a way that

$$\psi_k \to 0 \text{ as } k \to +\infty \quad \text{and } \psi_k \to b \text{ as } k \to -\infty. \tag{4.1}$$

The periodic potential, denoted by $W(\psi)$, will then have a minimum at $\psi = 0$. The simplest form for W would be sinusoidal:

$$W(\psi) = A\left(1 - \cos\frac{2\pi\psi}{b}\right). \tag{4.2}$$

The atoms above the slip-plane exert certain forces on one another, and the simplifying assumption is made that only the interactions between pairs of adjacent atoms are important and that these interactions can be represented by a potential energy $w(\psi_{k+1} - \psi_k)$ depending only on the relative separation of the two atoms concerned. Again $w(\psi)$ must have a minimum at $\psi = 0$, since this corresponds to a relative separation equal to b, the equilibrium spacing for a perfect crystal. The simplest form for w is thus a quadratic one:

$$w(\psi_{k+1} - \psi_k) = \text{const.} + \tfrac{1}{2}\alpha(\psi_{k+1} - \psi_k)^2. \tag{4.3}$$

The total energy of the plane of atoms is thus

$$U = \sum_{k=-\infty}^{\infty} [w(\psi_{k+1} - \psi_k) + W(\psi_k)]. \tag{4.4}$$

The equilibrium positions of the atoms can now be determined by minimizing U subject to the boundary conditions (4.1). Thus we obtain the set of equations, for all integers k,

$$-w'(\psi_{k+1} - \psi_k) + w'(\psi_k - \psi_{k-1}) + W'(\psi_k) = 0. \tag{4.5}$$

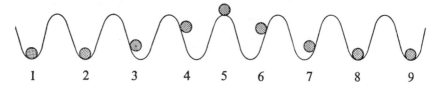

FIGURE 4.1

With the quadratic interaction of equation (4.3), this difference equation takes the form

$$\alpha(\psi_{k+1} - 2\psi_k + \psi_{k-1}) - W'(\psi_k) = 0. \tag{4.6}$$

An approximate solution can be obtained by replacing the second difference $(\psi_{k+1} - 2\psi_k + \psi_{k-1})$ by the second derivative $d^2\psi/dk^2$, when successive integrations of equation (4.6) give

$$\frac{\alpha}{2}\left(\frac{d\psi}{dk}\right)^2 = W, \tag{4.7}$$

$$\left(\frac{\alpha}{2}\right)^{1/2} \int_{b/2}^{\psi} \frac{d\psi'}{[W(\psi')]^{1/2}} = k_0 - k. \tag{4.8}$$

In equation (4.7) we have normalized W to have zero value at its minima, so that as $k \to \pm\infty$ both W and $d\psi/dk$ vanish. The constant of integration in equation (4.8) has been chosen so that $\psi = b/2$ when $k = k_0$, and we refer to the position k_0 as the centre of the dislocation. In Figure 4.1 the centre k_0 coincides with one of the maxima of W, although this need not be the case, at least within the present approximation of replacing the difference equation by a differential equation.

With the sinusoidal potential (4.2), the solution is

$$\psi = \frac{2b}{\pi} \tan^{-1}\left(\exp\frac{\pi(k_0 - k)}{l_0}\right), \tag{4.9}$$

which is sketched in Figure 4.2. The constant $l_0 = (b/2)(\alpha/A)^{1/2}$ characterizes the range of k values over which the atoms are very much disturbed from their potential wells, and $l_0 b$ is usually referred to as the *width* of the dislocation.

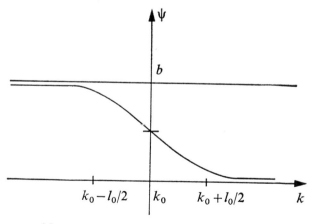

FIGURE 4.2

B *Estimation of the constants*

Various properties of the energy functions W and w can be related to the macroscopic elastic moduli. For example $W(\psi)$ specifies the change in energy per atom when one layer of atoms is sheared through a distance ψ relative to the layer below. For small ψ this corresponds to a macroscopic shear strain ψ/d, where d is the separation between the layers OC and ND in Figure 1.10(a), and hence to an energy per unit volume $\frac{1}{2}\mu(\psi/d)^2$, where μ is the elastic shear modulus. If we let c be the distance between atom layers perpendicular to the diagram of Figure 1.10, then we can equate $W(\psi)/bcd$ to this energy per unit volume for small ψ. In other words

$$W''(0) = \mu bc/d, \tag{4.10}$$

which for the particular case of equation (4.2) becomes

$$A = \mu b^3 c/4\pi^2 d. \tag{4.11}$$

In a similar way, $w(\Delta\psi)$ gives the change in energy per atom when a chain of atoms is stretched by an amount $\Delta\psi$ per interatomic spacing, that is when a longitudinal strain of amount $e_{xx} = \Delta\psi/b$ is imposed. For a chain forming part of a dislocation, it is most appropriate to compare with the macroscopic state of strain in which $e_{zz} = 0$ but $\sigma_{yy} = 0$ (with axes as in Fig. 3.2). Using equation (2.9), such a state corresponds to a transverse strain $e_{yy} = -[\nu/(1-\nu)]e_{xx}$, and hence to a strain energy per unit volume $\frac{1}{2}\lambda(e_{kk})^2 + \mu e_{ij}e_{ij} = [E/2(1-\nu^2)]e_{xx}^2$. Thus, for small $\Delta\psi$,

$$w(\Delta\psi) \sim (bcd)\,[E/2(1-\nu^2)]\,(\Delta\psi/b)^2,$$

or

$$w''(0) = [E/(1-\nu^2)]\,(cd/b). \tag{4.12}$$

This gives α directly in the case of equation (4.3).

With these values for A and α, the dislocation width becomes

$$l_0 b = (b^2/2)\,(\alpha/A)^{1/2} = \pi d[2/(1-\nu)]^{1/2}. \tag{4.13}$$

Thus the width is independent of the interatomic spacings b and c in the slip-plane itself and is directly proportional to the spacing between the slip-plane and adjacent parallel planes.

C *The energy of a Frenkel-Kontorova dislocation*

Within the present approximation of replacing the discrete set of k-values by a continuous range, the energy from equation (4.4) becomes

$$U \approx \int_{-\infty}^{\infty} \left[w\left(\frac{d\psi}{dk}\right) + W(\psi) \right] dk$$

$$= \int_{-\infty}^{\infty} \left[\frac{\alpha}{2}\left(\frac{d\psi}{dk}\right)^2 + W(\psi) \right] dk = \alpha \int_{-\infty}^{\infty} \left(\frac{d\psi}{dk}\right)^2 dk, \qquad (4.14)$$

where we have taken the special form of equation (4.3) and used equation (4.7). The solution (4.9) gives the explicit result

$$U = 4b(\alpha A)^{1/2}/\pi = (2\mu b^2/\pi^2)\,[2/(1-\nu)]^{1/2}. \qquad (4.15)$$

Although this energy is proportional to μb^2 as in the hollow-core model, it does not have the form of equation (3.23) – that is, in particular, it does not diverge as the specimen size increases. This feature must be regarded as arising from the essentially one-dimensional nature of the model, in which no account is taken of the volume distribution of strain energy. More important though for the present purposes, U has no dependence on the position of the centre k_0 of the dislocation, contrary to the expected periodic dependence. Such dependence has been averaged out by the continuous-range approximation.

In order to determine the fluctuation in energy as the dislocation moves through the potential background, we should return to the discrete set of equations (4.5) or (4.6) and calculate the energy from equation (4.4) without approximating the sum as an integral. Because of the difficulties in getting analytical results this way, Indenbom[2] suggested an intermediate approach which appears (see below) to give at least qualitatively correct results. He supposes that the discrete solution ψ_k is still given by equation (4.9) and proceeds to use it to evaluate U via equations (4.2), (4.3), and (4.4).

The solution ψ_k depends on the difference $k-k_0$, so that the sum in equation (4.4) can be written

$$\sum_{k=-\infty}^{\infty} f(k-k_0)$$

with $f(k-k_0)$ being some fairly complicated function. This sum is a periodic function of k_0 with period 1, and since it is also even in k_0 it can be represented by a Fourier cosine series. Hence

$$U = \sum_{m=0}^{\infty} B_m \cos 2\pi m k_0, \qquad (4.16)$$

where

$$B_m = 2 \int_0^1 \sum_{k=-\infty}^{\infty} f(k-k_0) \cos 2\pi m k_0 \, dk_0 \quad (m \neq 0), \qquad (4.17)$$

and B_0 is given by a similar formula without the 2 factor. Interchanging summation and integration, which is valid since $f(k-k_0)$ tends exponentially to zero for large $|k|$ for k_0 in $[0, 1]$, and introducing $k' = k-k_0$, we get

$$B_m = 2 \sum_{k=-\infty}^{\infty} \int_{k-1}^{k} f(k') \cos 2\pi mk' \, dk'$$

$$= 2 \int_{-\infty}^{\infty} f(k') \cos 2\pi mk' \, dk'. \tag{4.18}$$

Now from equations (4.2), (4.3), and (4.4),

$$f(k-k_0) = \frac{\alpha}{2} (\psi_{k+1} - \psi_k)^2 + 2A \sin^2(\pi\psi_k/b)$$

$$\approx \frac{\alpha}{2} \left(\psi'_{k+\frac{1}{2}} + \frac{1}{24} \psi'''_{k+\frac{1}{2}} + \dots \right)^2 + 2A \sin^2(\pi\psi_k/b), \tag{4.19}$$

using a Taylor approximation for ψ about the value $k+\frac{1}{2}$. Substituting the solution (4.9) then gives

$$f(k-k_0) \approx \frac{\alpha}{2} \left(\frac{b}{l_0} \right)^2 \operatorname{sech}^2 q_1 \left(1 - \frac{\pi^2}{24l_0^2} + \frac{\pi^2}{12l_0^2} \tanh^2 q_1 \right)^2 + 2A \operatorname{sech}^2 q, \tag{4.20}$$

where $q = \pi(k_0-k)/l_0$ and $q_1 = q - \pi/2l_0$. Combining equations (4.20) and (4.18) now enables the Fourier coefficients to be evaluated.

It turns out that each B_m has a factor cosech ($\pi m l_0$), and since for any physical case we expect the width bl_0 of the dislocation to be at least one or two lattice spacings, it is clear that only the smallest m values are going to contribute appreciably. These are given by

$$B_0 = \frac{2\alpha b^2}{\pi l_0} \left(1 - \frac{\varepsilon}{6} + \frac{7\varepsilon^2}{120} \right) \approx \frac{2\alpha b^2}{\pi l_0}, \tag{4.21}$$

$$B_1 = \alpha b^2 \operatorname{cosech}(\pi l_0) \left[\frac{\pi^2}{9} \left(1 - \frac{\pi^2}{120} \right) + \frac{\varepsilon}{3} - \frac{7\varepsilon^2}{60} \right] \approx 2\alpha b^2 e^{-\pi l_0}, \tag{4.22}$$

where $\varepsilon = \pi^2/12l_0^2$ and the approximations are certainly valid for $l_0 > 2$. Thus we obtain the energy of the dislocation

$$U \approx \frac{2\alpha b^2}{\pi l_0} (1 + \pi l_0 e^{-\pi l_0} \cos 2\pi k_0). \tag{4.23}$$

The term B_0 agrees with the energy in equation (4.15) calculated using an integral approximation. B_1 gives a sinusoidal variation of energy with period equal to one lattice spacing in which configurations of the type illustrated in Figure 4.1 are the unstable maxima and the stable positions are such that the centre of the dislocation coincides with one of the minima of the background potential. This energy fluctuation implies that a certain external stress is necessary to move the dislocation from one stable position to the next. If we define the force per unit length of dislocation line by equation (3.33) and compare with equation (3.36), we see that an external shear stress of magnitude

$$\sigma_{xy} = -\frac{1}{bc}\frac{\partial U}{\partial(bk_0)} = \frac{2\pi}{b^2 c}B_1 \sin 2\pi k_0$$

is necessary to overcome the energy barriers presented by the lattice. Maximizing over k_0, we obtain the resistance of the lattice to slip, usually called the *Peierls resistance*, as

$$\sigma_p = \frac{2\pi B_1}{b^2 c} = \frac{8\pi\mu}{1-\nu}\frac{d}{b}e^{-\pi l_0} \tag{4.24}$$

after using equations (4.22) and (4.12). Substituting for l_0 and taking $\nu = \frac{1}{3}$, we get a typical numerical result

$$\sigma_p/\mu \sim 40(d/b)\exp(-17d/b). \tag{4.25}$$

For a simple cubic crystal in which $d = b$ we get a ratio of approximately 4×10^{-6}, which is not an unreasonable value. Taking $d/b = (\frac{2}{3})^{1/2}$, which is the spacing ratio appropriate for face-centered cubic or hexagonal close-packed lattices, we get σ_p/μ to be about 6.5×10^{-5}.

The result (4.22) is smaller by a factor of roughly $\frac{2}{3}$ than the corresponding result obtained by Indenbom by almost the same method, because of an improvement of the approximations in the present development. On the question of approximation, it is worth pointing out that had we not included the third derivative term in equation (4.19) B_1 would have turned out to be zero and the fluctuations in U would have had a period of only half a lattice spacing with the much smaller amplitude B_2. This feature of the present calculation raises the question as to whether the Taylor series in equation (4.19) has been carried far enough. For ψ given by equation (4.9) the ratio of the next term involving a fifth derivative to the one involving the third derivative is of the order $(1/80)(\pi/l_0)^2$, so that the present approximation is certainly reasonable for $l_0 \gtrsim 3$.

As remarked earlier, it would be desirable to deal entirely with the discrete system of equations (4.6) and the expression (4.4) for U as a sum, without having resort to continuous-k methods. Numerical solutions to this problem have been obtained by Hobart[3] in the following way. The position ψ_0 of the atom $k = 0$ is fixed at some value slightly greater than $b/2$ so that the dislocation centre is at some position between the unstable equilibrium at $k = 0$ and the next stable equilibrium to the right. Then ψ_1 is selected arbitrarily and adjusted until the difference equations give a set of solutions ψ_k for $k \geq 2$, which monotonically decrease to zero as $k \to \infty$. By this technique the original two-point boundary-value problem, for the system (4.6) in which ψ_0 and $\lim_{k \to \infty} \psi_k = 0$ are specified, is changed into an initial-value problem in which ψ_0 and ψ_1 are specified, with ψ_1 being adjusted later to meet the boundary condition at infinity. For large k, ψ_k is small and, using equation (4.2), the system (4.6) is approximately $\psi_{k+1} - 2\psi_k + \psi_{k-1} = (A/\alpha)(2\pi/b)^2 \psi_k$, whose two independent solutions are $e^{\pm \lambda k}$ where λ is some real positive constant. Thus any error in ψ_1 introduces an exponentially increasing term $\delta e^{\lambda k}$ in ψ_k, which rapidly makes ψ_k either decrease below zero if $\delta < 0$ or begin to increase if $\delta > 0$. Because of this, the method used is sensitive to errors in the chosen value of ψ_1 and enables the correct value to be found quite rapidly.

A similar procedure is used to calculate the positions on the left by selecting and adjusting ψ_{-1}. In the resulting solution the $k = 0$ atom is not in equilibrium, of course, and the force necessary to hold it in position is

$$F(\psi_0) = \alpha(\psi_1 - 2\psi_0 + \psi_{-1}) - (2\pi A/b) \sin(2\pi\psi_0/b).$$

By integrating this force over values of ψ_0 between $b/2$ and the position of stable equilibrium (in which $\psi_0 = -\psi_1$) a value is obtained for the energy difference between the energy maximum and minimum, that is for $2B_1$. Thus the Peierls resistance can be calculated. Comparing with the result (4.24), Hobart found that his results agree for large widths with the indicated exponential dependence $\sigma_p \propto e^{-\pi/o}$, but the coefficient found earlier multiplying this factor is much too small, by a factor of 3.

Perhaps the most significant result emerging from these considerations is the high sensitivity of the Peierls resistance to the dislocation width and in turn to the lattice ratio (d/b). Because of this sensitivity slip is strongly favoured to take place on planes in which the packing of atoms is close and which are widely separated from neighbouring parallel planes, with the direction of the Burgers vector being that of closest packing within the plane of slip. Such behaviour is a well-established empirical feature.

c *Epitaxial films*

As a representation of a dislocation within the body of a medium, the Frenkel-Kontorova model is too idealized to be used as more than an indication of real behaviour. This is less true for dislocations close to a free surface which runs parallel to the slip-plane, in which case most of the strain will occur on just one side of the slip-plane, as the model supposes. One particular important case to which the model has been applied by Frank and van der Merwe,[4] and for which it is quite realistic, concerns a thin single-crystal film attached to the surface of a single-crystal substrate of a different material. Then for a thin enough film the positions of atoms in the substrate will be affected only a little by the presence of the film, so that their effect on the film can quite reasonably be taken as a periodic potential $W(\psi)$. The atoms of the film take up positions such as in Figure 4.1, in which the two influences of substrate potential and their own interactions are in equilibrium. For a monolayer this model is reasonably exact, but it can also be applied to layers several atoms thick, with the atoms of Figure 4.1 then representing the columns of atoms through the film thickness. As the thickness increases, however, a point is reached where the variation of strain through the thickness must be taken into account. This can be done using a modification of the Peierls-Nabarro model (§4.2)[5,6]

The problem of surface films is in reality two-dimensional. Frank and van der Merwe[4] have shown that, when the background potential and the atoms of the film both form rectangular lattices, it is possible to have edge dislocations in the two lattice directions which do not interact with one another (also two sets of screws are possible). For such configurations the one-dimensional model is applicable. There will exist, however, other solutions of the two-dimensional equations which cannot be obtained from the one-dimensional model.

The expression analogous to the combination of equations (4.2)–(4.4) for the potential energy of a chain of atoms in the present case is

$$U = \sum_{k=-\infty}^{\infty} \left[\frac{\alpha}{2} (\psi_{k+1} - \psi_k - \delta)^2 + A \left(1 - \cos \frac{2\pi \psi_k}{b} \right) \right]. \tag{4.26}$$

Here b is the interatomic spacing in the substrate, which determines the period of the background potential, and ψ_k is the displacement of the kth atom of the film relative to the kth minimum of the potential. Since the natural lattice spacing in the film will differ from that in the substrate, the interaction energy of pairs of atoms in the chain will have a minimum when $\psi_{k+1} - \psi_k = \delta$, the difference between the two natural spacings. Then, minimizing U, we obtain the equilibrium equations

$$\alpha(\psi_{k+1} - 2\psi_k + \psi_{k-1}) - (2\pi A/b) \sin(2\pi \psi_k/b) = 0 \tag{4.27}$$

or, replacing the difference by a derivative,

$$\alpha \, d^2\psi/dk^2 - (2\pi A/b) \sin(2\pi \psi/b) = 0. \tag{4.28}$$

The discrepancy δ does not appear in these equations, and thus has no influence on the solution for an infinite chain. This is not true, however, for a finite chain with, say, free ends. If the right-hand end of the chain occurs at $k = N$, the sum in equation (4.26) stops at $k = N-1$ for the first term and $k = N$ for the second, and the equilibrium equation for the Nth atom is

$$\alpha(\psi_N - \psi_{N-1} - \delta) - 2\pi(A/b) \sin(2\pi \psi_N/b) = 0. \tag{4.29}$$

This becomes simpler if we define ψ_{N+1} in such a way that equation (4.27) is satisfied for $k = N$, and then subtract equation (4.29) from it, to give

$$\psi_{N+1} - \psi_N - \delta = 0. \tag{4.30}$$

In the continuum approximation this becomes

$$d\psi/dk = \delta \quad \text{at a free end.} \tag{4.31}$$

(A somewhat different condition is obtained by approximating equation (4.29) directly. The extra term is small, however, if $l_0 = (b/2)(\alpha/A)^{1/2}$ is large, and this is a necessary condition for the validity of the continuum approximation. If the extra term is kept, the conclusions we shall draw remain essentially unchanged.)

The solution (4.9) of equation (4.28) can be generalized if we drop the constraint that $d\psi/dk \to 0$ as $k \to \pm\infty$, and we obtain (see Prob. 4.7)

$$\pm \frac{\pi(k-k_0)}{l_0 p} = \int_0^\phi \frac{d\theta}{(1-p^2 \sin^2\theta)^{1/2}} = F(\phi, p), \tag{4.32}$$

using the usual notation for elliptic integrals.[7] The constants k_0 and p arise as constants of integration, with the previous solution corresponding to $p = 1$, and we have set $\phi = \pi[(\psi/b) - \frac{1}{2}]$. The $\pm$ solutions refer to negative and positive dislocation solutions respectively.

Taking negative dislocations for definiteness, the solution is represented graphically in Figures 4.3 and 4.4. There is now a periodic array of dislocations with the distance B between dislocations being

$$B = \frac{2l_0 p b}{\pi} K(p), \tag{4.33}$$

where $K(p) = F(\pi/2, p)$ is the complete elliptic integral of the first kind. The energy per dislocation, from equation (4.26), is

$$U_D \approx \int_{k_0}^{k_0 + B} \left[\frac{\alpha}{2} \left(\frac{\partial \psi}{\partial k} - \delta \right)^2 + 2A \sin^2 \frac{\pi\psi}{b} \right] dk$$

$$= 2Al_0^2 \left(\frac{2}{\pi p l_0} [2E(p) - (1-p^2)K(p)] - \frac{2\delta}{b} + \frac{B\delta^2}{b^3} \right) \tag{4.34}$$

where $E(p)$ denotes the complete elliptic integral of the second kind.

For a finite chain, condition (4.31) must be satisfied at the ends, and substituting the solution (4.32) it follows that at a free end

$$\sin^2\phi = p^{-2} - l_0^2(\delta/b)^2. \tag{4.35}$$

Thus a solution for the finite chain can be found only if

$$(p^{-2} - 1)^{1/2} \le l_0(\delta/b) \le p^{-1}. \tag{4.36}$$

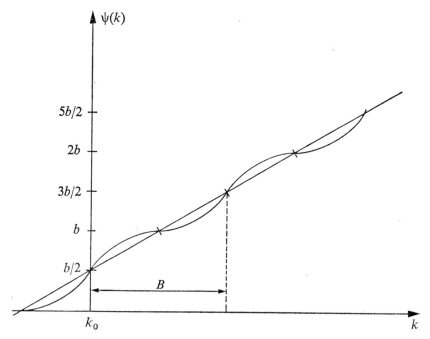

FIGURE 4.3

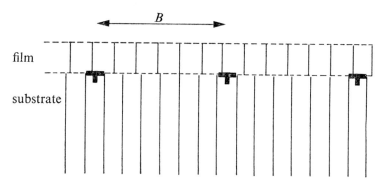

FIGURE 4.4

Now p^{-1} gives a rough measure of the density of dislocations – for small values of p, $K(p) \sim \pi/2$, and the distance between dislocations from equation (4.33) is approximately $p l_0 b$. Thus, if in the chain the dislocation density is so small that $p^{-1} < l_0(\delta/b)$, then more dislocations will be spontaneously generated in order to satisfy condition (4.36), whereas, if the density is so large that $(p^{-2} - 1)^{1/2} > l_0(\delta/b)$, then some of them will spontaneously run out from the ends.

It is a particularly important question whether the film can exist without any dislocations at all. In such a configuration, the crystal structure of the film will exactly match that of the face of the substrate to which it is attached, a phenomenon which is observed in practice and is known as *epitaxy*. Now from equation (4.33) $B \to \infty$ as $p \to 1$, so that this limit corresponds to a zero density of dislocations, and such a solution can exist with free ends provided $l_0(\delta/b) \leq 1$. This gives a maximum value for the relative discrepancy (δ/b) of the two natural lattices in order that an expitaxial film can exist.

Frank and van der Merwe[4] have estimated l_0 for a monolayer with certain assumptions about the interatomic forces and have concluded that $l_0 \sim 7$ when the film and substrate consist of the same material. This value is increased if the film is relatively more extensible or is more weakly bound to the substrate, and decreased if it is more rigid or more strongly bound. (Note also that the estimate (4.13) is of about this same order of magnitude.) Taking this value, we obtain a critical lattice discrepancy $\delta/b \sim 14$ per cent beyond which epitaxy cannot occur.

There is a further consideration that must be kept in mind, i.e. the calculation of the value of p which corresponds to the lowest total energy, since this value specifies the stable equilibrium configuration. For a given number of atoms in the chain, there will usually be several solutions of the type (4.32) satisfying the free end conditions (4.31). These correspond to different values of p within the range determined by conditions (4.36). We are here concerned with the particular one of these solutions whose energy is least. Combining equations (4.33) and (4.34) we obtain the energy per unit length of film, U_D/B, as a function of p. Minimizing with respect to p, and using

$$d[pK(p)]/dp = E(p)/(1-p^2), \qquad d[E(p)/p]/dp = -K(p)/p^2,$$

we obtain the condition that for minimum energy

$$l_0(\delta/b) = 2E(p)/\pi p. \qquad (4.37)$$

For $l_0(\delta/b) \geq 2/\pi$ this equation determines a value of p uniquely. For $l_0(\delta/b) < 2/\pi$ there is no turning value of the energy, but $d(U_D/B)/dp$ is negative for all p. The configuration with $p = 1$ then corresponds to the smallest possible energy, so that epitaxy is not only possible within this region, but actually is the stable state. For $l_0 \sim 7$ this state of affairs holds for lattice discrepancies up to about 9 per cent. Within the range $2/\pi < l_0(\delta/b) < 1$, although epitaxy is statically possible, it is only metastable and would be destroyed by thermal fluctuations at temperatures above absolute zero.

4.2
PEIERLS-NABARRO DISLOCATIONS

A *Edge dislocation*

The two most obvious criticisms that can be made against the Frenkel-Kontorova model are that it treats the two atom planes, one above and one below the slip-plane, in an asymmetric manner, and that the long-range stress-field in the two half-crystals is not included. A model of dislocations which removes these objections but is otherwise quite similar to the earlier one was first suggested by Peierls[8] and later developed by Nabarro.[9] In this model we can visualize an edge dislocation as being produced in the following way (see Fig. 4.5). The material is initially supposed to consist of two blocks, each of which can be treated as a continuum, one occupying the region $y > d/2$ and the other $y < -d/2$. The quantity d is here the separation of crystallographic layers in the y-direction and the thin space $-d/2 < y < d/2$ between the two halves of the body represents the space between the two planes of atoms on each side of the slip-plane, $y = 0$. The atoms on these two planes are drawn in Figure 4.5 in positions they would occupy if there were no forces of interaction between the two blocks. In this initial state, when neither of the two blocks is strained, the top one is wider than the bottom by an amount b, which equals the Burgers vector of the dislocation; we suppose the top block occupies the region $-L -b/4 < x < L+b/4$ in the x-direction and the lower block occupies $-L+b/4 < x < L-b/4$. The dislocation, centered at the origin, is then formed by applying equal and opposite shear tractions to the two sides $y = \pm d/2$ of the slip-plane in such a way that the top block is squeezed in by an amount $b/2$ and the bottom block expanded by $b/2$, so that both blocks occupy $-L < x < L$, and then switching on the interatomic forces across the slip-plane in order to hold the blocks in their distorted positions. The magnitude of these forces is determined by the relative positions of neighbouring pairs of atoms on the two sides of the slip-plane, which is determined by the displacement fields in the two blocks, which in turn are determined by the forces acting across the slip-plane. Hence there is a self-consistency requirement on these forces, or equivalently on the relative displacement of the two faces of the slip-plane, which enables the displacement to be found.

We let the displacement field after the dislocation is formed be $(u_x(x, y), u_y(x, y))$, which applies to both halves of the crystal, and we denote by $u_x(x, 0+)$ and $u_x(x, 0-)$ the limits onto the slip-plane, from above and below respectively, of the displacement component in the x-direction. From the symmetry of the problem, it is clear that $u_x(x, y)$ is an odd function of both x and y, and we use the notation $u(x) = u_x(x, 0+) = -u_x(x, 0-)$. The boundary conditions on the problem require that $u(x) \to -b/4$ as

E

$x \to L$ and $u(x) \to +b/4$ and $x \to -L$, so that $u(x)$ has the form sketched in Figure 4.6. (We have here allowed $L \to \infty$.)

First of all, let us suppose that $u(x)$ is a known function. Then the determination of the stress and displacement fields within each of the two blocks is just a straightforward boundary-value problem for an elastic half-space, with given tangential displacements $(\pm u(x))$ and given (zero) normal tractions. In the next chapter (§5.5B) we shall show how such problems can be solved by using linear arrays of dislocations, and obtain the general result that the Airy stress function is, for both $y > 0$ and $y < 0$,

$$\Psi(x, y) = \frac{\mu}{2\pi(1-\nu)} \int_{-\infty}^{\infty} y \ln[(x-x')^2 + y^2] u'(x') dx', \qquad (4.38)$$

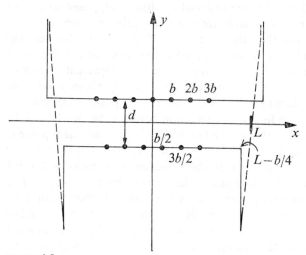

FIGURE 4.5

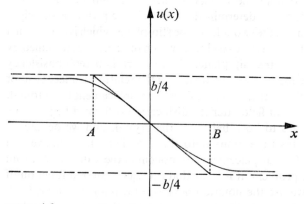

FIGURE 4.6

where $u'(x) = du/dx$ (cf. eqs. (5.100) and (5.101)). In particular, from these equations we obtain the limiting stress component:

$$\sigma_{xy}(x, 0) = -\frac{\mu}{\pi(1-\nu)} P \int_{-\infty}^{\infty} \frac{u'(x')dx'}{x-x'}, \qquad (4.39)$$

where the P indicates that the integral is a principal-value integral.

These tractions have to be supplied by the interatomic forces acting between the adjacent layers of atoms at $y = \pm d/2$. In the unstrained configurations, the atoms on $y = +d/2$ have positions $x = 0, \pm b, \pm 2b, \ldots$ while those on $y = -d/2$ have positions $x = \pm b/2, \pm 3b/2, \pm 5b/2, \ldots$ (Fig. 4.5). In the final configuration, these positions change by $u(x)$ on $y = d/2$ and by $-u(x)$ on $y = -d/2$, as shown in the enlarged diagram of Figure 4.7 for $x > 0$. The atom A_0 moves to A, through an amount $-u(x)$ to the left, while B_0 on the bottom layer moves to the right to B. The horizontal separation of A_0 and B_0 was $b/2$; that of A and B is $\phi(x) = b/2 + 2u(x)$ (note that $u(x) < 0$ for $x > 0$). For $x < 0$, the motions of atoms are exactly opposite those in $x > 0$, and the horizontal separation of neighbouring atoms is $\phi(x) = -b/2 + 2u(x)$.

The strain in the material at point x on the slip-plane is $e_{xy} = \phi(x)/2d$, and if Hooke's law were valid, this would correspond to shear stresses $\sigma_{xy}(x, 0) = 2\mu e_{xy} = \mu\phi(x)/d$. However, the strains are large – for $x \approx 0$, $e_{xy} \sim b/4d \sim \frac{1}{4}$ – so that the use of Hooke's law for the stresses becomes a questionable matter. In fact, at such large strains, it is relevant to use the periodicity of the lattice, which implies that σ_{xy} as a function of the separation ϕ must be periodic, with period b. The simplest periodic form would

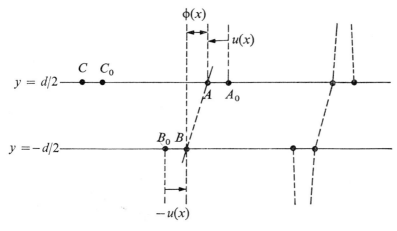

FIGURE 4.7

have $\sigma_{xy} \propto \sin(2\pi\phi/b)$, and since for small ϕ this has to reduce Hooke's law, it must have the form

$$\sigma_{xy} = \frac{\mu b}{2\pi d} \sin \frac{2\pi\phi}{b} = -\frac{\mu b}{2\pi d} \sin \frac{4\pi u}{b}. \tag{4.40}$$

Combining equations (4.39) and (4.40) then yields an integral equation for $u(x)$:

$$P \int_{-\infty}^{\infty} \frac{u'(x')dx'}{x-x'} = \frac{b(1-\nu)}{2d} \sin \frac{4\pi u(x)}{b}. \tag{4.41}$$

The solution of this equation, which satisfies the required conditions at infinity (cf. Fig. 4.6), is

$$u(x) = -\frac{b}{2\pi} \tan^{-1} \frac{x}{\zeta}, \tag{4.42}$$

where $\zeta = d/2(1-\nu)$. ζ is called the width of the Peierls-Nabarro dislocation as in the results (4.9) and (4.13) for the Frenkel-Kontorova model. (The correspondence of quantities between the two models is $x = b(k-k_0)$, $u = \psi/2 - b/4$, and $\zeta \sim bl_0/\pi$, although, as is clear, the two solutions are only in approximate agreement.) Combining equations (4.38) and (4.42) gives the stress-function

$$\Psi(x, y) = \frac{\mu b \zeta y}{4\pi^2(1-\nu)} \int_{-\infty}^{\infty} \frac{\ln[(x-x')^2+y^2]}{x'^2+\zeta^2} dx'$$

$$= -\frac{\mu b}{4\pi(1-\nu)} y \ln[x^2+(y\pm\zeta)^2], \tag{4.43}$$

where the $+$ sign applies for $y > 0$ and the $-$ sign for $y < 0$. From this, the following stresses are obtained:

$$\sigma_{xx}(x, y) = -\frac{\mu b}{2\pi(1-\nu)} \left(\frac{3y\pm 2\zeta}{D} - \frac{2y(y\pm\zeta)^2}{D^2} \right),$$

$$\sigma_{yy}(x, y) = -\frac{\mu b}{2\pi(1-\nu)} \left(\frac{y}{D} - \frac{2x^2 y}{D^2} \right), \tag{4.44}$$

$$\sigma_{xy}(x, y) = \frac{\mu b}{2\pi(1-\nu)} \left(\frac{x}{D} - \frac{2xy(y\pm\zeta)}{D^2} \right),$$

where $D = x^2+(y\pm\zeta)^2$. At large distances from the origin, these stresses approach those in equation (3.19), obtained from the hollow-core model.

B *The energy of a Peierls-Nabarro dislocation*

The energy of a Peierls-Nabarro dislocation has a similar form to that of the hollow dislocation of §3.2. Considering the first stage in the formation of the dislocation, namely the imposition of the shear tractions on the faces $y = \pm d/2$ of the two blocks, an amount of work

$$-\tfrac{1}{2} \int_{-\infty}^{\infty} \sigma_{xy}(x, 0) u(x) dx$$

is performed on each face. It can be seen from equations (4.44) that this integral diverges at infinity, just as in §3.2 the energy diverges for an infinite medium. We overcame this difficulty before by placing the dislocation at the centre of a cylinder of radius R whose outer surface is free, and we can do the same thing in the present case. Previously this required an extra term in the stress-function equal to

$$\Psi_1(x, y) = \frac{\mu b}{4\pi(1-\nu)} \frac{y(x^2+y^2)}{R^2} \tag{4.45}$$

(cf. eq. (3.13)). It is clear that in the present case this will also be the appropriate correction (for $\zeta \ll R$), since the stresses in equations (4.44) are almost identical with those of equation (3.16) except near the origin, and so will need the same stress-function to cancel them out on the cylinder $r = R$.

Combining the two stress-functions then gives

$$\sigma_{xy}(x, 0) = \frac{\mu b}{2\pi(1-\nu)} \left(\frac{x}{x^2+\zeta^2} - \frac{x}{R^2} \right) \tag{4.46}$$

The stress-function of equation (4.45) leads to displacements which vanish on $y = 0$. The energy contained in the two blocks is then, per unit length in the z-direction,

$$\frac{U}{L} = - \int_{-R}^{R} \sigma_{xy}(x, 0) u(x) dx = \frac{\mu b^2}{4\pi^2(1-\nu)} \int_{-R}^{R} \left(\frac{x}{x^2+\zeta^2} - \frac{x}{R^2} \right) \tan^{-1} \frac{x}{\zeta} dx$$

$$= \frac{\mu b^2}{4\pi(1-\nu)} \left(\ln \frac{R}{2\zeta} - \frac{1}{2} \right). \tag{4.47}$$

In addition, there is a certain amount of energy stored in the interatomic potentials across the slip-plane – that is, from the sinusoidal tractions of equation (4.40). This is often called the *misfit energy* and denoted by U_m. It turns out in fact[10] to have a value which is quite independent of the particular form of the relationship between σ_{xy} and ϕ which we assumed above to be given by equation (4.40). Let the relation be $\sigma_{xy} = \sigma_{xy}(\phi)$; then the energy per unit area of slip-plane at x is $\int_{0}^{\phi(x)} \sigma_{xy}(\phi) d\phi$, so that the total misfit energy per unit length in the z-direction is

$$\frac{U_m}{L} = \int_{-\infty}^{\infty} dx \int_{0}^{\phi(x)} \sigma_{xy}(\phi)d\phi = -\int_{-\infty}^{\infty} x\sigma_{xy}(x, 0)\frac{d\phi}{dx}\, dx \qquad (4.48)$$

after integrating by parts. But $\sigma_{xy}(x, 0)$ is related to $u(x)$ by equation (4.39), and $d\phi/dx = 2u'(x)$, so that

$$\frac{U_m}{L} = +\frac{2\mu}{\pi(1-\nu)} P \int\int_{-\infty}^{\infty} \frac{x}{x-x'} u'(x)u'(x')dx\, dx'.$$

Interchanging x and x' and adding, we get

$$\frac{U_m}{L} = \frac{\mu}{\pi(1-\nu)} \int\int_{-\infty}^{\infty} u'(x)u'(x')dx\, dx' = \frac{\mu b^2}{4\pi(1-\nu)}, \qquad (4.49)$$

since $u(\infty)-u(-\infty) = -b/2$. Adding this to equation (4.47) gives the total energy per unit length of a Peierls-Nabarro dislocation in a cylinder of radius $R\,(\gg\zeta)$ as

$$\frac{U^d}{L} = \frac{\mu b^2}{4\pi(1-\nu)}\left(\ln\frac{R}{2\zeta} + \frac{1}{2}\right). \qquad (4.50)$$

It is clear from these results that the Peierls-Nabarro model removes at least some of the weaknesses of the earlier theory. It is no longer necessary to suppose a hollow cylindrical core in order to avoid a singularity in stress at the dislocation line – the stresses of equations (4.44) are perfectly well behaved, if somewhat large, at $x = y = 0$. And the energy integral is convergent. The hollow-core model involves the introduction of a new parameter – the core radius – while in the present model the only parameters (b and d) are determined by the crystal structure. In fact, we can regard the Peierls-Nabarro theory as providing a value for r_0 since, if we compare the energy expressions (3.20) and (4.50), they become identical if we take

$$r_0 = 2\zeta/e^{3/2} = d/e^{3/2}(1-\nu). \qquad (4.51)$$

This is certainly of the right order of magnitude, although often inaccurate by a factor of 2 or more.

Despite its successes, it is possible to raise serious questions about the validity of this model. The assumption of a sinusoidal stress-displacement relation across the slip-plane is made more for mathematical simplicity than for physical applicability. Displacements of the atoms on the two planes $y = \pm d/2$ in the y-direction is ignored. Interatomic forces between, say, the planes $y = d/2$ and $y = -3d/2$ are not included – only nearest-neighbour forces are counted. And, perhaps most questionable of all, the slip-plane is treated by an odd mixture of continuum and discrete theories: it is regarded as being the continuous boundary of the two blocks $y > 0$ and $y < 0$, but as two discrete planes of atoms in computing the forces between the two blocks. The resulting stress gradients in the neighbourhood of the origin from equations (4.44) have very large values, of the order of μ/d, raising doubts about any theory which gives a continuum treatment of the core atoms. In spite of all this, however, the model constitutes a very valuable advance in the face of an immensely difficult problem.

c Screw dislocation[11,12]

It is possible to give a similar treatment of the screw dislocation. Again we start with two blocks $y > d/2$ and $y < -d/2$, which extend to infinity in the x-direction, the top one of which occupies the region $0 < z < L$ and the bottom $-b/2 < z < L-b/2$, so that the atoms on the two sides of the slip-plane (e.g. A_0 and B_0 in Fig. 4.8) are out of alignment in the z-direction by $b/2$. Then shear tractions in the z-direction are applied to the two faces $y = \pm d/2$ so as to cause an antiplane deformation, $u_z(x, y)$, with boundary values $u(x) = u_z(x, 0+) = -u_z(x, 0-)$. The asymptotic limits are

$$u(x) \to -b/4 \text{ as } x \to \infty \quad \text{and } u(x) \to b/4 \text{ as } x \to -\infty,$$

so that at $\pm \infty$ the atoms match across the slip-plane. Figure 4.8 illustrates the displacement of atoms for a typical plane $x = $ constant, $x > 0$.

The horizontal separation of atoms is again given by

$$\phi(x) = \begin{cases} b/2 + 2u(x) & (x > 0), \\ -b/2 + 2u(x) & (x < 0). \end{cases}$$

Just as in equation (4.40), this leads to shear tractions acting between the two planes, which we suppose to have the form

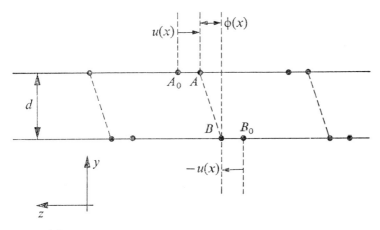

FIGURE 4.8

$$\sigma_{yz}(x, 0) = \frac{\mu b}{2\pi d} \sin \frac{2\pi\phi(x)}{b} = -\frac{\mu b}{2\pi d} \sin \frac{4\pi u(x)}{b}. \tag{4.52}$$

The antiplane displacement field $u_z(x, y)$ with boundary values $\pm u(x)$ on $y = 0\pm$ can be found using linear dislocation arrays (cf. §5.5A) and is given by

$$u_z(x, y) = -\frac{1}{\pi} \int_{-\infty}^{\infty} \tan^{-1}\left(\frac{y}{x-x'}\right) u'(x')dx'.$$

From this we obtain (cf. eqs. (5.92) and (5.95))

$$\sigma_{yz}(x, 0) = -\frac{\mu}{\pi} P \int_{-\infty}^{\infty} \frac{u'(x')}{x-x'} dx'. \tag{4.53}$$

Thus we obtain the following integral equation for $u(x)$:

$$P \int_{-\infty}^{\infty} \frac{u'(x')dx'}{x-x'} = \frac{b}{2d} \sin \frac{4\pi u(x)}{b}.$$

The solution of this equation which satisfies the appropriate conditions at infinity is

$$u(x) = -\frac{b}{2\pi} \tan^{-1} \frac{x}{\zeta}, \tag{4.54}$$

where the width ζ for a screw is equal to $d/2$. Then, using the above formula for $u(x)$, we can derive the stress components:

$$\sigma_{yz}(x, y) = \frac{\mu b\zeta}{2\pi^2} \int_{-\infty}^{\infty} \frac{(x-x')}{(x-x')^2+y^2} \frac{dx'}{x'^2+\zeta^2} = \frac{\mu b}{2\pi} \frac{x}{x^2+(y\pm\zeta)^2},$$

$$\sigma_{xz}(x, y) = \frac{\mu b\zeta}{2\pi^2} \int_{-\infty}^{\infty} \frac{y}{(x-x')^2+y^2} \frac{dx'}{x'^2+\zeta^2} = -\frac{\mu b}{2\pi} \frac{y\pm\zeta}{x^2+(y\pm\zeta)^2}, \tag{4.55}$$

where the $+$ sign applies for $y > 0$ and the $-$ sign for $y < 0$. As for the edge dislocation the stresses are bounded in the neighbourhood of the dislocation core.

We can calculate the energy of a screw dislocation, just as for an edge. The stresses of equation (4.55) satisfy approximate free-surface conditions on cylindrical surfaces when the radius $R \gg \zeta$, so that no modification is necessary for the case of a dislocation in a cylinder. The energy stored in causing the deformation fields in the two blocks is then

$$-\int_{-R}^{R} \sigma_{yz}(x, 0)u(x)dx,$$

which has the value $(\mu b^2/4\pi) \ln(R/2\zeta)$. The misfit energy per unit length is

$$\frac{U_m}{L} = \int_{-\infty}^{\infty} dx \int_{0}^{\phi(x)} \sigma_{yz}(\phi)d\phi$$

and, by an argument similar to that leading to equation (4.49), we can show that $U_m/L = \mu b^2/4\pi$. Thus the total energy per unit length is

$$\frac{U^d}{L} = \frac{\mu b^2}{4\pi}\left(\ln\frac{R}{2\zeta} + 1\right). \tag{4.56}$$

Comparing this with equation (3.4) gives a value for the core radius of

$$r_0 = 2\zeta/e = d/e. \tag{4.57}$$

D *Modified interaction*

The dislocation width $\zeta = d/2(1-\nu)$ predicted by Peierls and Nabarro is smaller than would be expected in practice, being only of the order of one lattice spacing. One reason for this is the assumption of a sinusoidal inter-action across the slip-plane which was chosen in equation (4.40). In order to overcome this difficulty, Foreman, Jaswon, and Wood[13] have considered a more general one-parameter family of interactions which lead to a variety of dislocation widths, depending on the value of the additional parameter. They proceed inversely to suppose a certain form for the displacement $u(x)$ and to calculate the shear tractions from equation (4.39). The class of solutions considered is

$$u(x) = -\frac{b}{2\pi}\left(\tan^{-1}\frac{x}{a\zeta} + \beta\frac{x\zeta}{x^2+a^2\zeta^2}\right), \tag{4.58}$$

where a is a free parameter and β will be determined later. Then from equation (4.39) we obtain, after carrying out the integration,

$$\sigma_{xy}(x, 0) = \frac{\mu b}{2\pi(1-\nu)}\left(\frac{x}{x^2+a^2\zeta^2} + 2\beta a\zeta^2\frac{x}{(x^2+a^2\zeta^2)^2}\right). \tag{4.59}$$

As $x \to \infty$, the relative displacement $\phi(x) = b/2+2u(x)$ of adjacent atoms (see Fig. 4.7) becomes small, and Hooke's law must be obeyed. Hence we must have $\sigma_{xy} \sim \mu\phi(x)/d$ as $x \to \infty$, and this requirement fixes $\beta = a-1$.

If we eliminate x between equations (4.58) and (4.59) we obtain the relation between σ_{xy} and $u(x)$ corresponding to this solution which replaces equation (4.40). Some numerical results are given in Figure 4.9.[13] It is convenient when evaluating these results to set $x/a\zeta = \cot \theta/2$, in which case, for $x > 0$,

$$\frac{2\pi\phi}{b} = \pi + \frac{4\pi u}{b} = \theta - \frac{a-1}{a}\sin\theta, \tag{4.60}$$

$$\frac{8\pi(1-\nu)\zeta}{\mu b}\sigma_{xy} = \frac{1}{a^2}[2(2a-1)\sin\theta-(a-1)\sin 2\theta]. \tag{4.61}$$

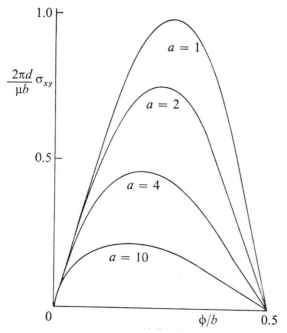

FIGURE 4.9

The maximum value of σ_{xy} represents the ideal shear strength of the lattice with the type of interaction considered. It occurs at $\theta = \pi/2$ (i.e. $\phi = b/4$) when $a = 1$ but moves nearer to $\phi = 0$ as a increases. For large a it occurs at $\theta \sim 0.62\pi$, that is at $\phi \sim 0.16b$. The value at the maximum of $(2\pi d/\mu b)\sigma_{xy}^{\max}$ is 1 for $a = 1$ and is approximately $2.23/a$ for large values of a.

The displacement function $u(x)$ in equation (4.58) has the same general form as in Figure 4.6, but the width of the curve increases as a increases. Perhaps the most convenient way of defining width for these more general cases is to construct the tangent at the origin and find the distance AB between its intersections with the asymptotes, $u = \pm b/4$ (see Fig. 4.6). Then we define the width as

$$\zeta(a) = AB/\pi,$$

with the factor π included to ensure that $\zeta(1) = \zeta$ for consistency with the earlier nomenclature. Then from equation (4.58) it is easily seen that

$$\zeta(a) = \zeta a^2/(2a-1).$$

It can be seen from these results[13] that the quantity $[4\pi(1-\nu)/\mu b]\sigma_{xy}^{\max}\,\zeta(a)$ equals 1 for $a = 1$ and tends to 1.11 as $a \to \infty$, that is it is virtually independent of a, giving a convenient relationship between dislocation widths and the shear strength of the crystal.

Foreman, Jaswon, and Wood compare their results with the observed form of dislocations in certain bubble rafts.[14] They find that the curve $a = 4$ in Figure 4.9 gives the best fit to the observed stress-strain behaviour of the rafts considered, and the solution with $a = 4$ in equation (4.58) fits reasonably well the observed distribution of displacement when the bubble raft is dislocated.

4.3
PEIERLS-NABARRO RESISTANCE

A *Atomistic calculation of energies*

Peierls[8] and Nabarro[9,10] extended their model to include asymmetrical configurations, that is configurations in which the centre of the dislocation does not coincide with a lattice site. A sequence of such configurations is shown in Figure 4.10, in which the centre translates through a distance equal to half a lattice spacing. The energy of the dislocation can again be found, and its variation with the position of the dislocation centre gives a measure of the resistance which the crystal structure imposes on the motion of the dislocation. Let the energy per unit length be $U(\xi)/L$ when the centre is at ξ. Then $U(\xi)$ must certainly be periodic, with period b, and we would expect intuitively that it should be a minimum when $\xi = 0$ (that is in the symmetrical arrangement of Fig. 4.10(a)) and should attain a maximum between $\xi = 0$ and $\xi = b$. We can define, by analogy with equation (2.92), a resistive stress per unit length

$$\sigma_p = \frac{1}{b}\left(\frac{\partial U(\xi)/L}{\partial \xi}\right)_{max} \tag{4.62}$$

(maximized over ξ), so that an external shear stress σ_{xy} greater than σ_p is necessary in order to move the dislocation out of its stable position. Thus σ_p represents one contribution to the yield stress of a material, though usually it is by no means the most important one.

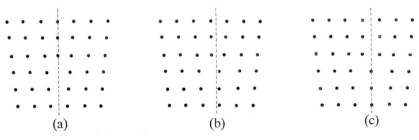

(a)   (b)   (c)

FIGURE 4.10

In order to calculate $U(\xi)$, we consider first the state $\xi = 0$ and recalculate the misfit energy from an atomistic viewpoint. Consider three atoms, as drawn in Figure 4.7, which in the unstrained state before the two crystal blocks were joined occupied the positions A_0, B_0, and C_0, having co-ordinates $(nb, d/2)$, $((n-\frac{1}{2})b, -d/2)$, and $((n-1)b, d/2)$, respectively. When the dislocation is formed, these atoms move to positions A, B, and C, and we define $\psi_+(n)$ and $\psi_-(n)$ to be the horizontal separations between A and B and between B and C respectively. Then in terms of the displacement field $u(x) = u_x(x, 0+)$, we have

$$\psi_+(n) = b/2 + u(nb) + u((n-\tfrac{1}{2})b),$$

$$\psi_-(n) = -b/2 + u((n-1)b) + u((n-\tfrac{1}{2})b). \tag{4.63}$$

These two functions give a more accurate value of horizontal separation than the function $\phi(x)$ used earlier. We shall, however, use the Taylor approximations

$$\psi_+(n) \approx b/2 + 2u((n-\tfrac{1}{4})b),$$

$$\psi_-(n) \approx -b/2 + 2u((n-\tfrac{3}{4})b), \tag{4.64}$$

in which case

$$\phi((n-\tfrac{1}{4})b) \approx \psi_+(n) \quad (n \geq 1), \qquad \phi((n-\tfrac{3}{4})b) \approx \psi_-(n) \quad (n \leq 0). \tag{4.65}$$

Now the misfit energy is the total energy of interaction between atoms on the two planes $y = \pm d/2$, and we assume that such interaction takes the form of two-body forces between pairs of atoms with an associated potential energy per unit length in the z-direction $W(\phi)$ for a pair of atom rows whose horizontal separation is equal to ϕ. Then, counting only nearest-neighbour interactions, the total energy of misfit is, per unit length,

$$U_m/L = \sum_{n=-\infty}^{\infty} \{W[\psi_+(n)] + W[\psi_-(n)]\}. \tag{4.66}$$

The interaction potential $W(\phi)$ must be chosen in a way consistent with the assumed form of the shear tractions acting between the two blocks (cf. eq. (4.40)). The force acting, say, on the atom B in Figure 4.7 is $\partial(U_m/L)/\partial u$ evaluated at $x = (n-\frac{1}{2})b$, which equals $W'[\psi_+(n)] + W'[\psi_-(n)]$, and since the separation of the atoms is on the average b, we must set this equal to $b\sigma_{xy}$. Remembering equation (4.40), we then have

$$W'[\psi_+(n)] + W'[\psi_-(n)] = \frac{\mu b^2}{2\pi d} \sin \frac{2\pi\psi_+(n)}{b} \quad (n \geq 1),$$

$$= \frac{\mu b^2}{2\pi d} \sin \frac{2\pi\psi_-(n)}{b} \quad (n \leq 0). \tag{4.67}$$

These conditions do not determine $W(\phi)$ without additional assumptions, and there are two 'simplest' cases we can consider, which represent extremes in the range of possibilities. On the one hand, we could assume that each atom on, say, $y = -d/2$ interacts with only the single nearest neighbour, so that $W'[\psi_-(n)] = 0$ for $n \geq 1$ and $W'[\psi_+(n)] = 0$ for $n \leq 0$. Such an assumption would lead to inaccuracies in the core region, where there is not very much difference in length between the BC bond and the BA bond (Fig. 4.7). On the other hand, we could take the solution

$$W(\phi) = \frac{\mu b^3}{8\pi^2 d}\left(1 - \cos\frac{2\pi\phi}{b}\right) \tag{4.68}$$

(note that $\psi_-(n) = \psi_+(n) - b$ to first order in b/ζ) in which the two types of bond contribute equally to the energy of misfit. Although this improves the treatment of the core, it introduces errors at large distances, where the interactions must certainly be dominated by the nearest neighbour. We shall consider this second possibility first since it corresponds more closely with the original results of Peierls and Nabarro, returning later to the other alternative.

Combining equations (4.64) and (4.68) with the solution for $u(x)$ obtained earlier in equation (4.42) we obtain

$$W[\psi_+(n)] = \frac{\mu b \zeta^2}{4\pi^2 d}\frac{1}{(n-\frac{1}{4})^2+q^2},$$

$$W[\psi_-(n)] = \frac{\mu b \zeta^2}{4\pi^2 d}\frac{1}{(n-\frac{3}{4})^2+q^2}, \tag{4.69}$$

where $q = \zeta/b$. The sum in equation (4.66) can now be evaluated using the result[15]

$$\sum_{n=-\infty}^{\infty}\frac{1}{(n-\alpha)^2+q^2} = \frac{\pi}{q}\frac{\sinh 2q\pi}{\cosh 2q\pi - \cos 2\alpha\pi}. \tag{4.70}$$

The final result is

$$U_m/L = (\mu b^2 \zeta/2\pi d)\tanh 2q\pi. \tag{4.71}$$

For $q = \zeta/b \gtrsim \frac{2}{3}$ there is virtually no difference between this result and equation (4.49). Since for very narrow dislocations, with ζ less than b, the whole model becomes questionable, with its continuum treatment of the dislocation core, the two methods of calculating U_m/L do not differ within the range of significance of the theory.

B *Non-symmetric configurations*

We consider now the situation in which the dislocation centre has moved to some point ξ, which does not in general coincide with a lattice site. The Peierls-Nabarro model, treating as it does the two half-spaces $y > 0$ and $y < 0$ as elastic continua, makes no stipulation that the centre should be at a lattice site. If the centre moves, it simply translates the deformation fields with it: the boundary displacement

$$u_1(x) \equiv u(x-\xi) = -(b/2\pi) \tan^{-1}[(x-\xi)/\zeta]$$

is still a solution of the integral equation (4.41) which summarizes the model. This feature, of course, constitutes a substantial weakness of the model, since it seems highly likely that, at least in the vicinity of the core, some changes in the atomic configurations will occur in moving from symmetric to asymmetric configurations. In particular, variations in the atomic displacements normal to the slip-plane, which the model completely ignores, can be expected to be an important factor.

One immediate consequence of the translation of the displacements is that there is no change in the elastic energy in the two half-spaces (cf. eq. (4.47)), the only energy change occurring being the misfit energy. But this, using equation (4.66), becomes after translation

$$\frac{U_m(\xi)}{L} = \sum_{n=-\infty}^{\infty} \{W[\psi_+(n-\xi/b)] + W[\psi_-(n-\xi/b)]\} \tag{4.72}$$

$$= \frac{\mu b \zeta^2}{4\pi^2 d} \sum_{n=-\infty}^{\infty} \left(\frac{1}{(n-\frac{1}{4}-\xi/b)^2 + q^2} + \frac{1}{(n-\frac{3}{4}-\xi/b)^2 + q^2} \right) \tag{4.73}$$

after using equations (4.69). The summations can be performed using equation (4.70), and the final result is

$$\frac{U_m(\xi)}{L} = \frac{\mu b^2 \zeta}{4\pi d} \tanh 2q\pi \left[\left(1 + \frac{\sin 2\pi\xi/b}{\cosh 2q\pi} \right)^{-1} + \left(1 - \frac{\sin 2\pi\xi/b}{\cosh 2q\pi} \right)^{-1} \right]$$

$$\approx \frac{\mu b^2}{4\pi(1-\nu)} - \frac{\mu b^2}{2\pi(1-\nu)} \exp\left(-\frac{4\pi\zeta}{b} \right) \cos \frac{4\pi\xi}{b}, \tag{4.74}$$

where the approximation is good for $\zeta \gtrsim b$. From equation (4.62) the Peierls stress is then seen to be

$$\sigma_p = [2\mu/(1-\nu)] \exp(-4\pi\zeta/b). \tag{4.75}$$

These results are roughly as expected. The energy is periodic with the symmetric configuration $\xi = 0$ being stable, and the energy fluctuations are exponentially sensitive to the dislocation width, just as in the Frenkel-Kontorova model. The numerical form for a typical value of ν,

$$\sigma_p/\mu = 3 \exp(-9d/b), \tag{4.76}$$

should be compared with the earlier result (4.25). Both results have roughly the same order of magnitude, with the present being somewhat larger. A comparison with experimental values is not simple owing to the difficulty of separating the lattice contribution to the yield stress from other contributions, in particular the effect of impurities. However, it is clear that[16] the result of equation (4.75) is certainly too high in general, and possibly the predictions of both models are somewhat too high. In this context, it is perhaps worth while illustrating the difficulties inherent in any attempt to calculate σ_p. The value of σ_p, as indicated by experiment, is not greater than $10^{-4}\mu$, and probably as a rule considerably less. The difference in energy between the positions $\xi = 0$ and $\xi = b/2$ cannot exceed $\sigma_p b^2/2$ (cf. eq. (4.62)) and this is smaller than U_m/L itself by a factor of about 10^{-4}, or less. Thus we are trying to calculate a very small effect relative to the dominant terms in the energy. Considerable care over the application of approximations is essential in any such calculation.

Equation (4.74) does contain one surprise, namely that U_m has period $b/2$ rather than simply b. One possibility is that this has arisen through a too brusque treatment of the Taylor approximation between equation (4.63) and (4.64), and this seems particularly likely after a similar experience with the Frenkel-Kontorova model.[2] However, this is not the case, and it can be shown (see Prob. 4.13) that the exact energy $U_m(\xi)$ in equation (4.72) is always periodic with period $b/2$ when the potential W satisfies $W(\phi) = W(\phi+b)$, which is true of equation (4.68). Thus it is the assumption leading to this equation which is the source of our unusual result, and in the following section we shall investigate the alternative possibility that we mentioned at the time of making it.

c Nearest-neighbour interaction model

If we suppose that only ψ_+-type bonds contribute to the misfit energy on the right of the dislocation centre, while ψ_--type contribute on the left, the appropriate solution of equation (4.67) is

$$W(\phi) = \frac{\mu b^3}{4\pi^2 d}\left(1-\cos\frac{2\pi\phi}{b}\right). \tag{4.77}$$

Furthermore, the misfit energy is, in place of equation (4.72),

$$\frac{U_m(\xi)}{L} = \sum_{n=1}^{\infty} W[\psi_+(n-\xi/b)] + \sum_{n=-\infty}^{0} W[\psi_-(n-\xi/b)]. \tag{4.78}$$

Using the Taylor approximations (4.64) for $\psi_\pm$ and equation (4.42) for $u(x)$, this becomes

$$\frac{U_m(\xi)}{L} = \frac{\mu b \zeta^2}{2\pi^2 d}\left(\sum_{n=1}^{\infty}\frac{1}{(n-\frac{1}{4}-\xi/b)^2+q^2} + \sum_{n=-\infty}^{0}\frac{1}{(n-\frac{3}{4}-\xi/b)^2+q^2}\right). \tag{4.79}$$

It is important to realize that these equations give $U_m(\xi)$ only within the range $-\frac{1}{2} < \xi/b < \frac{1}{2}$, since outside this interval the atoms nearest the core switch over from a ψ_+ to a ψ_- dominant bond or vice versa. This discrete behaviour emphasizes the limitation of the present model as opposed to the smooth transition of bonding present in the previous one. However equation (4.78) gives a more realistic representation of regions distant from the centre.

For large q (that is, for wide dislocations) the sums in equation (4.79) can be approximated by integrals, using

$$\sum_{n=1}^{\infty} f(n) \sim \int_{1/2}^{\infty} f(x)dx + \frac{1}{24}f'(\tfrac{1}{2}) + O(f'''). \tag{4.80}$$

The result is

$$\frac{U_m(\xi)}{L} \sim \frac{\mu b^2 \zeta}{2\pi d}\left[1 - \frac{1}{2\pi q} - \frac{1}{32\pi q^3} + \frac{1}{2\pi q^3}\left(\frac{\xi}{b}\right)^2 + O(q^{-5})\right]. \tag{4.81}$$

Thus again the symmetric position $\xi = 0$ corresponds to an energy minimum with an energy that equals to first order the value calculated in the last section (cf. eq. (4.74)). However, the variation with ξ is no longer even approximately sinusoidal, and the coefficient of variation is quite different from the previous one. For example, using equation (4.62) to calculate σ_p, the maximum occurs at $\xi = b/2$, which is the limit of applicability of equation (4.81), and we obtain, setting $q = \zeta/b$,

$$\sigma_p/\mu = b^3/4\pi^2 d\zeta^2. \tag{4.82}$$

Although σ_p decreases as the width ζ increases, it no longer shows the exponential dependence of the last section. Also the predicted value of σ_p is too large unless ζ is of the order of magnitude $10b$, which is wider than a physically relevant dislocation would be, and certainly would not fit the requirement that $\zeta = d/2(1-\nu)$ for any realistic d/b ratio. Since the approximations made are good for ζ down to about $3b$, we have to conclude that the present nearest-neighbour model itself overestimates σ_p quite considerably.

D Further discussion

We have seen in both of the previous sections, in which two quite different views of the interactions across the slip-plane were adopted, that the calculations tend to overestimate the Peierls resistance. A number of suggestions have been made as to the reason for this. The most direct arises from the observation that in both of the results (4.75) and (4.82) σ_p is sensitively dependent on the width ζ of the dislocation, so that the overestimation of σ_p and the underestimation of ζ are both consequences of the same flaw in the model. Hence a calculation of σ_p for the more general model of Foreman,

Jaswon, and Wood[13] (see §4.2D), in which the dislocation width can assume values larger than ζ, should produce a correspondingly reduced result for the Peierls resistance. As we shall see, this does indeed turn out to be the case.

In the previous calculations we found the interatomic potentials in equation (4.66) by equating $W'[\psi_+(n)] + W'[\psi_-(n)]$ to the shear stress $b\sigma_{xy}(x, 0)$ at the point in question. For the generalized model, we have equations (4.60) and (4.61) determining σ_{xy} and $\phi \equiv \psi_+$ in terms of θ, where $\cot \theta/2 = x/a\zeta$, and points $x > 0$ only are considered. Then using $\psi_- \approx \psi_+ - b$ (cf. eq. (4.64)), we obtain the conditions that for $x > 0$

$$W'(\phi) + W'(\phi-b) = \frac{\mu b^2}{8\pi(1-\nu)\zeta a^2} [2(2a-1) \sin \theta - (a-1) \sin 2\theta], \quad (4.83)$$

where

$$\frac{2\pi\phi}{b} = \theta - \frac{a-1}{a} \sin \theta. \quad (4.84)$$

As before, there are two simple solutions of equations (4.83) and (4.84), one in which the two neighbour interactions contribute equally and another in which only the single nearest neighbour contributes. We consider these in turn.

Supposing first that $W'(\phi) = W'(\phi-b)$ and integrating equation (4.83), we obtain, after replacing $\cot \theta/2$ by $x/a\zeta$,

$$W[\phi(x)] = \frac{\mu b^3 \zeta}{8\pi^2(1-\nu)a} \left(\frac{a}{x^2+a^2\zeta^2} + \frac{(a^2-1)a^2\zeta^2}{(x^2+a^2\zeta^2)^2} + \frac{4(a-1)^2a^4\zeta^4}{3(x^2+a^2\zeta^2)^3} \right). \quad (4.85)$$

The total misfit energy can now be obtained from equation (4.72) where, by equation (4.64), we must replace $\psi_+(n-\xi/b)$ by $\phi(nb-\frac{1}{4}b-\xi)$ and $\psi_-(n-\xi/b)$ by $\phi(nb-\frac{3}{4}b-\xi)$ in the two sums. The summations can be performed using equation (4.70) and corresponding results which can be obtained by differentiating that equation with respect to q, and we arrive at an expression for $U_m(\xi)$ which in its leading terms is precisely the same as equation (4.74) except for an extra factor a in the exponent. Thus, corresponding to equation (4.75),

$$\sigma_p = [2\mu/(1-\nu)] \exp(-4\pi a\zeta/b). \quad (4.86)$$

In view of the appearance of a in the exponent, the predicted value of σ_p can be drastically decreased by a comparatively small increase in a from the value $a = 1$ of the Peierls-Nabarro model.

A similar if less dramatic result follows for the single nearest-neighbour model. Here we assume that in equation (4.83) only the first term on the left is non-zero for $x > 0$, and the solution (4.85) for $W(\phi)$ is doubled. The misfit energy is then given by equation (4.78), where $\psi_\pm(n-\xi/b)$ are to be replaced in terms of ϕ as in the last paragraph. In place of equation (4.79), we obtain

$$\frac{U_m(\xi)}{L} = \frac{\mu b}{4\pi^2(1-\nu)}\left[\sum_{n=1}^{\infty}\left(\frac{1}{D_+}+\frac{a^2-1}{a}\frac{r^2}{D_+^2}+\frac{4(a-1)^2}{3a}\frac{r^4}{D_+^3}\right)\right.$$
$$\left.+\sum_{n=-\infty}^{0}\left(\frac{1}{D_-}+\frac{a^2-1}{a}\frac{r^2}{D_-^2}+\frac{4(a-1)^2}{3a}\frac{r^4}{D_-^3}\right)\right], \quad (4.87)$$

where $r = a\zeta/b$ and $D_\pm = (n-\tfrac{1}{2}\pm\tfrac{1}{4}-\xi/b)^2+r^2$. Using equation (4.80) to approximate the sums for large values of r, we obtain a result for $U_m(\xi)$ whose leading terms are

$$\frac{U_m(\xi)}{L} \sim \frac{\mu b^2}{4\pi(1-\nu)}\left[1-\frac{1}{2\pi r}+\frac{(2a-1)(3a-2)}{a^2}\frac{1}{2\pi r^3}\left(\frac{\xi}{b}\right)^2\right].$$

Hence, maximizing $\partial U_m/\partial\xi$, we obtain, in place of equation (4.82),

$$\sigma_p/\mu = [b^3/4\pi^2 d\zeta^2]\,[(2a-1)(3a-2)/a^5]. \quad (4.88)$$

Again σ_p can be greatly reduced by choosing a large value for a.

Another effect which will result in a reduction in the yield stress from the calculated value of σ_p arises from the fact that the dislocation lines in an actual crystal will not lie parallel to the lattice directions, but will be distributed in all directions, most of which will be inclined with respect to the lattice. The energy hills presented by the lattice will tend to make the inclined dislocation adopt a zig-zag configuration (Fig. 4.11) with short segments over the hills and the greatest lengths possible along the Peierls valleys. (The short segments, one lattice spacing in length, are usually referred to as *kinks*.) On the other hand, the total elastic energy of the dislocation is increased by increasing the length of the line, and this effect tends to

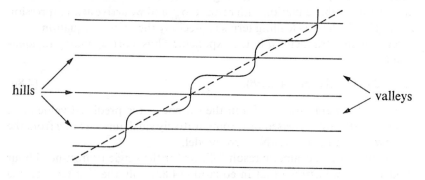

FIGURE 4.11

eliminate the kinks, leaving a straight dislocation line. Of these two compet-
ing effects, the second is, in general, dominant, since the energy associated
with an increase in length will be of the same order of magnitude as the
energy per unit length of original dislocation line, i.e. $\mu b^2/4\pi$, while the
energy increase per unit length caused by eliminating the kinks can be only
of the order of $\sigma_p b^2$, which in most cases will be much smaller. A more
detailed calculation by Mott and Nabarro[17,10] has shown that the dislocation will deviate from a straight line only by an angle of about 10^{-2} to 10^{-3}.

In relating the computed value of σ_p to the yield stress, it is assumed that
the dislocation line in equilibrium rests along one of the energy valleys. If
the line is inclined to the valleys, with the consequence that much of its
length lies across the brows of the hills even in the equilibrium configuration,
then the average stress needed to displace the line parallel to itself is greatly
reduced from the maximum resistance σ_p presented by the steepest part of
any one hill.

The misfit energy $U_m(\xi)$ was found in equation (4.49) on the assumption
that the slip-plane could be treated as a continuum, in which case it is
independent of ξ, and was calculated from equation (4.72) on the assumption
that the particles bordering the slip-plane were discrete points. Dietze[18] has
pointed out that the actual situation lies somewhere between these two: the
atoms on each side of the slip-plane are, in fact, distributions of interacting
matter occupying volumes whose diameters are comparable to the lattice
spacing. Thus we can expect the true lattice resistance to lie between the
values zero and σ_p which result from the two extreme assumptions. Treating
the case of a Gaussian distribution of matter in each atom, of width s,
Dietze shows that the resistance has an additional factor $\exp(-ks^2)$, where
k is a constant. As s increases from 0 to ∞ the transition from mass-point
model to continuum model occurs.

4.4

INTERCRYSTALLINE BOUNDARIES

A *Interface between two different crystals*

In §4.1D we described the application of the Frenkel-Kontorova model to
the problems of surface films on substrates of different material, and we
mentioned then that, when the thickness of the film exceeds a few atoms,
that model becomes inappropriate, and it is necessary to take into account
the non-uniformity of strain through the thickness of the film. Van der
Merwe[5,6,19] has shown how this can be done using an extension of the
Peierls-Nabarro model. We shall consider the extreme case in which the
thickness of the 'film' is infinite, so that the situation is very similar to
Figure 4.5, except that the two blocks consist of different materials with

different lattice sizes. A more complete investigation[6, 19] shows, in fact, that once the thickness of the film exceeds the quantity p defined in equation (4.89) it has very little effect on the total energy, and the results for an infinite film closely correspond to those of the more realistic finite film.

Naming the two materials A and B in $y > 0$ and $y < 0$ respectively, we denote by μ_a and μ_b the two shear moduli, by ν_a and ν_b the two Poisson ratios, and by a and b the two lattice spacings in the x-direction. We take $a > b$, and we shall consider the case when there exists an integer P such that $Pa = (P+1)b$. Then we define c and p by the conditions

$$p = Pa = (P+1)b = (P+\tfrac{1}{2})c, \tag{4.89}$$

so that c represents a lattice spacing intermediate between those of the two materials, and p gives the length of one period of misfit between the two lattices. In Figure 4.12 is drawn the reference configuration of the two planes of atoms on either side of the interface, where the two atoms A_0 and B_0 are exactly aligned and this alignment of atoms is repeated every distance p. The horizontal separation, $U(x)$, of two atoms A_k and B_k located near $x = kc$ increases from zero to c as x changes from zero to p (taking $p/c \gg 1$), so that in the reference state $U(x) = cx/p$. When the interactions between the two planes of atoms are switched on, deformations are induced in the two blocks, giving x-components of displacement on the boundary $y = 0$ equal to, say, $u_a(x)$ and $u_b(x)$. The resulting total misalignment is therefore

$$U(x) = \frac{cx}{p} + u_a(x) - u_b(x) \tag{4.90}$$

in the final state.

The tractions acting across the interface at x will depend on $U(x)$, and we shall assume, as in equation (4.40), that

$$\sigma_{xy}(x, 0) = \frac{\mu_0}{2\pi} \sin \frac{2\pi U(x)}{c}. \tag{4.91}$$

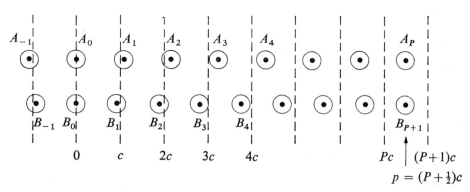

FIGURE 4.12

This equation, when combined with equation (4.90) satisfies the requirement that all quantities should be periodic in x with period p, as is clear from Figure 4.12, and also leads to shear tractions which change sign at the point of symmetry, $x = p/2$. It does, however, represent only one particular case, and although we can expect the resulting predictions to be qualitatively correct, they will differ in detail from what is observed, just as with the Peierls-Nabarro model itself. The constant μ_0 measures the shear strength of the bonding between the two crystals.

Again equation (4.39) applies to the elastic boundary-value problems in the two half-spaces, so that we have the system of equations

$$-\frac{\mu_a}{\pi(1-\nu_a)} P \int_{-\infty}^{\infty} \frac{u_a'(x')dx'}{x-x'} = \frac{\mu_b}{\pi(1-\nu_b)} P \int_{-\infty}^{\infty} \frac{u_b'(x')dx'}{x-x'}$$

$$= \frac{\mu_0}{2\pi} \sin \frac{2\pi}{p} \left(x + \frac{p}{c} u_a(x) - \frac{p}{c} u_b(x) \right). \quad (4.92)$$

Since $u_a(0) = u_b(0) = 0$, the first two parts imply that

$$[\mu_a/(1-\nu_a)]u_a(x) = -[\mu_b/(1-\nu_b)]u_b(x) \equiv (c\lambda/p)u(x), \quad (4.93)$$

say, where

$$\lambda^{-1} = (1-\nu_a)/\mu_a + (1-\nu_b)/\mu_b. \quad (4.94)$$

Then the remaining equation, when expressed in terms of $u(x)$, takes the form

$$P \int_{-\infty}^{\infty} \frac{u'(x')dx'}{x'-x} = \pi\beta^{-1} \sin \frac{2\pi}{p} [x+u(x)], \quad (4.95)$$

where

$$\beta = (2\pi\lambda/\mu_0) (c/p). \quad (4.96)$$

The constant β is small if c/p is small, that is if there is little difference between the spacings of the two lattices, and if μ_0 is large, that is if there is strong bonding between the two crystals.

The solution of this equation (or rather a related integral equation which can be derived from it – see Prob. 4.19) was originally found by van der Merwe as a power series in β^{-1}, which it turned out could be summed in closed form. The resulting solution can be written as

$$u(x) = \frac{p}{\pi} \cot^{-1} \left(\beta^{-1}[1+(1+\beta^2)^{1/2}] \cot \frac{\pi x}{p} \right) - x, \quad (4.97)$$

which can readily be verified by substitution. The misfit distance $U(x) = c[x+u(x)]/p$ has the form sketched in Figure 4.13. Rather than the linear variation of misfit which was present before the interaction across the

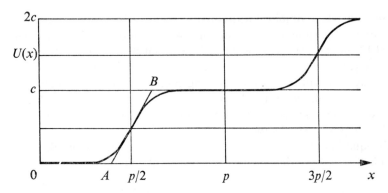

FIGURE 4.13

interface was switched on, the misfit is seen to be concentrated in the form of dislocations around the points $x = p/2$, $3p/2$, etc. Using the tangent at $p/2$ to determine the width of the dislocation at that point, we find a value $x_B - x_A = c\lambda/\mu_0$. This width will usually be of the same order as c, since we can expect that μ_0 will be of the same order as the shear moduli in many cases, whereas the distance between the dislocations is the much larger quantity p.

B *Interface energy*

The calculation of the energy associated with an interface is closely similar to that already carried out for the Peierls-Nabarro dislocation. We consider the two contributions, elastic strain energy in the two half-crystals and misfit energy corresponding to the bonding forces across the interface, calculating each for a strip $-p/2 < x < p/2$ (which suffices owing to the periodicity of the situation). From equation (4.91), the surface density of misfit energy at x is $(\mu_0 c/4\pi^2)\{1 - \cos[2\pi U(x)/c]\}$, so that the total misfit energy per unit area of interface is

$$U_m = p^{-1}(\mu_0 c/4\pi^2) \int_{-p/2}^{p/2} \{1 - \cos[2\pi U(x)/c]\}\, dx.$$

Substituting from equation (4.97) and performing the integration, we find that

$$U_m = (\mu_0 c/4\pi^2)\,[1 + \beta - (1 + \beta^2)^{1/2}]. \qquad (4.98)$$

The elastic energy stored between $x = \pm p/2$ is given by

$$-\frac{1}{2}\int_{-p/2}^{p/2} \sigma_{xy}(x, 0)\,[u_a(x) - u_b(x)]dx = -\frac{c}{2p}\int_{-p/2}^{p/2} \sigma_{xy}(x, 0)\, u\,(x)dx.$$

(The contributions from the planes $x = \pm p/2$ cancel because of the periodicity.) Obtaining σ_{xy} from equation (4.91) with $u(x)$ given in equation (4.97), the integration can be performed to give an elastic energy per unit area of interface equal to

$$U_e = -(\mu_0 c/4\pi^2)\, \beta \ln[2\beta(1+\beta^2)^{1/2} - 2\beta^2]. \qquad (4.99)$$

If we combine the two results (4.98) and (4.99), we get the total energy per unit area as a function of the parameter β. As β increases from 0 to large values, which corresponds to an increase in the difference between the lattice sizes from zero to large relative amounts, the total interfacial energy $(U_m + U_e)$ increases monotonically from 0 to a limiting value of $\mu c/4\pi^2$. When $\mu_a = \mu_b = \mu_0$, the energy reaches a level of about 85 per cent of the limit by the time the quantity c/p, which measures the misfit between the two lattices, becomes 0.4.[5]

As in §4.1D we are interested in the question of whether epitaxy can occur, that is whether a surface film can exist in a state in which its lattice spacing is constrained to match that of the substrate. Although the misfit energy is minimized in such a configuration, the strain energy in the film is increased, and the balance of these two effects determines whether epitaxy can occur. In the epitaxial state the film undergoes a uniform longitudinal strain of magnitude $e_{xx} = (a-b)/a \approx c/p$. The strain energy density, using the results of the discussion preceding equation (4.12), is $[\mu_a/(1-\nu_a)]e_{xx}^2$ so that the energy per unit area of interface is

$$U_{ep} = [\mu_a/(1-\nu_a)]\, h(c/p)^2, \qquad (4.100)$$

where h is the film thickness. Then we can say that epitaxy is the stable state if $U_{ep} < U_e + U_m$, and otherwise the dislocated configuration is stable.

For a given film thickness h, this condition determines a maximum value for the misfit parameter (c/p) which permits a stable epitaxial state. Combining equations (4.98), (4.99), and (4.100) this maximum misfit, when $\beta \ll 1$, is the solution of the equation

$$\ln \frac{e\mu_0 p}{4\pi\lambda c} = \frac{2\pi\mu_a h}{\lambda p(1-\nu_a)}.$$

To illustrate the type of behaviour implied by this condition, we quote the following numerical results[19] for the case when $\nu_a = \nu_b = 0.3$ and $\mu_a = \mu_b = \mu_0$: for $h = a$, the critical value of c/p is 0.1, which closely corresponds to the results of the Frenkel-Kontorova model; for $h = 5a$, $(c/p)_{crit} = 0.03$; for $h = 20a$, $(c/p)_{crit} = 0.01$.

c *Other interfaces*

1. Misfit in two directions
The work of the previous two sections can readily be extended to model the

situation of an interface between two crystals with lattice discrepancies in two directions. Figure 4.12 again represents the atoms in two rows in the x-direction in their reference positions, and there is now a similar diagram to represent rows in the z-direction. As well as equation (4.90), which expresses the misalignment of two atoms in the x-direction, there is the corresponding equation

$$W(z) = (c'z/p') + w_a(z) - w_b(z)$$

expressing the misalignment in the z-direction in terms of w_a and w_b, the two z-components of the displacement field, and a parameter c'/p' which is derived from the misfit of the lattice vectors in the z-direction as c/p was derived previously for the x-direction. If we now assume that $\sigma_{xy}(x, 0, z)$ is still given by equation (4.91), while the boundary value of σ_{yz} on $y = 0$ is given by a similar condition in terms of $W(z)$, $\sigma_{yz}(x, 0, z) = (\mu_0'/2\pi)$ $\times \sin[2\pi W(z)/c']$, then the problem is solved by the superposition of two solutions of the kind found earlier. One of the solutions is plane strain perpendicular to z and corresponds to an array of edge dislocations of spacing p with lines parallel to z, and the other is plane strain perpendicular to x and corresponds to an array of edges of spacing p' and lines parallel to x. The two energies will be simply additive.

The assumption that the two shear tractions σ_{xy} and σ_{yz} are determined independently by the misfits in their respective directions is hard to justify, and provides a limitation to the applicability of the model. The assumed relations are certainly consistent with the independence implied by Hooke's law, which we can reasonably expect to hold when the strains across the interface are small. However, there are regions, near the interfacial dis-locations, in which the strains are large, and in which the non-linear stress-strain relations such as equation (4.91) can be expected to involve also some cross-relation between the two sets of components. Nevertheless, the model provides a useful first approximation to the situation.

2. Screw boundary
Let us suppose that the two materials A and B on either side of the interface are identical substances in which the two-dimensional lattices of atoms in the interface are oblique rather than rectangular. The atom rows in the z-direction are taken to be parallel in A and B, but we suppose that the second set of atom rows makes equal and opposite angles (of $\theta/2$) with the x-direction in the two materials. A plan view of these two sets of atom rows is shown in the dashed lines of Figure 4.14, where the z-rows have been omitted for clarity. This interface is a simple example of a *twin boundary*, that is a boundary between two crystals which are mirror images of one another, the mirror plane being in this case the z-plane.

In their reference positions, corresponding atoms on the two planes are

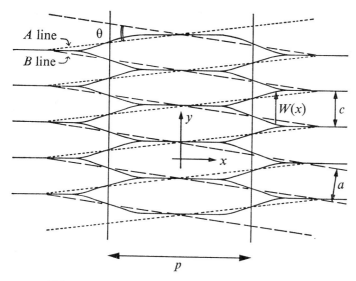

FIGURE 4.14

displaced in the z-direction relative to one another by an amount which increases linearly with x. More precisely, $W(x) = cx/p$, where $p = \frac{1}{2}a$ cosec $\theta/2$ and $c = a$ sec $\theta/2$ (a denotes the perpendicular spacings of the oblique lattice lines). When the interactions across the interface are switched on, deformation fields occur in each crystal, and the relative displacement is changed to

$$W(x) = cx/p + w_a - w_b, \tag{4.101}$$

where w_a and w_b denote the z-components of displacement on $y = 0$ in the two crystals. Then we assume that the binding forces across the interface can be represented by a shear traction σ_{yz} given in terms of W by an equation analogous to equation (4.91):

$$\sigma_{yz}(x, 0) = (\bar{\mu}/2\pi) \sin[2\pi W(x)/c]. \tag{4.102}$$

For the resulting state of antiplane strain the traction in, say, the upper half-crystal is given by equation (4.53) with u replaced by w_a. For identical materials, $w_b = -w_a$, so that we end up with the integral equation

$$P \int_{-\infty}^{\infty} \frac{w_a'(x')dx'}{x' - x} = \frac{\bar{\mu}}{2\mu} \sin \frac{2\pi}{p} \left(x + \frac{2p}{c} w_a(x) \right)$$

after substituting for σ_{yz} from equation (4.102). This equation is exactly the same as equation (4.95) if we set $u = 2pw/c$ and $\beta = \pi\mu c/\bar{\mu}p$ so that we can take over the earlier solution directly. As before, the misalignment of atoms across the interface is concentrated in regions near the points $x = \pm p/2$,

$\pm 3p/2$, etc., the width of the regions being $c\mu/2\bar{\mu}$. This is illustrated by the solid lines in Figure 4.14, which represent the displaced positions of dashed lattice lines. In the final state, the atom rows in A and B almost coincide over most of the interface, with the exceptional regions forming an array of screw dislocations whose separation is p and whose Burgers vectors are equal to c (cf. Fig. 4.13).

3. Twist boundary

The most important application of the previous example concerns the *twist boundary*, which is an interface between two identical crystals which, in their reference configurations, are simply rotated with respect to one another about a direction normal to the interface. Such an interface is of interest as one of the simplest kinds of grain boundary in a polycrystal. We shall consider only the case when the two-dimensional arrays of atoms of A and B on the interface form square lattices, each with spacing a. Now we have seen that a periodic array of screw dislocations with lines parallel to the z-axis, Burgers vectors $c = a \sec \theta/2$, and separation $p = \frac{1}{2}a \operatorname{cosec} \theta/2$ accounts for the situation in which the z-rows of atoms in the two crystals are parallel and the x-rows are inclined at an angle θ to one another. It follows that a similar array with lines parallel to the x-direction would allow the two sets of z-rows to be inclined at θ to one another while keeping the x-rows parallel. Thus we are naturally led to consider the two arrays together as a model of the twist boundary, obtaining what is termed a *cross-grid* of screws.

At the purely formal level, such a cross-grid will be a solution of the problem provided equations (4.101) and (4.102) are still valid and provided we assume a similar pair of equations for the x-component of relative displacement of corresponding atoms, U, and the shear traction σ_{xy} on $y = 0$. Thus we need

$$U(z) = cz/p + u_a - u_b,$$

$$\sigma_{xy}(y = 0) = (\bar{\mu}/2\pi) \sin[2\pi U(z)/c].$$

If these assumptions are made, then it follows immediately that the deformation field is simply a superposition of two antiplane deformations, one in the z-direction and the other in the x-direction, and there is no interference between them. In particular, the interfacial energy is the sum of the energies of the two separate states (see Prob. 4.22).

There are two criticisms to be made of these assumptions. First, there is the same point that we made in example 1, namely that in a non-linear situation such as we are considering at present we should expect some dependence of σ_{yz} on U as well as on W and of σ_{xy} on W as well as on U. Secondly, the model allows for interaction across the interface only between rows which are parallel when $\theta = 0$. As θ increases, we can expect that interactions between, say, x-rows in A and z-rows in B will become increasingly important, and will, in fact, become dominant when θ exceeds $\pi/4$. For both of these reasons, the model can be regarded as reliable only for small values of θ.

4. Tilt boundary

A second type of symmetric grain boundary occurs when the two crystals are rotated relative to one another about an axis lying in the interface – the so-called *tilt boundary*. Considering only simple cubic crystals with rotation about one of the cube axes, the situation is illustrated in Figure 4.15, in which the interface is again $y = 0$, with the two crystals being tilted through angles of $\pm \theta/2$ about the z-axis. In the reference configuration illustrated, the separation of atoms on either side of the interface in the y-direction increases linearly with x: $V(x) = cx/p$, where $c = a \sec \theta/2$ and $p = \frac{1}{2}a \csc \theta/2$. Thus, including the displacement components,

$$V(x) = cx/p + v_a(x) - v_b(x),$$

where $v_{a,b}(x)$ are the limiting values on $y = 0$ of the y-components of displacement in A and B. The forces across the interface will be principally normal in this case, changing from compressive to tensile and back again with period p in x. Thus the simplest form they can take is

$$\sigma_{yy}(x, 0) = (\mu_1/2\pi) \sin[2\pi V(x)/c],$$

which certainly has the correct periodicity in x, but, like all our earlier such sinusoidal interactions, cannot be expected to do more than approximate to reality.

Once we have made this basic assumption, the solution of the problem follows exactly as before, since, treating the two half-crystals as an isotropic elastic continuum, we have an identical integral relation to equation (4.39) connecting $\sigma_{yy}(x, 0)$ with $v_a(x)$ (cf. eqs. (5.106) and (5.109)). Consequently

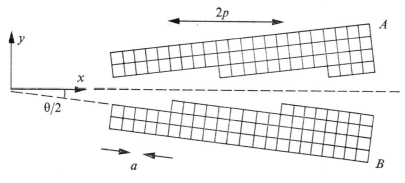

FIGURE 4.15

the solution is again given by equation (4.97), where $u(x)$ is now to be defined as $[pV(x)/c] - x$, and $\beta = [\pi\mu/\mu_1(1-\nu)](c/p) = [2\pi\mu/\mu_1(1-\nu)]\tan\theta/2$. Figure 4.13 represents the separation $V(x)$ itself. The interpretation of the solution in this case is as an array of edge dislocations, with lines in the z-direction and Burgers vectors of magnitude c in the y-direction. The dislocations have separations equal to p and widths equal to $c\mu/2\mu_1(1-\nu)$.

Further discussion of crystalline interfaces is contained in the review article by Amelinckx and Dekeyser.[20]

4.5

ATOMISTIC CALCULATIONS

It would be highly desirable to calculate the equilibrium configuration of the atoms around a dislocation using a balance-of-forces equation for each atom, together with a knowledge, possibly empirical, of the forces of inter-action between pairs of atoms. However, even waiving the question of the role of quantum mechanics in such a calculation, it presents great difficulties in terms of the algebraic complexities involved. Nevertheless, some attempts have been made along these lines and we shall briefly describe two of them.[21-23]

Both calculations concern alkali halides because these materials can be well approximated as an array of positive and negative ions with simple electrostatic forces between them, with in addition short-range repulsive forces arising from the ion cores. The equilibrium crystalline arrangement can be obtained from a simple cubic lattice by filling the lattice positions alternately with positive and negative ions (see Figure 4.16 in which shaded and unshaded circles denote the two types of ion). It is known from experimental evidence that slip occurs predominantly in the $\langle 110 \rangle$ directions on $\{1\bar{1}0\}$ planes in these materials, and we shall consider here only the case of pure screw dislocations which can produce such slip, Other dislocations producing the same slip as well as other slip-planes are considered by Huntingdon, Dickey, and Thomson.[21,22] We take explicitly a dislocation whose Burgers vector and line direction lie parallel to [110], which we denote by Oz' (Fig. 4.16), which can glide in the z-direction on the plane $(1\bar{1}0)$. Axes Ox' and Oy' are taken in the directions shown (i.e. $x' = z$ and Oy' is perpendicular to Ox' and Oz').

Figure 4.17 shows the arrangement of atoms looking along Oz', with small and large circles denoting atoms on successive planes, spaced by an amount $b/2$ in the z'-direction ($b = a\sqrt{2}$, with a being the cube dimension). A screw dislocation lying along Oz', with Burgers vector b, then has a variety of configurations according to the position of its centre along the x'-direction, with two extremes occurring when the centre is at positions such as O_1 and O_2 in Figure 4.17. The one of these two positions which leads to

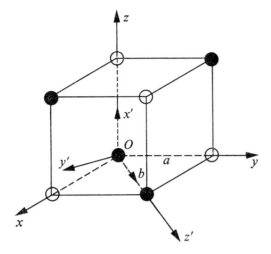

FIGURE 4.16

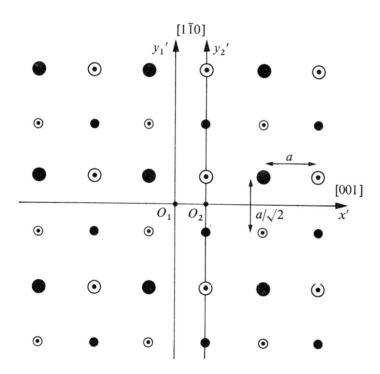

FIGURE 4.17

the least total energy corresponds to the stable configuration for the dislocation, with the difference in energy measuring the barrier to dislocation
motion, and hence giving an estimate of the Peierls stress (at least provided
no intermediate configuration has an even higher energy). We refer to the
two configurations as I and II.

In Maradudin's calculation,[23] it is assumed that atoms are displaced only
in the z'-direction, and we denote by $w_{m,n}$ the displacement of the row of
atoms which in Figure 4.17 occupies the position $(ma, na/\sqrt{2})$. n has values
$\pm\frac{1}{2}, \pm\frac{3}{2}, \ldots$, while m has the same set of values in configuration I and the
set $0, \pm 1, \pm 2, \ldots$ in configuration II. It is then assumed that only interaction
energies between nearest-neighbour atom rows need be taken into account,
and that these can be approximated by quadratic terms, much as in equation
(4.3) for the Frenkel-Kontorova model. Thus the total energy is

$$U = \sideset{}{'}\sum_{m,n} [\tfrac{1}{2}A(w_{m+1,n} - w_{m,n})^2 + \tfrac{1}{2}B(w_{m,n+1} - w_{m,n})^2]$$
$$+ \sum_{m>0} \tfrac{1}{2}B(w_{m,1/2} - w_{m,-1/2} - b)^2,$$

where in the first sum the interactions accounted for by the second sum are
to be omitted. These latter terms allow for the relative displacement
through one Burgers vector of atoms on the half $x' > 0$ of the slip-plane.

Minimizing the energy leads to a set of difference equations for the $w_{m,n}$
which Maradudin solves explicitly for the two configurations (in II there is
the additional symmetry property that $w_{0,1/2} = -w_{0,-1/2} = b/4$). At
large distances from the centre, the displacement solution approaches that
of continuum elasticity theory, and near the centre some departures occur,
although these turn out to be relatively small, particularly in configuration
I. The energies in the two configurations can be fitted to expressions such as
equation (3.4) and so lead to predicted values for the core radius, r_0. Using
values of the parameters A and B appropriate to sodium chloride,
Maradudin finds $r_0 = 0.13b$ in I and $0.85 \times 10^{-3}b$ in II. The second of these
values is much smaller than expected, but the first is not all that different
from the prediction of the Peierls-Nabarro model (cf. eq. (4.57)). The
corresponding difference in energy between the two configurations is also
much greater than measurements of yield stresses would lead us to believe,
being in fact of the same order of magnitude as the energies themselves.

One source of error which contributes to the surprising results of
Maradudin's model for configuration II is the exclusion of displacements in
the $x'y'$-plane. In the calculation of Huntingdon, Dickey, and Thomson,[22]
such displacements are allowed, and they are found to be significant,
particularly the y' displacements of the two rows $m = 0, n = \pm\frac{1}{2}$ in configuration II. These authors also use more realistic expressions for the
interaction potentials between pairs of atom rows in sodium chloride. Their

procedure is to set, as a zeroth approximation, the atoms in positions indicated by continuum elasticity, and then to adjust their positions progressively outwards from the dislocation centre, minimizing the interaction energy at each step. This procedure is followed out to distances enclosing up to 24 atom rows. Calling this maximum distance r_1, the resulting minimum total energy of the rows within $r < r_1$ is fitted to an expression of the type $A \ln(r_1/r_0)$, where A is the coefficient obtained from continuum elasticity (i.e. A would be $\mu b^2/4\pi$ for an isotropic material). The resulting core radius turns out to be $\sim 0.5b$ in both configurations – there is, in fact, relatively little difference in this improved calculation between the energies of the two configurations, although I appears still to be slightly the more stable. A somewhat surprising outcome of this analysis is that the dislocation is spread out more in the y'-direction than in the x'-direction, which might imply that the Peierls resistance is lower for motion in the former direction, that is [$1\bar{1}0$]. This would conflict with observation if it did turn out to be the case, but the question must remain open since values for the two energy barriers are not calculated.

PROBLEMS

1 Find the general Fourier coefficient B_m from equation (4.18) and verify that $B_m \ll B_1$ for $m \geqslant 2$.

2 Investigate the effects on B_1 of including the next higher term (fifth derivative) in the Taylor approximation in equation (4.19). Find a limit on l_0 which ensures that the additional term is small.

3 The background potential in equation (4.2) can be improved by including the next harmonic:

$$W(\psi) = A\{1 - \cos(2\pi\psi/b) + k[1 - \cos(4\pi\psi/b)]\}.$$

Derive limits on k which ensure that W has minima only at the lattice points $\psi = nb$. How are equations (4.11) and (4.13) changed? Obtain the solution analogous to equation (4.9).

4 The background of equation (4.2) can be approximated by a sequence of parabolic wells, so that for $k > k_0$, $W(\psi) \approx (2\pi^2 A/b^2)\psi^2$. Find the solution ψ_k in this case, making an appropriate modification for $k < k_0$.

5 Obtain the analogues of equations (4.5) and (4.6) when interactions such as $w(\psi_{k+2} - \psi_k)$ between second nearest neighbours are taken into account. Using the parabolic model of Problem 4 find the solution ψ_k and show that if $l_0 \gg 1$, $\psi_k \approx (b/2) \exp[-\pi(k - k_0)/l_0\sqrt{5}]$ for $k > k_0$. Is this a reasonable way of including second-neighbour forces?

6 Suggest an alternative numerical method for solving the difference equations (4.27) subject to the conditions that ψ_0 and $\lim_{k\to\infty} \psi_k = 0$ are given.

7 Obtain the general solution (4.32) of equation (4.28) and derive equations (4.33) and (4.34).

8 Show that p can always be found such that conditions (4.36) are satisfied.

9 Find the condition analogous to equation (4.31) that would hold at the boundary edge between two surface films laid on the same substrate.

10 Can the Frenkel-Kontorova model be used to describe screw dislocations?

11 To describe the the two-dimensional extension of the Frenkel-Kontorova
    model[14] necessary for a complete description of expitaxial films, we
    consider a square lattice, labelling the potential wells by two indices (k, l),
    and let $(\psi_{k, l}, \phi_{k, l})$ be the displacement components in the two lattice
    directions of the atom (k, l) in the film relative to the well (k, l). In place of
    equation (4.26), we take

$$U = \sum_{k,l = -\infty}^{\infty} \left[\tfrac{1}{2}\alpha(\psi_{k+1, l} - \psi_{k, l} - \delta)^2 + \tfrac{1}{2}\alpha(\phi_{k, l+1} - \phi_{k, l} - \delta)^2 \right.$$
$$+ \tfrac{1}{2}\beta(\psi_{k, l+1} - \psi_{k, l} + \phi_{k+1, l} - \phi_{k, l})^2 + \tfrac{1}{2}\gamma(\psi_{k+1, l} - \psi_{k, l} - \delta)(\phi_{k, l+1} - \phi_{k, l} - \delta)$$
$$\left. + A\left(2 - \cos\frac{2\pi\psi_{k, l}}{b} - \cos\frac{2\pi\phi_{k, l}}{b}\right) \right].$$

Interpret the constants β and γ. Obtain the equations of motion and show that
solutions in which $\psi_{k, l}$ is independent of l and $\phi_{k, l}$ of k, or vice versa, are
possible. Interpret these two types of solution. Making the usual difference
to derivative replacement, show that solutions are also possible in which
$\psi_{k, l} = \pm\phi_{k, l}$. Find these explicitly when $\psi_{k, l} = \psi(s)$, where $s = k\cos\theta$
$+ l\sin\theta$ with θ constant, and interpret the result.

12 Derive the result (4.49) for the misfit energy directly using the explicit
    solutions for the Peierls-Nabarro displacement field and shear stress.

13 Combining equations (4.72) and (4.63), show that $U_m(\xi + b/2) = U_m(\xi)$
    whenever $W(\phi) = W(\phi + b)$, so that the misfit energy has period $b/2$ in such
    cases.

14 In obtaining equations (4.69), the approximation (4.64) was used in equation
    (4.68). Suppose the next non-zero term in the Taylor approximation of
    equations (4.63) is included. What are the resulting values of $W[\psi \pm (n)]$?
    Show that the total misfit energy in equation (4.71) then has a correction
    factor $(1 + 1/64q^2 \ldots)$ where the omitted terms involve factors $e^{-2\pi q}$ or
    smaller $(q = \zeta/b)$.

15 Use the Fourier series technique of equations (4.16)–(4.18) to analyse the
    expression (4.73). Show that the first three terms in its Fourier series
    reproduce the result (4.74).

16 Use the method of Problem 15 when $U_m(\xi)$ is given by equation (4.79) to
    show that the Fourier coefficients of $U_m(\xi)/L$ are given by

$$B_m = \frac{2\mu b\zeta^2}{\pi^2 d} \int_{\frac{1}{2}}^{\infty} \frac{\cos 2\pi mk}{(k - \frac{1}{4})^2 + q^2} \, dk \quad (m \neq 0).$$

Show that for large q, $B_0 \sim (\mu b^2\zeta/2\pi d)(1 - 1/2\pi q)$ and $B_m \sim (-1)^m\mu b^5/$
$4\pi^4 m^2 d\zeta^2$ $(m \neq 0)$. Compare the amplitude $2B_1$ of the first harmonic with the
amplitude of equation (4.81).

17 Derive the results (4.86) and (4.88).

18 If in the state represented by Figure 4.5 the atoms on $y = \pm d/2$ are exactly
    in phase, then in any state of dislocation the horizontal separation of
    adjacent atoms, $\phi(x)$, equals $2u(x)$ (provided that $u(x)$ nowhere exceeds $b/4$).
    Obtain the equation analogous to (4.40) and (4.41). Show that

$$u(x) = \frac{b}{2\pi}\cot^{-1}\left(\frac{x^2\sin\theta}{4\zeta^2} - \cot\theta\right) + \frac{b\theta}{8\pi}$$

satisfies the new equations, provided an additional stress equal to $(\mu b/2\pi d)$
$\sin\theta/2$ is included in equation (4.40). By sketching $u(x)$, show that it

represents a dislocation pair, whose separation is $2\zeta/\theta$ for small values of θ, superimposed on a uniform shear of the lattice.[10] Why is this shear necessary?

19 Use the fact that $u(x)$ has period p to reduce the left-hand side of equation (4.95) to an integral over the interval $(0, p)$. (You should find a kernel involving a cotangent.)

20 Verify that equation (4.97) satisfies equation (4.95). (A contour method is perhaps most suitable for evaluating the integral.)

21 By allowing $p \to \infty$ in §4.4A, retrieve the results of the Peierls-Nabarro model.

22 Show that the interfacial energy (per unit area) of a twist boundary is $(\bar{\mu}c/2\pi^2)\ \{1+\beta-(1+\beta^2)^{1/2}-\beta\ln[2\beta(1+\beta^2)^{1/2}-2\beta^2]\}$, where $\beta = \pi\mu c/\bar{\mu}p = (2\pi\mu/\bar{\mu})\tan\theta/2$.

23 Show that the interfacial energy per unit area of a symmetric tilt boundary is equal to one-half the expression of Problem 22 with $\bar{\mu}$ replaced by μ_1 and β defined by equations (4.94) and (4.96).

24 Using equations (5.93), (5.94), (5.100), and (5.107), derive the stress-fields of the various boundaries considered in §4.4. (For the results, see reference 5.) All of the stress components have a factor $\exp(-2\pi y/p)$ in $y > 0$.

REFERENCES

1 Frenkel, J., and Kontorova, T., Phys. Z. Sowjet **13** (1938), 1

2 Indenbom, V.L., Sov. Phys. Cryst. **3** (1958), 193

3 Hobart, R., J. Appl. Phys. **36** (1965), 1944

4 Frank, F.C., and van der Merwe, J.H., Proc. Phys. Soc. **A198** (1949), 205, 216

5 van der Merwe, J.H., Proc. Phys. Soc. **A63** (1950), 616

6 van der Merwe, J.H., J. Appl. Phys. **34** (1963), 117

7 Abramowitz, M., and Stegun, I.A. (eds.), *Handbook of mathematical functions* (New York: Dover, 1965)

8 Peierls, R.E., Proc. Phys. Soc. **A52** (1940), 34

9 Nabarro, F.R.N., Proc. Phys. Soc. **A59** (1947), 256

10 Nabarro, F.R.N., Adv. Phys. **1** (1952), 360

11 Eshelby, J.D., Phil. Mag. **40** (1949), 903

12 Leibfried, G., and Dietze, H.D., Z. Phys. **126** (1949), 790

13 Foreman, A.J., Jaswon, M.A., and Wood, J.K., Proc. Phys. Soc. **A64** (1951), 156

14 Bragg, W.L., and Nye, J.F., Proc. Roy. Soc. **A190** (1947), 474

15 Cottrell, A.H., *Dislocations and plastic flow in crystals* (Oxford, 1953), p. 98

16 Cottrell, A.H., Progr. Metal Phys. **1** (1949), 91

17 Mott, N.F., and Nabarro, F.R.N., *Bristol conference on strength of solids* (London: Physical Society, 1948), p. 1

18 Dietze, H.D., Z. Phys. **132** (1952), 107

19 van der Merwe, J.H., in *Single crystal films*, ed. M.H. Francombe and H. Sato (London: Macmillan, 1964), p. 139

20 Amelinckx, S., and Dekeyser, W., Solid State Phys. **8** (1959), 327

21 Huntingdon, H.B., Phys. Rev. **59** (1941), 942

22 Huntingdon, H.B., Dickey, J.E., and Thomson, R., Phys. Rev. **100** (1955), 1117

23 Maradudin, A.A., J. Phys. Chem. Solids **9** (1958), 1

F

5

LINEAR DISLOCATION
ARRAYS AND CRACKS

5.1
HEAD-LOUAT SOLUTIONS

A *Equilibrium equations for an array*

Consider a row of coplanar and parallel straight screw dislocations, lying, say, in the plane $y = 0$ with their lines and Burgers vectors parallel to the z-axis, in an infinite isotropic medium. Suppose there are n dislocations in the row with x-coordinates $x_1, x_2, \ldots, x_n$ (Fig. 5.1) and Burgers vectors $b_1, b_2, \ldots, b_n$ (which would in most cases equal $\pm b$, the smallest possible Burgers vector). Such an array would be produced, for example, by a Frank-Read source lying in the plane $y = 0$. The sequence of concentric loops which would be produced by the operation of a source of this kind becomes more and more like a series of parallel straight dislocations as it moves farther from the source. This effect is emphasized by the differing mobilities of edge and screw dislocations, which often lead to one pair of sides of the loop moving away relatively quickly, leaving the slower sides as long and straight lines forming a row of the above type. We shall be concerned with the equilibrium configurations of such linear arrays under the influence of certain prescribed external stresses, taking account of the

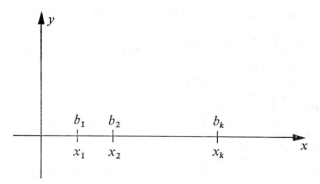

FIGURE 5.1

mutual interactions of dislocations in the array. An example, perhaps the most important, of the type of physical situation of interest here is that of the dislocation pile-up. In this case, there is a barrier on the plane $y = 0$ (for instance a grain boundary) which prevents the dislocations from moving past it, and under the influence of a suitable external shear stress, they will pile up against it. Large stresses therefore arise in the neighbourhood of the barrier, whose magnitude is an important factor in certain phenomena.

The shear stress at the position x_k of the kth dislocation due to all the other dislocations is, from equation (3.2),

$$\sigma_{yz}^{(k)} = \sum_{l \neq k} \frac{\mu}{2\pi} \frac{b_l}{x_k - x_l}. \tag{5.1}$$

This will cause a force per unit length equal to $b_k \sigma_{yz}^{(k)}$, and the kth dislocation will be in equilibrium if this force together with any additional forces, which we write as $b_k P_k$, is equal to zero. P_k contains the external shear stress $\sigma_{yz}(x_k)$ as well as any internal dislocation forces such as the Peierls-Nabarro resistance and the short-range stress-fields of barriers. Thus we have the equilibrium equations

$$\sum_{l \neq k} \frac{\mu}{2\pi} \frac{b_l}{x_k - x_l} + P_k = 0 \quad (k = 1, 2, \ldots, n). \tag{5.2}$$

The solution of these equations for the equilibrium positions (x_k) in general is difficult. Eshelby, Frank, and Nabarro[1] have obtained closed-form solutions for several problems of this kind, showing, for example, that for a simple pile-up of n dislocations under uniform stress with one of the dislocations locked at $x = 0$ the positions of the other $n-1$ are given, in appropriate units, by the zeros of L_n', the derivative of the nth Laguerre polynomial. Other authors have used computers to derive numerical solutions of equations (5.2). However, a much simpler method, first suggested by Leibfried,[2] which leads to a soluble system of equations for any coplanar array, and which appears to be valid provided there are about five or more dislocations involved in the array,[3] is to make a continuum approximation right from the beginning. Instead of dealing with a row of discrete dislocations, we suppose that the slip-plane $y = 0$ contains a smeared-out sheet of dislocation density, with an amount $bf(x)dx$ of Burgers vector between x and $x+dx$. Then the shear stress at some point x on the array due to all the dislocation except that in the neighbourhood of x itself is

$$\int_{-\infty}^{x-\varepsilon} + \int_{x+\varepsilon}^{\infty} \frac{\mu b \, f(x')dx'}{2\pi} \frac{}{x - x'},$$

where we have excluded an interval $(x-\varepsilon, x+\varepsilon)$ just as we exclude $l = k$ in the sum in equation (5.1). Taking the limit as $\varepsilon \to 0$, we obtain a Cauchy principal-value integral,

$$P \int_{-\infty}^{\infty} \frac{\mu b}{2\pi} \frac{f(x')dx'}{x-x'} \,. \tag{5.3}$$

If now the external shear stress at x is $P(x)$, the equilibrium equations take the form of a singular integral equation

$$\frac{\mu b}{2\pi} P \int_{-\infty}^{\infty} \frac{f(x')dx'}{x-x'} + P(x) = 0. \tag{5.4}$$

We have written the limits as $\pm\infty$, although in most cases the array would be bounded and the limits finite. Equation (5.4) must hold at all points x on the array (i.e. at which $f(x) \neq 0$).

Singular integral equations have received considerable attention from a number of researchers over the last fifty years and a great deal is known about their solutions.[4,5,6] In the appendix to this chapter we give a brief account of the theory to the extent that it concerns us here. Using the notation $F(x) = (2\pi/\mu b)P(x)$, the general solution of equation (5.4) when the integration range is actually the finite range (a, b) is given by equation (5.A.17) where D is an arbitrary constant. If $f(x)$ is required to be bounded at $x = a$, the solution is equation (5.A.18), and if $f(x)$ is bounded at both $x = a$ and $x = b$, a solution exists only if condition (5.A.19) is met, in which case the solution is given by equation (5.A.20). More generally, when $f(x)$ is non-zero in several disjoint regions, the solution is described in equations (5.A.24) and (5.A.25). In the following section we shall give four of the examples solved by Head and Louat[7] using this technique of a continuum approximation.

B *Examples*

1. n positive screws are locked between two barriers at $x = \pm a$, and are in equilibrium under their own mutual repulsions, with no external forces acting. Then $P(x)$ and $F(x)$ are zero, and from equation (5.A.17) with $b = -a$

$$f(x) = D_1/(a^2 - x^2)^{1/2}. \tag{5.5}$$

Thus there is no possibility of avoiding the unboundedness of $f(x)$ in the neighbourhoods of the two blocks. The constant D_1 is determined by the requirement that the total Burgers vector between $-a$ and a be nb:

$$n = \int_{-a}^{a} f(x)dx = \pi D_1.$$

2. Two positive dislocations are locked at $x = \pm a$ and n more occupy the region between, being in equilibrium under their own mutual repulsions and the repulsion of the two locked dislocations. We use the continuum approximation of the n free dislocations, and treat the shear stress of the two locked ones as an external force, so that

$$P(x) = \frac{\mu b}{2\pi}\left(\frac{1}{x+a} + \frac{1}{x-a}\right) \text{ and } F(x) = \left(\frac{1}{x+a} + \frac{1}{x-a}\right). \tag{5.6}$$

The two fixed dislocations prevent any of the free ones from approaching too closely to $x = \pm a$, and it is reasonable to suppose that $f(x) = 0$ in some region $b < |x| < a$. Taking $f(x)$ to be bounded at $x = \pm b$, the solution from equation (5.A.20) is

$$f(x) = -\frac{1}{\pi^2}(b^2 - x^2)^{1/2} P \int_{-b}^{b} \frac{1}{(b^2 - x'^2)^{1/2}}\left(\frac{1}{x'+a} + \frac{1}{x'-a}\right)\frac{dx'}{x'-x}$$

$$= \frac{2a}{\pi(a^2-b^2)^{1/2}}\frac{(b^2-x^2)^{1/2}}{a^2-x^2}. \tag{5.7}$$

Condition (5.A.19) is automatically satisfied since $F(x)$ is an odd function of x, and b is determined from the condition

$$\int_{-b}^{b} f(x)dx = n$$

as

$$b = a[n(n+4)]^{1/2}/(n+2).$$

For large values of n, $b \approx a$, and the solution for $f(x)$ becomes almost identical with that in the first example.

3. n positive screw dislocations lying in the region $x > 0$ are driven to the left by a constant external shear stress $\sigma_{zy} = -\sigma$. They are prevented from moving beyond $x = 0$ by a barrier there, and are forced to pile up against it. We suppose that $f(x)$ is non-zero over some range $0 < x < a$, where a gives a measure of the width of the pile-up on the slip-plane, and that $f(x)$ is bounded at $x = a$ but unbounded at $x = 0$. Then the appropriate form of the solution is equation (5.A.18) with $F(x) = -2\pi\sigma/\mu b$:

$$f(x) = -\frac{1}{\pi^2}\left(\frac{a-x}{x}\right)^{1/2} P \int_{0}^{a}\left(\frac{x'}{a-x'}\right)^{1/2}\left(-\frac{2\pi\sigma}{\mu b}\right)\frac{dx'}{x'-x} = \frac{2\sigma}{\mu b}\left(\frac{a-x}{x}\right)^{1/2}. \tag{5.8}$$

The width a is determined by the number of dislocations:

$$n = \int_{0}^{a} f(x)dx = \pi\sigma a/\mu b \rightarrow a = \mu n b/\pi\sigma. \tag{5.9}$$

It is of interest to calculate the shear stress in the neighbourhood of the tip of the pile-up, and in particular on $y = 0$ for $x \leqslant 0$. Using formula (5.3),

$$\sigma_{zy}(x, 0) = -\sigma + \frac{\mu b}{2\pi} \int_0^a \frac{f(x')dx'}{x-x'},$$

where the first term represents the external stress. Substituting from equation (5.8) we obtain

$$\sigma_{zy}(x, 0) = -\sigma[(x-a)/x]^{1/2} \quad \text{for } x > a \text{ or } x < 0. \tag{5.10}$$

Thus there is a square-root singularity in stress (a *stress concentration*) at the tip of a pile-up.

4. A somewhat more elaborate case is the double pile-up against two barriers. Here there are barriers at $x = \pm a$, and at $x = 0$ there is a Frank-Read source which produces equal numbers of positive and negative dislocations. Under the action of an external shear stress $\sigma_{yz} = \sigma$, the positive dislocations move to the right and pile up at $x = a$, while the negative dislocations move to the left and pile up at $x = -a$ (Fig. 5.2). The source ceases to operate when the shear stress on it is reduced to zero by the stress-fields of the two piled-up arrays.

If $f(x)$ is allowed to be unbounded at $x = \pm a$, the solution is given by equation (5.A.17):

$$f(x) = -\frac{1}{\pi^2} \frac{1}{(a^2-x^2)^{1/2}} P \int_{-a}^a (a^2-x'^2)^{1/2} \left(\frac{2\pi\sigma}{\mu b}\right) \frac{dx'}{x'-x} + \frac{D_1}{(a^2-x^2)^{1/2}}$$

$$= \frac{2\sigma}{\mu b} \frac{x}{(a^2-x^2)^{1/2}} + \frac{D_1}{(a^2-x^2)^{1/2}}. \tag{5.11}$$

Now the source generates equal numbers of positive and negative dislocations, so that $f(x)$ must be odd about $x = 0$, and hence $D_1 = 0$. The number of dislocation pairs generated by the source is equal to

$$\int_0^a f(x)dx = 2\sigma a/\mu b.$$

Again we can calculate the shear stress in the neighbourhoods of the two pile-ups, and obtain

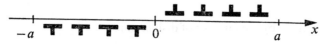

$$-a \qquad\qquad\qquad 0 \qquad\qquad\qquad a \qquad x$$

FIGURE 5.2

$$\sigma_{zy}(x, 0) = \sigma + \frac{\mu b}{2\pi} \int_{-a}^{a} \frac{f(x')dx'}{x-x'} = \frac{\sigma|x|}{(x^2-a^2)^{1/2}} \cdot \tag{5.12}$$

As expected, there is again a square-root stress singularity.

c Edge dislocation arrays

If the array in Figure 5.1 consists of edge dislocations with Burgers vectors in the x-direction, the equilibrium requirement is that the stress component σ_{xy} vanish at the position of each dislocation. Using equations (3.16), (3.19), we have in place of equation (5.2):

$$\sum_{l \neq k} \frac{\mu}{2\pi(1-\nu)} \frac{b_l}{x_k-x_l} + P_k = 0,$$

where P_k is the external stress component σ_{xy} at x_k. In the continuum approximation, the equilibrium equations take the same form as equation (5.4) but with μ replaced by $\mu/(1-\nu)$. Thus all of the above results and examples can be applied to the edge case simply by making this replacement and by putting σ_{xy} for σ_{zy}.

5.2
ELASTIC SHEAR CRACKS

A Plane shear crack

One of the most useful by-products of the theory of linear dislocation arrays is its application to elastic and elastoplastic crack problems. Solutions to many of these problems can be obtained by other methods, and in some cases have been known since before 1920. However, the application of dislocation methods has in recent years provided some new insights. We consider first the plane strain deformation of an infinite body with a crack in the form of an infinite plane strip, under an externally applied load which approaches pure shear at infinity. We take the deformation to be plane perpendicular to z, and the crack to occupy the region $y = 0$, $-a < x < a$ (Fig. 5.3.). The stress-field approaches a constant shear ($\sigma_{xy} = \sigma$) at infinity. Then the boundary conditions satisfied by this boundary-value problem of classical elasticity are:

on $y = 0$, $|x| < a$: $\sigma_{yy} = \sigma_{xy} = 0$;

as $(x, y) \to \infty$, $\sigma_{xx}, \sigma_{yy} \to 0$ and $\sigma_{xy} \to \sigma$. $\tag{5.13}$

But all of these conditions are satisfied by the stress field of the double pile-up example in the last section (§5.1B, example 4), provided we use edge dislocations rather than screws, and hence we can invoke the uniqueness theorem to claim that this stress-field solves the crack problem. The linear density of dislocations in that problem is given by (cf. eq. (5.11) and §5.1c)

$$f(x) = \frac{2(1-\nu)\sigma}{\mu b}\frac{x}{(a^2-x^2)^{1/2}} \tag{5.14}$$

and, combining this with the stress-field in equation (3.16), we get for example

$$\sigma_{xx}(x, y) = \int_{-a}^{a} -\frac{\mu b f(x')dx'}{2\pi(1-\nu)}\frac{y[3(x-x')^2+y^2]}{[(x-x')^2+y^2]^2}$$

$$= -\frac{\sigma y}{\pi}\int_{-a}^{a}\frac{3(x-x')^2+y^2}{[(x-x')^2+y^2]^2}\frac{x'}{(a^2-x'^2)^{1/2}}\,dx'. \tag{5.15}$$

The other stress components follow similarly, and are

$$\sigma_{yy}(x, y) = \frac{\sigma y}{\pi}\int_{-a}^{a}\frac{(x-x')^2-y^2}{[(x-x')^2+y^2]^2}\frac{x'}{(a^2-x'^2)^{1/2}}\,dx', \tag{5.16}$$

$$\sigma_{xy}(x, y) = \sigma + \frac{\sigma}{\pi}\int_{-a}^{a}\frac{(x-x')[(x-x')^2-y^2]}{[(x-x')^2+y^2]^2}\frac{x'}{(a^2-x'^2)^{1/2}}\,dx'. \tag{5.17}$$

Conditions (5.13) at infinity are all obviously satisfied by these stresses, and the requirement that σ_{xy} should vanish on the crack was just the condition for dislocation equilibrium which we imposed in the last section in order to obtain the solution (5.14). This leaves the condition that σ_{yy} should vanish on the crack, which is not quite immediate from equation (5.16) since

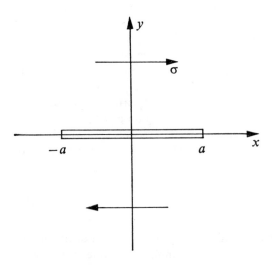

FIGURE 5.3

setting $y = 0$ in the integral produces a divergence. However, by explicitly evaluating the integral in equation (5.16), we obtain

$$\sigma_{yy}(x, y) = (\sigma a^2 y/R^3) \sin(3\theta). \tag{5.18}$$

Here R and θ are determined from the equation

$$[a^2 - (x+iy)^2]^{1/2} = Re^{i\theta}, \tag{5.19}$$

with $\theta = 0$ on $|x| < a$, $y = 0+$ and θ defined by analytic continuation elsewhere. Clearly $\sigma_{yy}(x, 0)$ vanishes, in fact, for all values of x.

The integrals in equations (5.15)–(5.17) can conveniently be evaluated by contour methods, which we apply for σ_{xy}. Equation (5.17) can be rewritten

$$\sigma_{xy}(x, y) = \sigma + \frac{\sigma}{2\pi} \oint_\Gamma \frac{(x-\zeta)\,[(x-\zeta)^2 - y^2]}{[(x-\zeta)^2 + y^2]^2}\, \frac{\zeta d\zeta}{(a^2 - \zeta^2)^{1/2}},$$

where Γ is a contour around the crack (Fig. 5.4) and $(a^2 - \zeta^2)^{1/2}$ is the analytic branch of this function which takes real and positive boundary values on the top face of the crack ($|x| < a$, $y = 0+$). Using Cauchy's theorem, we can replace the integral around Γ by one around Γ_R, a large

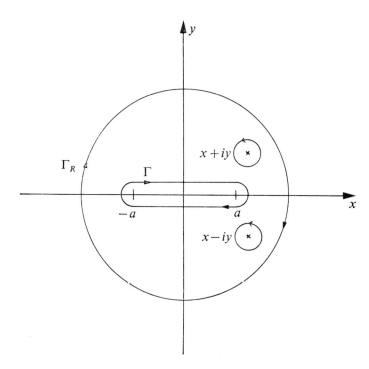

FIGURE 5.4

circle, and residues from the two poles at $x \pm iy$. For $|\zeta|$ large, $(a^2 - \zeta^2)^{1/2} \sim -i\zeta$, so that

$$\oint_{\Gamma_R} \sim \oint_{\Gamma_R} \frac{d\zeta}{i\zeta} = -2\pi,$$

and the contribution from Γ_R just cancels the constant σ-term. The residues at $\zeta = x \pm iy$ are

$$\frac{-(x \pm iy)}{2[a^2 - (x \pm iy)^2]^{1/2}} \mp \frac{iya^2}{2[a^2 - (x \pm iy)^2]^{3/2}},$$

so that

$$\sigma_{xy}(x, y) = \frac{i\sigma}{2} \left(\frac{(x+iy)^3 - xa^2}{[a^2 - (x+iy)^2]^{3/2}} + \frac{(x-iy)^3 - xa^2}{[a^2 - (x-iy)^2]^{3/2}} \right). \tag{5.20}$$

Making use of equation (5.19) we can express this in the real form

$$\sigma_{xy} = \sigma \left(-\frac{x}{R} \sin \theta + \frac{y}{R} \cos \theta - \frac{a^2 y}{R^3} \cos 3\theta \right). \tag{5.21}$$

Similar evaluation of the other stress components leads to equation (5.18) and

$$\sigma_{xx}(x, y) = -\sigma \left(\frac{2x}{R} \cos \theta + \frac{2y}{R} \sin \theta + \frac{a^2 y}{R^3} \sin 3\theta \right). \tag{5.22}$$

Just as for the pile-up itself, the crack produces singularities in the stress-field near its tips. Considering the neighbourhood of $x = a$, and writing $x + iy = a + \rho e^{i\alpha}$, so that $R \approx (2a\rho)^{1/2}$ and $\theta \approx \alpha/2 - \pi/2$, we obtain

$$\sigma_{xy} \sim \sigma \left(\frac{a}{2\rho} \right)^{1/2} \left(\cos \frac{\alpha}{2} - \frac{1}{2} \sin \alpha \sin \frac{3\alpha}{2} \right), \tag{5.23}$$

$$\sigma_{xx} \sim -\sigma \left(\frac{a}{2\rho} \right)^{1/2} \left(2 \sin \frac{\alpha}{2} + \frac{1}{2} \sin \alpha \cos \frac{3\alpha}{2} \right), \tag{5.24}$$

$$\sigma_{yy} \sim \sigma \left(\frac{a}{2\rho} \right)^{1/2} \frac{1}{2} \sin \alpha \cos \frac{3\alpha}{2} \tag{5.25}$$

(cf. eq. (5.12)). As a measure of the strength of the stress concentrations induced at the tip of a crack, it is conventional to introduce a number of quantities called *stress-intensity factors*. With the geometry of the present section, these are defined for the tip $x = a$ as

$$K_\mathrm{I} = \lim_{x \to a} (x-a)^{1/2} \sigma_{yy}(x, 0), \tag{5.26}$$

$$K_\mathrm{II} = \lim_{x \to a} (x-a)^{1/2} \sigma_{xy}(x, 0), \tag{5.27}$$

$$K_\mathrm{III} = \lim_{x \to a} (x-a)^{1/2} \sigma_{yz}(x, 0). \tag{5.28}$$

For the present case of a crack under plane shear, only K_{II} is non-zero; it is given by $K_{\text{II}} = \sigma(a/2)^{1/2}$. (Note that some authors use stress-intensity factors which differ from these by a constant.)

The relative displacement of the two crack faces is equal to the total Burgers vector between the point under consideration and either of the crack-tips:

$$\Delta u(x) = \int_x^a bf(x)dx = \frac{2(1-\nu)\sigma}{\mu}(a^2 - x^2)^{1/2}, \tag{5.29}$$

which of course is independent of b.

The energy of the cracked medium may be obtained from the general work of §3.3A, and is most usefully measured by ΔU^{tot} as given in equation (3.30). The state $\{\mathbf{u}^0, \boldsymbol{\sigma}^0\}$ is the deformation that would be produced in the medium subjected to the same boundary conditions but with the crack absent. For the present plane shear crack, $\boldsymbol{\sigma}^0$ is a uniform state of shear in which $\sigma_{xy}{}^0 = \sigma$ and the other $\sigma_{ij}{}^0 = 0$. Then ΔU^{tot} gives the change in energy of both the medium and the agents supplying the external tractions when the crack is formed in the medium which has already been subjected to this uniform shear. From equation (3.30), using the displacement in equation (5.29), we have

$$\Delta U^{\text{tot}} = -\frac{(1-\nu)\sigma^2}{\mu}\int_{-a}^a (a^2 - x^2)^{1/2}\, dx = \frac{-\pi(1-\nu)}{2\mu}\sigma^2 a^2. \tag{5.30}$$

This is the energy per unit distance in the z-direction.

B Antiplane shear crack

The corresponding elastic boundary-value problem for a crack loaded in antiplane shear satisfies the boundary conditions:

$$\text{on } y = 0, |x| < a : \sigma_{yz} = 0,$$
$$\text{as } (x, y) \to \infty, \sigma_{xz} \to 0, \sigma_{yz} \to \sigma. \tag{5.31}$$

These conditions are all immediately satisfied by the double pile-up problem of §5.1B using screw dislocations. The equilibrium density of screws turned out to be $f(x) = (2\sigma/\mu b)x/(a^2 - x^2)^{1/2}$, so that using equation (3.2) we obtain for the stress-field

$$\sigma_{xz}(x, y) = -\frac{\sigma y}{\pi}\int_{-a}^a \frac{1}{(x-x')^2 + y^2}\frac{x'}{(a^2 - x'^2)^{1/2}}\, dx', \tag{5.32}$$

$$\sigma_{yz}(x, y) = \sigma + \frac{\sigma}{\pi}\int_{-a}^a \frac{x-x'}{(x-x')^2 + y^2}\frac{x'}{(a^2 - x'^2)^{1/2}}\, dx'. \tag{5.33}$$

Evaluating these integrals by the method of section A, we get, with the notation of equation (5.19),

$$\sigma_{xz}(x, y) = -\sigma \, \mathrm{Re} \, \frac{x+iy}{[a^2-(x+iy)^2]^{1/2}} = -\frac{\sigma}{R}(x \cos \theta + y \sin \theta), \qquad (5.34)$$

$$\sigma_{yz}(x, y) = \sigma \, \mathrm{Im} \, \frac{x+iy}{[a^2-(x+iy)^2]^{1/2}} = \frac{\sigma}{R}(-x \sin \theta + y \cos \theta). \qquad (5.35)$$

Near the crack-tip, we again write $x = a + \rho \cos \alpha$, $y = \rho \sin \alpha$, so that for small ρ the dominant terms in the stresses are

$$\sigma_{xz} \sim -\sigma \left(\frac{a}{2\rho}\right)^{1/2} \sin \frac{\alpha}{2}, \qquad \sigma_{yz} \sim \sigma \left(\frac{a}{2\rho}\right)^{1/2} \cos \frac{\alpha}{2}, \qquad (5.36)$$

agreeing with equation (5.12) for $\alpha = 0$. Then, using the definitions of the stress-intensity factors in equation (5.28), we see that the antiplane shear crack corresponds to non-zero K_{III}, which has the value $K_{\mathrm{III}} = \sigma(a/2)^{1/2}$.

The relative displacement of the two crack faces is, in the z-direction,

$$\Delta u(x) = \int_x^a bf(x)dx = \frac{2\sigma}{\mu}(a^2-x^2)^{1/2}. \qquad (5.37)$$

Using equation (3.30) with $\mathbf{t}^0 = (0, 0, -\sigma)$ gives the change in energy on forming the crack, for unit distance in the z-direction,

$$\Delta U^{\mathrm{tot}} = \frac{1}{2}\int_{-a}^a -\frac{2\sigma^2}{\mu}(a^2-x^2)^{1/2}dx = -\frac{\pi}{2\mu}\sigma^2 a^2. \qquad (5.38)$$

c Shear cracks with non-uniform loading

The dislocation method can be applied not only to cracks under uniform loading at infinity but to cases of arbitrary non-uniform applied stress distributions. For the sake of definiteness we shall first consider the case of an antiplane crack under arbitrary internal loading. Then the boundary conditions for the elastic problem are:

$$\sigma_{xz}, \sigma_{yz} \rightarrow 0 \text{ at infinity};$$
$$\text{on } y = 0, |x| < a, \sigma_{yz} = p(x), \qquad (5.39)$$

where $p(x)$ is a prescribed loading function.

Now suppose that we replace the crack by a linear array of screw dislocations with density function $f(x)$. Then the shear stress σ_{yz} due to the dislocations in the region $y = 0$, $|x| < a$ is

$$\frac{\mu b}{2\pi}\int_{-a}^a \frac{f(x')dx'}{x-x'},$$

and we select the density function in such a way that this is identical with $p(x)$. Then the stress-field of the dislocation array satisfies all of the conditions (5.39), and hence must solve the elastic problem.

$f(x)$ satisfies equation (5.A.7) with integration range $(-a, a)$ and $F(x) = -(2\pi/\mu b)p(x)$ so that the solution, given by equation (5.A.17), is

$$f(x) = -\frac{1}{\pi^2} \frac{1}{(a^2-x^2)^{1/2}} P \int_{-a}^{a} (a^2-x'^2)^{1/2} \left(-\frac{2\pi}{\mu b} p(x')\right) \frac{dx'}{x'-x}. \qquad (5.40)$$

The constant D has to be zero since

$$\int_{-a}^{a} f(x)dx = 0$$

(because the relative displacement of the two crack faces must vanish for $x = \pm a$) and this condition is met by equation (5.40). From equation (3.2),

$$\sigma_{yz}(x, y)+i\sigma_{xz}(x, y) = \frac{\mu b}{2\pi} \int_{-a}^{a} \frac{1}{(x-x'')+iy} f(x'')dx''$$

$$= \frac{1}{\pi^2} \int_{-a}^{a} (a^2-x'^2)^{1/2} p(x')dx' P \int_{-a}^{a} \frac{1}{(x+iy-x'')} \frac{1}{(x'-x'')} \frac{dx''}{(a^2-x''^2)^{1/2}}, \qquad (5.41)$$

after using equation (5.40) and interchanging the orders of integration. Using partial fractions to evaluate the second integral gives, with the notation (5.19),

$$\sigma_{yz}+i\sigma_{xz} = -\frac{i}{\pi R} e^{-i\theta} \int_{-a}^{a} \frac{(a^2-x'^2)^{1/2} p(x')dx'}{x'-(x+iy)}. \qquad (5.42)$$

Near $x = a$ we write $x = a+\rho e^{i\theta}$ and obtain the dominant stresses in the form

$$\sigma_{yz} \sim K_{\mathrm{III}} \cos \tfrac{1}{2}\alpha/\sqrt{\rho}, \qquad \sigma_{xz} \sim -K_{\mathrm{III}} \sin \tfrac{1}{2}\alpha/\sqrt{\rho}, \qquad (5.43)$$

where the stress-intensity factor is given by

$$K_{\mathrm{III}} = -\frac{1}{\pi(2a)^{1/2}} \int_{-a}^{a} \left(\frac{a+x'}{a-x'}\right)^{1/2} p(x')dx'. \qquad (5.44)$$

For $p(x') = -\sigma$, these formulae agree with equation (5.36) obtained earlier.

The elastic energy stored in the deformed medium is

$$U = -\tfrac{1}{2} \int_{-a}^{a} p(x)\Delta u(x)dx \qquad (5.45)$$

per unit distance in the z-direction, since the applied traction on the face $y = 0+$ is $-p(x)$. The relative displacement of the two crack faces is

$$\Delta u(x) = \int_{x}^{a} bf(x'')dx''$$

and substituting from equation (5.40) gives

$$U = -\frac{1}{\pi\mu} \int_{-a}^{a} p(x)dx \int_{x}^{a} \frac{dx''}{(a^2-x''^2)^{1/2}} \int_{-a}^{a} (a^2-x'^2)^{1/2} \frac{p(x')dx'}{x'-x''}$$

$$= \frac{1}{\pi\mu} \int_{-a}^{a} p(x)dx \int_{-a}^{a} p(x')dx' \ln \left| \frac{\left(\dfrac{a-x}{a+x}\right)^{1/2} + \left(\dfrac{a-x'}{a+x'}\right)^{1/2}}{\left(\dfrac{a-x}{a+x}\right)^{1/2} - \left(\dfrac{a-x'}{a+x'}\right)^{1/2}} \right| \tag{5.46}$$

after performing the integration over x''.

For the analogous problem in plane strain, the internally loaded shear crack has the boundary conditions

$$\sigma_{xx}, \sigma_{yy}, \sigma_{xy} \to 0 \text{ at infinity,}$$

$$\sigma_{yy} = 0, \sigma_{xy} = p(x) \text{ on } y = 0, |x| < a. \tag{5.47}$$

The corresponding edge-dislocation problem has density again given by equation (5.40) but with μ replaced by $\mu/(1-\nu)$. Then using formulae (3.16) we obtain the stresses in the form

$$\sigma_{ij}(x, y) = \frac{1}{\pi^2} \int_{-a}^{a} (a^2-x'^2)^{1/2} p(x')dx' \cdot$$

$$P \int_{-a}^{a} \frac{K_{ij}(x'')}{[(x-x'')^2+y^2]^2(x'-x'')} \frac{dx''}{(a^2-x''^2)^{1/2}},$$

where $K_{xx} = -y[3(x-x'')^2+y^2]$, $K_{yy} = y[(x-x'')^2-y^2]$, and $K_{xy} = (x-x'')[(x-x'')^2-y^2]$. The integrations over x'' can be performed by replacing the principal-value integral by half the contour integral in Figure 5.5 (note that the two half-residues at x' cancel because of the different signs of the square root on the two sides of the real line from $-a$ to $+a$), and then expanding the contour to pick up the residues at $x+iy$ and $x-iy$. The results are as follows, where R and θ are defined in equation (5.19):

$$\sigma_{xx}(x, y)+\sigma_{yy}(x, y) = \frac{2}{\pi R} \int_{-a}^{a} \frac{(x-x') \cos\theta - y \sin\theta}{(x-x')^2+y^2} (a^2-x'^2)^{1/2} p(x')dx',$$

$$\tag{5.48}$$

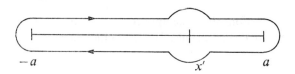

FIGURE 5.5

$\sigma_{yy}(x, y)$

$$= \frac{y}{\pi R} \int_{-a}^{a} \frac{[(x-x')^2 - y^2]\sin\theta + 2y(x-x')\cos\theta}{[(x-x')^2 + y^2]^2} (a^2 - x'^2)^{1/2} p(x')dx'$$

$$- \frac{y}{\pi R^3} \int_{-a}^{a} \frac{(x^2 + y^2 - xx')\sin 3\theta + x'y\cos 3\theta}{(x-x')^2 + y^2} (a^2 - x'^2)^{1/2} p(x')dx,$$

$$\tag{5.49}$$

$\sigma_{xy}(x, y)$

$$= \frac{y}{\pi R} \int_{-a}^{a} \frac{[(x-x')^2 - y^2]\cos\theta - 2y(x-x')\sin\theta}{[(x-x')^2 + y^2]^2} (a^2 - x'^2)^{1/2} p(x')dx'$$

$$+ \frac{y}{\pi R^3} \int_{-a}^{a} \frac{-(x^2 + y^2 - xx')\cos 3\theta + x'y\sin 3\theta}{(x-x')^2 + y^2} (a^2 - x'^2)^{1/2} p(x')dx'$$

$$+ \frac{1}{\pi R} \int_{-a}^{a} \frac{(x-x')\sin\theta + y\cos\theta}{(x-x')^2 + y^2} (a^2 - x'^2)^{1/2} p(x')dx'. \tag{5.50}$$

Considering a point directly in front of the crack-tip ($x = a+\rho, y = 0$ where $\rho \ll a$), we have σ_{xx} and σ_{yy} both zero, while the shear stress again has a square-root singularity. Writing the dominant term as $\sigma_{xy} \sim K_{II}/\sqrt{\rho}$ (cf. eq. (5.27)), we have

$$K_{II} = -\frac{1}{\pi(2a)^{1/2}} \int_{-a}^{a} \left(\frac{a+x'}{a-x'}\right)^{1/2} p(x')dx', \tag{5.51}$$

which is the same formula as in the antiplane case. Again setting $p(x) = -\sigma$ we return to the result $K_{II} = \sigma(a/2)^{1/2}$, which we had before for uniform stressing.

The elastic energy is again given by equation (5.45) but with $\Delta u(x)$ the relative displacement of the two crack faces in the x-direction. Equation (5.46) then follows, but with μ replaced by $\mu/(1-\nu)$.

5.3

ELASTOPLASTIC SHEAR CRACKS

A BCS model

In the last section we showed that the elastic boundary-value problem for a crack occupying the region $y = 0$, $|x| < a$, $-\infty < z < \infty$ and subject to plane or antiplane shear stresses at infinity is equivalent to a double pile-up of dislocations in the region $-a < x < a$ caused by a source at $x = 0$. It turned out from the solution thus obtained that certain components of stress have square-root singularities in the vicinity of the crack-tips, $x = \pm a$,

$y = 0$. Hence there will be a certain zone around each tip in which the stresses are above the yield limit and will activate dislocation sources, thus causing regions of plasticity in the material just beyond the crack-tips. The plastic flow in these regions continues until the stress singularity is removed, either through blunting of the crack or by the stresses of the dislocations created during the plasticity.

The stress-fields of elastic cracks can be obtained by the methods of classical elasticity,[8-10] so that the dislocation method of solution described in §5.2, while perhaps simpler, does not provide any further information than these older methods. The situation is, however, quite different for elastoplastic cracks. The use of the mathematical theory of plasticity has yielded an exact solution only for the antiplane shear crack problem,[11] although some approximate solutions (using deformation theory of plasticity) have also been obtained for cracks in tension.[12,13] Consequently, the dislocation array model, which has been extended to elastoplastic cracks by Bilby, Cottrell, and Swinden[14] (BCS), is in this case capable of furnishing new and useful results. It should, however, be emphasized right at the start that, although the dislocation array approach gives an exact solution to the elastic crack problem, its extension to the plastic problem represents a substantial idealization from the complexities of the real situation. It is not easy to offer convincing reasons a priori to justify the BCS model, and perhaps its strongest support comes from the close agreement between its predictions and the classical results of Hult and McClintock[11] for the antiplane shear case.

Consider a crack occupying the region $|x| < c$, $y = 0$ in an infinite medium which is subject to shear stress $\sigma_{xy} \to \sigma$ at infinity. Instead of supposing the dislocations created by the source at $x = y = 0$ to be blocked at $x = \pm c$ as in Figure 5.2, we now allow them to run into the material in the regions $|x| > c$ (Fig. 5.6), thus causing plastic shear in these regions. We suppose that the motion of the dislocations through the material is opposed

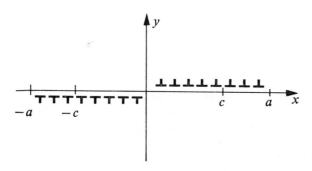

FIGURE 5.6

by a frictional resistance $b\sigma_1$ per unit length of dislocation line, with σ_1 being the yield stress, and that they are able to penetrate as far as the points $x = \pm a$ beyond the crack-tips. Then the equilibrium equation for the array, with density $f(x)$, is from equation (5.4)

$$\frac{\mu b}{2\pi(1-\nu)} P \int_{-a}^{a} \frac{f(x')dx'}{x-x'} = \begin{cases} -\sigma & \text{for } -c < x < c, \\ -(\sigma-\sigma_1) & \text{for } c < |x| < a. \end{cases} \tag{5.52}$$

Since there are no barriers at $x = \pm a$, it is to be expected that $f(x)$ will be bounded there, and the solution of the integral equation (5.52) (or (5.A.7)) is then given by equation (5.A.20), provided condition (5.A.19) holds. This condition is

$$\int_{-c}^{c} \frac{\sigma dx'}{(a^2-x'^2)^{1/2}} + \int_{-a}^{-c} + \int_{c}^{a} \frac{(\sigma-\sigma_1)dx'}{(a^2-x'^2)^{1/2}} = 0,$$

i.e.

$$c/a = \cos(\pi\sigma/2\sigma_1). \tag{5.53}$$

This equation determines the size of the plastic zones, $a-c$, in terms of the applied stress. The solution for the density is then

$$f(x) = +\frac{1}{\pi^2}(a^2-x^2)^{1/2}\left(-\frac{2\pi(1-\nu)}{\mu b}\right)\left(\sigma P \int_{-c}^{c} \frac{1}{(a^2-x'^2)^{1/2}}\frac{dx'}{x'-x}\right.$$

$$\left.+(\sigma-\sigma_1)P\int_{-a}^{-c} + \int_{c}^{a} \frac{1}{(a^2-x'^2)^{1/2}}\frac{dx'}{x'-x}\right)$$

$$= \frac{2(1-\nu)\sigma_1}{\pi\mu b}\ln\left|\frac{x(a^2-c^2)^{1/2}+c(a^2-x^2)^{1/2}}{x(a^2-c^2)^{1/2}-c(a^2-x^2)^{1/2}}\right|. \tag{5.54}$$

This function is sketched in Figure 5.7. $f(x)$ is still unbounded at the crack-tips, but the square-root singularity of equation (5.14) is now replaced by a logarithmic singularity, which is too weak to give rise to a stress concentration. (However, the unboundedness of $f(x)$ does destroy the mathematical arguments used in the Appendix to derive the solutions of singular integral equations. This shortcoming can be corrected by redefining the right-hand side of equation (5.52) so as to make it continuous, for example by setting it equal to $-\sigma+\sigma_1(|x|-c)/\varepsilon$ in some small region $c < |x| < c+\varepsilon$. This has a negligible effect on $f(x)$ except within a distance of order ε from the crack-tips, in which region it removes the infinity.)

The total plastic displacement at the crack-tip, Φ, is equal to the sum of the Burgers vectors of all the dislocations which have passed through $x = c$, i.e.

$$\Phi = \int_{c}^{a} bf(x)dx = \frac{4(1-\nu)\sigma_1 c}{\pi\mu}\ln\frac{a}{c}. \tag{5.55}$$

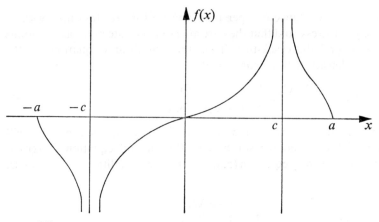

FIGURE 5.7

It is of interest to examine the approximate values of the various quantities for small stresses ($\sigma \ll \sigma_1$). From equations (5.53) and (5.55) we get

$$\frac{a-c}{c} \approx \frac{\pi^2 \sigma^2}{8\sigma_1{}^2}, \qquad \Phi \approx \frac{\pi(1-\nu)}{2\mu\sigma_1} \sigma^2 c. \tag{5.56}$$

Bearing in mind the value of the stress-intensity factor K_{II} for an elastic crack of length $2c$, these formulae can be written in the alternative form

$$(a-c) \approx \pi^2 K_{II}{}^2 / 4\sigma_1{}^2, \qquad \Phi \approx \pi(1-\nu) K_{II}{}^2 / \mu\sigma_1. \tag{5.57}$$

If we proceed right to the limit $\sigma_1 \to \infty$, we obtain again the results of §5.2A: $a \to c$ and

$$f(x) \sim \frac{2(1-\nu)}{\pi\mu b} \frac{2x}{c(a^2-x^2)^{1/2}} \sigma_1 (a^2-c^2)^{1/2} \sim \frac{2\sigma(1-\nu)}{\mu b} \frac{x}{(c^2-x^2)^{1/2}}.$$

The BCS model can equally well be applied to antiplane shear cracks under a stress $\sigma_{yz} \to \sigma$ at infinity. All we need to do is to use screw dislocations instead of edges, and all of the above formulae hold with all the $1-\nu$ factors removed and K_{II} replaced by K_{III}.

We have referred throughout to σ_1 as the 'yield stress,' and it is necessary to qualify this somewhat. At low stresses the plastic zones $c < |x| < a$ will be contained within a single grain in a polycrystal, and σ_1 should be associated with the frictional stress opposing the motion of dislocations across a grain. It contains contributions from the intrinsic lattice resistance (Peierls-Nabarro stress), from impurities and precipitates, and from other immobile dislocations which must be crossed by the mobile ones. At higher stresses, the plastic zones will cross several grains, and the largest contribution to σ_1 will usually arise from the resistive effect of grain boundaries. We can then set σ_1 equal to the macroscopic yield limit of the polycrystal.

B *Energy in the BCS model*

For essentially the same reasons as for elastic cracks, we are concerned with the change in energy of the system (body plus external stressing mechanisms) on forming a crack in an elastoplastic medium. We start from the deformation $\mathbf{u}^0$, which as usual corresponds to the same external boundary conditions applied to an uncracked medium (cf. §3.3A). For the BCS crack, this deformation consists of a uniform shear $\sigma_{xy}{}^0 = \sigma$. Now suppose that the tractions on the region $|x| < a$ are changed so that in $|x| < c$, $\sigma_{xy} = 0$, while in the plastic zones $c < |x| < a$, $\sigma_{xy} = \sigma_1$. Then the resulting state will correspond exactly to the BCS crack, since equation (5.52) implies that the crack satisfies these boundary conditions.

In terms of the notation of conditions (3.22), the BCS crack has tractions

$$
\mathbf{t}^c = \begin{cases} 0 & \text{for } -c < x < c, \\ (-\sigma_1, 0, 0) & \text{for } c < |x| < a. \end{cases}
\tag{5.58}
$$

(Note that on the upper surface the normal points in the $-y$ direction.) The corresponding energy change upon forming the crack can then be obtained from equation (3.30), using the displacement discontinuity

$$
\Delta u(x) = b \int_x^a f(x)dx.
$$

After integrating by parts, we obtain

$$
\begin{aligned}
\Delta U^{\text{tot}} &= \tfrac{1}{2} \int_{-a}^a (-\sigma)\Delta u(x)dx + \tfrac{1}{2} \int_{c<|x|<a} (-\sigma_1)\Delta u(x)dx \\
&= b\sigma \int_o^a xf(x)dx + b\sigma_1 \int_c^a xf(x)dx - bc\sigma_1 \int_c^a f(x)dx.
\end{aligned}
\tag{5.59}
$$

Substituting from equations (5.54) and (5.55) and performing the integrations gives

$$
\Delta U^{\text{tot}} = \frac{2(1-\nu)\sigma\sigma_1}{\mu} c^2 \left(\frac{a^2}{c^2} - 1\right)^{1/2} - \frac{4(1-\nu)\sigma_1{}^2 c^2}{\pi\mu} \ln\frac{a}{c}.
\tag{5.60}
$$

At low stresses this approximates to

$$
\Delta U^{\text{tot}} = -\frac{\pi(1-\nu)\sigma^2 c^2}{2\mu}\left[1 + O\left(\frac{\sigma^2}{\sigma_1{}^2}\right)\right].
$$

The first term here is the energy of a purely elastic shear crack (see equation (5.33)).

In the foregoing we have calculated the decrease in elastic energy on forming the crack. But because of the dislocation motion against the resistance σ_1, there is a certain frictional dissipation which must be allowed for in any energy calculation. The amount of energy dissipated is

$$U^D = 2 \int_c^a \sigma_1(x-c)bf(x)dx \tag{5.61}$$

since the dislocations at x move a distance $x-c$ against σ_1. Substituting for $f(x)$, this turns out to be

$$U^D = \frac{2(1-\nu)\sigma\sigma_1}{\mu} c^2 \left(\frac{a^2}{c^2} - 1\right)^{1/2} - \frac{8(1-\nu)\sigma_1^2 c^2}{\pi\mu} \ln\frac{a}{c}. \tag{5.62}$$

At low stresses this is smaller than ΔU^{tot} by a factor of order σ^2/σ_1^2.

If we are considering the process of formation of a crack, it may be relevant to consider the quantity $-\Delta U^{tot} - U^D$ as the amount of energy released. However, often we are more interested in the growth of a crack which already exists in a state of stress with appropriate plastic zones at its tips. Then we need to calculate the rate of dissipation when the crack length changes from c to $c+dc$, i.e.

$$D = 2 \int_c^a \sigma_1(x-c)\, b\, \frac{\partial}{\partial c} f(x)dx.$$

Now from equation (5.54), for a fixed value of σ, $f(x)$ is a function of x/c, so that $\partial f(x)/\partial c = -(x/c)f'(x)$, and

$$D = 2 \int_c^a \sigma_1(x-c)\, b\left(-\frac{x}{c}\right)f'(x)dx. \tag{5.63}$$

Evaluating this, we obtain

$$D = -d\Delta U^{tot}/dc, \tag{5.64}$$

with ΔU^{tot} given in equation (5.60). Thus we conclude that for a growing BCS crack there is no release of energy – the decrease in potential energy is exactly balanced by the dissipation.

It is perhaps unnecessary to point out that all of this discussion can be applied to describe the plastic zones at the tips of an antiplane shear crack. The only change is that $1-\nu$ is replaced by 1, σ_{xy} by σ_{yz}, and K_{II} by K_{III} in the formulae. In the antiplane case a solution of this problem has been derived[11] from the mathematical theory of plasticity, and it is possible to make a comparison between this and the BCS model. There is one obvious difference in that the plastic zone in the plasticity theory is not concentrated on the line of the crack as in the BCS model, but has a non-zero area which just touches the crack-tip. However, the two quantities of most interest, namely the maximum extent of the plastic zone in front of the crack and the plastic displacement Φ at the crack-tip, are in very close agreement for the two theories.[14]

c *Shear crack in a finite plate*[15]

Consider the situation, illustrated in Figure 5.8, of an infinite row of cracks lying on the plane $y = 0$, all of length $2c$, and centred at $x = 0$, $\pm 2h$, $\pm 4h$, ... The medium is deformed in antiplane strain with a shear stress $\sigma_{yz} \to \sigma$ at infinity. We suppose that the material is elastoplastic with yield stress σ_1, and in conformity with the BCS model we represent the plastic zones at the ends of all the cracks by planar distributions of screw dislocations of lengths $a - c$. The screws to the right of each crack are positive, and those to the left are negative. As in previous cases, we shall represent the cracks themselves also as linear arrays of dislocations. Then it is clear that the distribution of screw dislocations is antisymmetric about the planes $x = \pm h$, so that, by the image construction of §3.5, the stresses σ_{xz} vanish on these two planes. Therefore the stress distribution obtained from the infinite row of cracks also corresponds to the problem of a single crack, $-c < x < c$, in a finite plate, $-h < x < h$, with the surfaces $x = \pm h$ being traction-free. But, even further, it also solves the problem of an edge-crack $0 < x < c$ in a finite plate $0 < x < h$, since the dislocation distribution is also odd about $x = 0$.

The equilibrium equation for the dislocation density $f(x)$ which describes the row of cracks is similar to equation (5.52):

$$\frac{\mu b}{2\pi} P \int_{-\infty}^{\infty} \frac{f(x')dx'}{x-x'}$$

$$= \begin{cases} -\sigma & \text{for } 2nh-c < x < 2nh+c, n = 0, \pm 1, \pm 2, \ldots, \\ -(\sigma-\sigma_1) & \text{for } c < |x-2nh| < a, n = 0, \pm 1, \pm 2, \ldots. \end{cases} \qquad (5.65)$$

Splitting the left-hand side into contributions from each of the cracks and using the fact that $f(x')$ is identical for each of these contributions, i.e. $f(x+2nh) = f(x)$, we obtain

$$\frac{2\pi}{\mu b} \text{ LHS} = P \int_{-a}^{a} \sum_{n=-\infty}^{\infty} \frac{1}{x-x'-2nh} f(x')dx'.$$

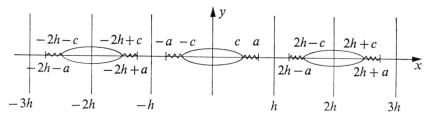

FIGURE 5.8

If we restrict x to lie between $-a$ and a and use the result

$$\sum_{n=1}^{\infty} \frac{1}{z^2-n^2} = -\frac{1}{2z^2} + \frac{\pi}{2z} \cot \pi z \quad (|z| < 1),$$

we get

$$\frac{2\pi}{\mu b} \text{LHS} = \frac{\pi}{2h} P \int_{-a}^{a} \cot \frac{\pi(x-x')}{2h} f(x')dx'$$

$$= \frac{\pi}{4h} P \int_{-a}^{a} \left(\cot \frac{\pi(x-x')}{2h} - \cot \frac{\pi(x+x')}{2h} \right) f(x')dx'$$

$$\text{(since } f(x') = -f(-x'))$$

$$= \frac{\pi}{2h} P \int_{-a}^{a} \frac{\sin(\pi x'/h)}{\cos(\pi x'/h) - \cos(\pi x/h)} f(x')dx'$$

$$= \frac{\pi}{2h} P \int_{-a}^{a} \frac{\sin(\pi x'/2h)\cos(\pi x'/2h)}{\sin^2(\pi x/2h) - \sin^2(\pi x'/2h)} f(x')dx'$$

$$= \frac{\pi}{2h} P \int_{-a}^{a} \frac{\cos(\pi x'/2h)}{\sin(\pi x/2h) - \sin(\pi x'/2h)} f(x')dx', \tag{5.66}$$

where we have used partial fractions and put $x' \to -x''$ in one of the terms.
We now introduce new variables as follows:

$$x_1 = \sin(\pi x/2h), \qquad x_1' = \sin(\pi x'/2h),$$

$$a_1 = \sin(\pi a/2h), \qquad c_1 = \sin(\pi c/2h). \tag{5.67}$$

Then the integral equation for $f(x)$ becomes

$$P \int_{-a_1}^{a_1} \frac{f_1(x_1')}{x_1 - x_1'} dx_1' = \frac{2\pi}{\mu b} \cdot \begin{cases} -\sigma & \text{for } -c_1 < x_1 < c_1, \\ -(\sigma - \sigma_1) & \text{for } c_1 < |x_1| < a_1, \end{cases} \tag{5.68}$$

where

$$f_1(x_1) = f_1\left(\sin \frac{\pi x}{2h} \right) \equiv f(x).$$

But this is precisely the same as equation (5.52) so that we can take over the results of the original BCS theory immediately. In particular from equations (5.53) and (5.54)

$$\frac{c_1}{a_1} = \frac{\sin(\pi c/2h)}{\sin(\pi a/2h)} = \cos \frac{\pi \sigma}{2\sigma_1} \tag{5.69}$$

and

$$f_1(x_1) = f(x) = \frac{2\sigma_1}{\pi\mu b} \ln \left| \frac{x_1(a_1{}^2 - c_1{}^2)^{1/2} + c_1(a_1{}^2 - x_1{}^2)^{1/2}}{x_1(a_1{}^2 - c_1{}^2)^{1/2} - c_1(a_1{}^2 - x_1{}^2)^{1/2}} \right|. \tag{5.70}$$

The crack-tip displacement is

$$\Phi = b \int_c^a f(x)dx = \frac{2bh}{\pi} \int_{c_1}^{a_1} f_1(x_1) \frac{dx_1}{(1 - x_1{}^2)^{1/2}}$$

$$= \frac{4\sigma_1 h}{\pi^2 \mu} \tan\frac{\pi a}{2h} \int_0^{\theta_0} \frac{\sin\theta \, d\theta}{[1 + \sin^2\theta \tan^2(\pi a/2h)]^{1/2}} \ln\left| \frac{\sin(\theta_0 + \theta)}{\sin(\theta_0 - \theta)} \right|, \tag{5.71}$$

where $\theta_0 = \pi\sigma/2\sigma_1$ and we have set $x_1 = a_1 \cos\theta$.

The low-stress approximations $(\sigma \ll \sigma_1)$ are

$$\frac{a - c}{c} \approx \frac{\pi^2\sigma^2}{8\sigma_1{}^2} \frac{\tan(\pi c/2h)}{\pi c/2h}, \tag{5.72}$$

$$\Phi \approx \frac{\sigma^2}{\mu\sigma_1} h \tan(\pi c/2h). \tag{5.73}$$

As $h \to \infty$, these reproduce the results of equation (5.56) (with $1 - \nu \to 1$) for a single crack in an infinite plate.

We can also retrieve the solution for a crack in a purely elastic finite plate by allowing $\sigma_1 \to \infty$. In this limit, we obtain from equation (5.70), using equation (5.72),

$$f(x) = \frac{2\sigma}{\mu b} \frac{\sin(\pi x/2h)}{[\sin^2(\pi c/2h) - \sin^2(\pi x/2h)]^{1/2}}. \tag{5.74}$$

The corresponding shear stress directly in front of the crack, on the plane $y = 0$, is obtained by summing up the effects of all the image cracks in Figure 5.8:

$$\sigma_{yz}(x, 0) = \sigma + \frac{\mu b}{2\pi} \int_{-\infty}^{\infty} \frac{f(x')dx'}{x - x'} = \sigma + \frac{\mu b}{2\pi} \frac{\pi}{2h} \int_{-a_1}^{a_1} \frac{f_1(x_1')dx_1'}{x_1 - x_1'},$$

where we have used the same line of argument as between equations (5.65) and (5.66). Then since, in equation (5.74),

$$f_1(x_1) = \frac{2\sigma}{\mu b} \frac{x_1}{(c_1{}^2 - x_1{}^2)^{1/2}},$$

we end up with

$$\sigma_{yz}(x, 0) = \frac{\sigma x_1}{(x_1{}^2 - c_1{}^2)^{1/2}} = \frac{\sigma \sin(\pi x/2h)}{[\sin^2(\pi x/2h) - \sin^2(\pi c/2h)]^{1/2}}. \tag{5.75}$$

The corresponding stress-intensity factor is

$$K_{III} = \lim_{x \to c+} (x-c)^{1/2}\sigma_{yz}(x, 0) = \sigma \left(\frac{h}{\pi} \tan \frac{\pi c}{2h}\right)^{1/2}. \qquad (5.76)$$

It is interesting to note that the results for the plastic crack at low stress levels can be written in a compact form in terms of the elastic stress-intensity factor which corresponds to the same applied stress. Thus equations (5.72) and (5.73) take the form

$$a-c \approx \pi^2 K_{III}^2/4\sigma_1^2, \qquad \Phi \sim \pi K_{III}^2/\mu\sigma_1, \qquad (5.77)$$

and the whole dependence of the plastic-zone size and the crack-tip displacement on the parameters σ, c, and h is contained in the one factor K_{III}^2. This is perhaps not surprising: at low stresses, the plastic zone will lie entirely within a region where the corresponding elastic stresses are dominated by the singular term $K_{III}/(x-c)^{1/2}$, and so its features will be determined by this term alone.

5.4
TENSILE CRACKS

A *Uniform loading*

In the dislocation pile-ups we considered in §5.1, the only allowed motion of the dislocations was glide in the plane of the pile-up, and we obtained the equilibrium equations by setting the total glide force on each dislocation equal to zero, or equal to the frictional resistance when this occurs. For edge dislocations, it is also possible to consider climb pile-ups, such as in Figure 5.9, where it is supposed that the relevant motion is climb, in the x-direction. Equilibrium of the dislocations requires that the total climb force should vanish, which implies, from equation (3.36), that $-\sigma_{yy}$ at the position of each dislocation should be zero, or should equal the climb resistance if this is non-zero. Now in fact the climb resistance can in most cases assume very large values, so that more or less any climb array can be maintained in equilibrium. The equilibrium positions of the dislocations, with the possible exception of one or two near the ends of a finite array, cannot usually be

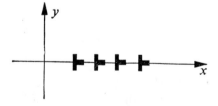

FIGURE 5.9

determined by the methods of §5.1. Moreover, the physical significance of climb pile-ups is much less than that of glide pile-ups, which, as we shall see in the next chapter, are central to some approaches to fracture initiation and to the propagation of yield through a polycrystal (see Fig. 1.25).

However, like glide arrays, climb arrays can be used to solve crack problems, this time for cracks under forces which are normal to the crack faces. Consider, as before, a source of edge pairs at $x = 0$ which under the action of an applied tension $\sigma_{yy} = \sigma$ produces edges which move away to $x > 0$ or $x < 0$ according to their sign (Fig. 5.10). We suppose that there are blocks at $x = \pm a$. Then supposing the climb resistance is zero, and replacing the discrete array by a continuum with density $f(x)$, equilibrium requires that $\sigma_{yy}(x, 0) = 0$ in $-a < x < a$. From the stresses in equation (3.16) we then obtain

$$\sigma_{yy}(x, 0) = -\frac{\mu b}{2\pi(1-\nu)} \int_{-a}^{a} \frac{f(x')dx'}{x-x'} + \sigma = 0. \tag{5.78}$$

This is the same equation as in example 4 of §5.1B apart from a factor $-(1-\nu)$, and we can take over the solution directly:

$$f(x) = -\frac{2(1-\nu)\sigma}{\mu b} \frac{x}{(a^2-x^2)^{1/2}}. \tag{5.79}$$

Just as in the shear case, this double pile-up satisfies all of the boundary conditions of the related crack problem (that $\sigma_{xy} \to 0$ and $\sigma_{yy} \to \sigma$ at infinity and that $\sigma_{xy} = \sigma_{yy} = 0$ on $y = 0$, $|x| < a$). In the region $x \gtrsim a$ just beyond the crack-tip, the tensile stress component is

$$\sigma_{yy}(x, 0) = \sigma - \frac{\mu b}{2\pi(1-\nu)} \int_{-a}^{a} \frac{f(x')dx'}{x-x'} = \frac{\sigma x}{(x^2-a^2)^{1/2}}. \tag{5.80}$$

Thus the stress-intensity factor K_I is non-zero (see eq. (5.26)) and equals $\sigma(a/2)^{1/2}$. The crack-opening displacement is

$$\Delta u(x) = -b \int_{x}^{a} f(x')dx' = \frac{2(1-\nu)\sigma}{\mu} (a^2-x^2)^{1/2}, \tag{5.81}$$

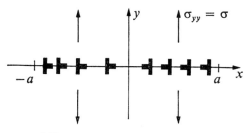

FIGURE 5.10

so that the crack assumes an elliptical shape. Finally the change in energy on making the crack can be derived from equation (3.30). Here $(\Delta u_i) = (0, \Delta u, 0)$, $(t_i^0) = (0, -\sigma, 0)$, and $\mathbf{t}^e = \mathbf{0}$ since the crack surface is free, so that

$$\Delta U^{\text{tot}} = \frac{1}{2} \int_{-a}^{a} -\frac{2(1-\nu)\sigma^2}{\mu} (a^2 - x^2)^{1/2} dx = -\frac{\pi(1-\nu)}{2\mu} \sigma^2 a^2 \qquad (5.82)$$

per unit distance in the z-direction.

Using equations (3.16) and (5.79) we can write down the complete stress-field for the tensile crack as

$$\sigma_{yy}(x, y) = \sigma + \frac{\sigma}{\pi} \int_{-a}^{a} \frac{(x-x')\,[(x-x')^2 + 3y^2]}{[(x-x')^2 + y^2]^2} \frac{x'}{(a^2 - x'^2)^{1/2}} dx',$$

$$\sigma_{xx}(x, y) = +\frac{\sigma}{\pi} \int_{-a}^{a} \frac{(x-x')\,[(x-x')^2 - y^2]}{[(x-x')^2 + y^2]^2} \frac{x'}{(a^2 - x'^2)^{1/2}} dx',$$

$$\sigma_{xy}(x, y) = +\frac{\sigma y}{\pi} \int_{-a}^{a} \frac{(x-x')^2 - y^2}{[(x-x')^2 + y^2]^2} \frac{x'}{(a^2 - x'^2)^{1/2}} dx'.$$

Following the method of §5.2A, these expressions can be evaluated to give

$$\sigma_{yy}(x, y) = -\sigma \left(\frac{x}{R} \sin\theta - \frac{y}{R} \cos\theta + \frac{a^2 y}{R^3} \cos 3\theta \right),$$

$$\sigma_{xx}(x, y) = -\sigma \left(1 + \frac{x}{R} \sin\theta - \frac{y}{R} \cos\theta - \frac{a^2 y}{R^3} \cos 3\theta \right), \qquad (5.83)$$

$$\sigma_{xy}(x, y) = \sigma \frac{a^2 y}{R^3} \sin 3\theta,$$

where R and θ are defined in equation (5.19). Again we can evaluate the singular terms near the crack-tip by setting $x = a + \rho \cos\alpha$, $y = \rho \sin\alpha$ and obtain for $\rho \ll a$

$$\sigma_{xx} \sim \sigma \left(\frac{a}{2\rho} \right)^{1/2} \left(\cos\frac{\alpha}{2} - \frac{1}{2} \sin\frac{3\alpha}{2} \sin\alpha \right),$$

$$\sigma_{yy} \sim \sigma \left(\frac{a}{2\rho} \right)^{1/2} \left(\cos\frac{\alpha}{2} + \frac{1}{2} \sin\frac{3\alpha}{2} \sin\alpha \right), \qquad (5.84)$$

$$\sigma_{xy} \sim \frac{1}{2} \sigma \left(\frac{a}{2\rho} \right)^{1/2} \sin\alpha \cos\frac{3\alpha}{2}.$$

B *Non-uniform loading*

Just as in the shear case, cracks under non-uniform normal loading on their inner faces can be solved using dislocation arrays. It is necessary, however, to add the proviso that the loading should be such that the crack is nowhere closed (i.e. $\Delta u(x) \geq 0$ for all x between $-a$ and $+a$.) The elastic boundary-value problem must then satisfy

$$\sigma_{xx}, \sigma_{xy}, \sigma_{yy} \to 0 \text{ at infinity,}$$

$$\sigma_{yy} = p(x), \sigma_{xy} = 0 \text{ on } y = 0, |x| < a. \tag{5.85}$$

Replacing the crack by a continuous array of dislocations in a climb pile-up, blocked at $x = \pm a$ (Fig. 5.10), we choose the density function $f(x)$, to be such that the stress component σ_{yy} due to the dislocations is equal to the given $p(x)$:

$$-\frac{\mu b}{2\pi(1-\nu)} P \int_{-a}^{a} \frac{f(x')dx'}{x-x'} = p(x). \tag{5.86}$$

Since by symmetry the shear stress of all the dislocations in Figure 5.10 vanishes on $y = 0$, we can expect the stress-field of the array to satisfy all the conditions (5.85).

The solution of equation (5.86) which satisfies

$$\int_{-a}^{a} f(x)dx = 0$$

is

$$f(x) = -\frac{2(1-\nu)}{\pi\mu b} \frac{1}{(a^2-x^2)^{1/2}} P \int_{-a}^{a} (a^2-x'^2)^{1/2} \frac{p(x')dx'}{x'-x}. \tag{5.87}$$

Then the stress-field is

$$\sigma_{ij}(x, y) =$$

$$\frac{1}{\pi^2} \int_{-a}^{a} (a^2-x'^2)^{1/2} p(x')dx' P \int_{-a}^{a} \frac{H_{ij}(x'')}{[(x-x'')^2+y^2]^2(x'-x'')} \frac{dx''}{(a^2-x''^2)^{1/2}}, \tag{5.88}$$

where $H_{xx} = (x-x'') [(x-x'')^2 - y^2]$, $H_{yy} = (x-x'') [(x-x'')^2 + 3y^2]$, $H_{xy} = y[(x-x'')^2 - y^2]$. The integrals involved here are similar to those in the shear case, and in fact $\sigma_{xx}(x, y)$ and $\sigma_{xy}(x, y)$ are given respectively by the right-hand sides of equations (5.50) and (5.49). The remaining result is that

$$\sigma_{xx}(x, y)+\sigma_{yy}(x, y) = \frac{2}{\pi R} \int_{-a}^{a} \frac{(x-x') \sin \theta + y \cos \theta}{(x-x')^2 + y^2} (a^2-x'^2)^{1/2}p(x')dx', \tag{5.89}$$

where R and θ are defined in equation (5.19).

On $y = 0$, these results reduce, using the fact that $\theta = -\pi/2$ there, to

$$\sigma_{xy} = 0,$$

$$\sigma_{xx}(x, 0) = \sigma_{yy}(x, 0) = -\frac{1}{\pi R} \int_{-a}^{a} \frac{(a^2 - x'^2)^{1/2} p(x')}{x - x'} \, dx', \qquad (5.90)$$

which gives a stress-intensity factor at $x = a$ of

$$K_I = -\frac{1}{\pi \sqrt{(2a)}} \int_{-a}^{a} \left(\frac{a+x'}{a-x'}\right)^{1/2} p(x') dx'. \qquad (5.91)$$

The elastic energy is again given by equation (5.46) but with μ replaced by $\mu/(1-\nu)$.

c Plasticity and tensile cracks

In §5.3 we discussed the BCS model of a shear crack in which the plastic zone at the crack-tips was accounted for by allowing the piled-up dislocation array to move into the material beyond the tips against a frictional resistance arising from the yield stress. It is possible to apply similar considerations in the tensile case, allowing the pile-up in Figure 5.10 to climb into the material against a resistance, σ_1. We would then obtain formally the same results as the BCS model for the length of the plastic zone and the crack-tip displacement.

There is, however, some difficulty in giving a physical interpretation to such a model. If σ_1 is taken to be the climb resistance, then this is so much larger than the resistance to glide in most cases that the crack-tip displacement Φ would be much smaller for a tensile crack than for a shear crack (cf. eq. (5.57)), which does not fit the general observations. On the other hand σ_1 cannot be identified with the frictional resistance to dislocation glide, since this is a shear stress and the crack produces a zero shear stress in its own plane. Mathematically, the model can be made meaningful by supposing that the criterion for plastic flow takes the form $\sigma_{yy} = \sigma_1$, involving the tensile stress in the plane of the crack rather than the shear stress. Such a view would leave undefined the physical mechanisms responsible for the yielding. A tensile-yield stress model of this kind was first given by Dugdale,[16] in fact before the BCS model was developed, and was extended by Goodier and Field.[17]

For an elastic shear crack, the plane of the crack is the plane of maximum shear stress in the material near the crack-tips, and it is natural to expect that, whatever the shape of the plastic zone, its maximum extent from the crack-tip should be in this plane. This expectation is supported by both experimental and theoretical results: the calculations of Hult and McClintock,[11] for example, predict that at low stresses the plastic zone should be a circle through the crack-tip with centre on the plane of the crack.

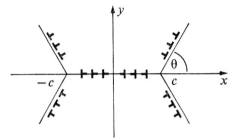

FIGURE 5.11

Because of this, it is perhaps not surprising that the BCS model, which squashes the plastic zone down onto this single plane, should give reasonable results. In the tensile case, however, the shear stress σ_{xy} vanishes on the plane of the crack, and the commonly observed plastic zone has a very small extension on this plane. The direction of maximum extent is usually inclined at a large angle, often about 70°, to the crack direction, and it is clearly quite unreasonable to model such a zone by a single plane. Nevertheless, cases are observed, particularly in conditions approximating plane stress rather than plane strain, with long narrow plastic zones, almost coplanar with the crack, and the Dugdale model would presumably be applicable to these. It is also found[18] that the plastic zone size predicted by that model is in approximate agreement with observations on tensile cracks.

The dislocation-array model has been extended to the tensile case in what is a physically more meaningful way by Bilby and Swinden.[19] They allow dislocation motion on two equally inclined planes at each crack-tip, so that the array of Figure 5.11 is obtained. The shear stress on the inclined planes is equated to the yield stress, while on the crack itself σ_{yy} is required to vanish. The system of equations can now only be solved numerically. Bilby and Swinden have done this for $\theta = 45°$, and compare the results so obtained with those obtained by taking over the formulae of §5.3A directly and replacing σ_1 by $2\sigma_1$. It turns out that the projection onto the x-axis of the length of the inclined arrays in Figure 5.11 is almost identical with $a - c$ from equation (5.53), while the normal crack-tip displacement is smaller by a factor of about $\sqrt{2}$ for the inclined-plane case.

5.5
ELASTIC BOUNDARY-VALUE PROBLEMS FOR A HALF-SPACE

A Antiplane strain

The theory of linear dislocation arrays can be used to construct in a straightforward way the solutions to certain plane and antiplane strain boundary-value problems for a half-space.[20] These include the general traction boundary-value problem and certain mixed problems. In this section we shall consider the antiplane case.

The antiplane displacement boundary-value problem for the half-space $y > 0$ has $u_z(x, y)$, $\sigma_{xz}(x, y)$, and $\sigma_{yz}(x, y)$ as the only non-zero displacement and stress components with boundary condition $u_z(x, 0+) = u(x)$. Consider in addition the corresponding antiplane deformation field $u_z(x, y)$ for $y < 0$ which satisfies $u_z(x, 0-) = -u(x)$. Then, by symmetry, $u_z(x, -y) = -u_z(x, y)$, and it follows from this by differentiation that $\sigma_{xz}(x, -y) = -\sigma_{xz}(x, y)$ and $\sigma_{yz}(x, -y) = \sigma_{yz}(x, y)$. In particular, σ_{yz} is continuous across $y = 0$, so that the two deformed half-spaces can be glued together and will hold one another in their deformed configurations. The resulting state of internal stress of the whole space $(-\infty < y < \infty)$ can then be regarded as generated by a sheet of screw dislocations lying on the plane $y = 0$. The Burgers vector of these dislocations is related to the discontinuity in displacement between the two half-spaces. In fact, if $bf(x)dx$ is the Burgers vector between x and $x+dx$, then (cf. Fig. 1.1(d) for the signs)

$$bf(x) = -2du/dx. \tag{5.92}$$

Using equation (3.2) we can now write down the stress-fields immediately:

$$\sigma_{xz}(x, y) = -\frac{\mu}{2\pi} \int_{-\infty}^{\infty} \frac{y}{(x-x')^2+y^2} bf(x')dx' = +\frac{\mu}{\pi} \int_{-\infty}^{\infty} \frac{yu'(x')}{(x-x')^2+y^2} dx', \tag{5.93}$$

$$\sigma_{yz}(x, y) = \frac{\mu}{2\pi} \int_{-\infty}^{\infty} \frac{x-x'}{(x-x')^2+y^2} bf(x')dx' = -\frac{\mu}{\pi} \int_{-\infty}^{\infty} \frac{(x-x')u'(x')}{(x-x')^2+y^2} dx'. \tag{5.94}$$

This solution of the displacement boundary-value problem can be used to derive a solution of the traction problem, in which $u(x)$ is not specified but rather $\sigma_{yz}(x, 0) = t(x)$ is given. From equations (5.93) and (5.94), by splitting the integrand into partial fractions and using the Plemelj formulae (eqs. (5.A.2), (5.A.3)), we have

$$\sigma_{xz}(x, 0+) = -\tfrac{1}{2}\mu bf(x), \qquad \sigma_{yz}(x, 0+) = -\frac{\mu b}{2\pi} P \int_{-\infty}^{\infty} \frac{f(x')dx'}{x'-x} = t(x). \tag{5.95}$$

Then the solution of the second equation is obtained from equation (5.A.6) as

$$bf(x) = \frac{2}{\pi\mu} P \int_{-\infty}^{\infty} \frac{t(x'')dx''}{x''-x}. \tag{5.96}$$

Equations (5.92) and (5.95a) now yield immediately the boundary values of the other stress component and the displacement. The full stress-field can be obtained by substituting from equation (5.96) into equations (5.93) and (5.94). Inverting the orders of the integrations involved in these expressions, the one over x' can be performed to give

$$\sigma_{xz}(x, y) = -\frac{1}{\pi} \int_{-\infty}^{\infty} \frac{x'' - x}{(x'' - x)^2 + y^2} t(x'')dx'',$$

$$\sigma_{yz}(x, y) = \frac{1}{\pi} \int_{-\infty}^{\infty} \frac{y}{(x'' - x)^2 + y^2} t(x'')dx''.$$

(5.97)

Finally the displacement field from which these stresses are derivable is

$$u_z(x, y) = \frac{1}{2\pi\mu} \int_{-\infty}^{\infty} \ln[(x - x'')^2 + y^2]t(x'')dx''.$$

(5.98)

In this way we have achieved a complete solution of the traction boundary-value problem.

B *Plane shear boundary-value problems*

Consider first the mixed problem in plane strain for the half-space $y > 0$ with boundary conditions

$$u_x(x, 0+) = u(x), \qquad \sigma_{yy}(x, 0+) = 0,$$

and let the solution be $u_x(x, y)$, $u_y(x, y)$, $\sigma_{ij}(x, y)$ ($y > 0$). We use the same symbols for the displacement and stress components in $y < 0$ which satisfy the boundary conditions

$$u_x(x, 0-) = -u(x), \qquad \sigma_{yy}(x, 0-) = 0.$$

Then by symmetry $u_x(x, -y) = -u_x(x, y)$ and $u_y(x, -y) = u_y(x, y)$ so that σ_{xx} and σ_{yy} are odd functions of y and σ_{xy} is an even function. It follows then that $\sigma_{yy}(x, 0-) = 0$ and $\sigma_{xy}(x, 0-) = \sigma_{xy}(x, 0+)$, so that the two halves $y > 0$ and $y < 0$ can again be joined together to form a deformed state of the whole region $-\infty < y < \infty$. The resulting state of internal stress can be regarded as generated by a sheet of edge dislocations with Burgers vector in the x-direction and density

$$bf(x) = -2du/dx.$$

(5.99)

From equation (3.16) we then obtain the stress-field as

$$\sigma_{ij}(x, y) = \frac{\mu}{2\pi(1-\nu)} \int_{-\infty}^{\infty} \frac{K_{ij}}{[(x-x')^2 + y^2]^2} bf(x')dx'$$

$$= -\frac{\mu}{\pi(1-\nu)} \int_{-\infty}^{\infty} \frac{K_{ij}}{[(x-x')^2 + y^2]^2} u'(x')dx',$$

(5.100)

where

$$K_{xx} = -y[3(x-x')^2+y^2], \quad K_{yy} = y[(x-x')^2-y^2],$$
$$K_{xy} = (x-x')[(x-x')^2-y^2].$$
(5.101)

The corresponding traction problem would have boundary conditions

$$\sigma_{yy}(x, 0) = 0, \qquad \sigma_{xy}(x, 0) = s(x),$$

and from equation (5.100) the second of these requires that

$$\frac{\mu b}{2\pi(1-\nu)} P \int_{-\infty}^{\infty} \frac{f(x')dx'}{x-x'} = s(x).$$
(5.102)

The solution from equation (5.A.6) is

$$bf(x) = \frac{2(1-\nu)}{\pi\mu} P \int_{-\infty}^{\infty} \frac{s(x'')dx''}{x''-x}.$$
(5.103)

Then substituting this into equation (5.100) and performing the x'-integration gives the complete stress-field as

$$\sigma_{xx}(x, y) = \frac{2}{\pi} \int_{-\infty}^{\infty} \frac{(x-x'')^3}{[(x-x'')^2+y^2]^2} s(x'')dx'',$$

$$\sigma_{yy}(x, y) = \frac{2}{\pi} \int_{-\infty}^{\infty} \frac{(x-x'')y^2}{[(x-x'')^2+y^2]^2} s(x'')dx'',$$
(5.104)

$$\sigma_{xy}(x, y) = \frac{2}{\pi} \int_{-\infty}^{\infty} \frac{(x-x'')^2y}{[(x-x'')^2+y^2]^2} s(x'')dx''.$$

The corresponding displacements, after using equations (2.11) and integrating, are seen to be, apart from an arbitrary rigid displacement,

$$u_x(x, y) = \frac{1}{2\pi\mu} \int_{-\infty}^{\infty} \left((1-\nu)\ln[(x-x'')^2+y^2] + \frac{y^2}{(x-x'')^2+y^2} \right) s(x'')dx'',$$

$$u_y(x, y) = \frac{1}{2\pi\mu} \int_{-\infty}^{\infty} \left((1-2\nu)\tan^{-1}\frac{y}{x-x''} - \frac{(x-x'')y}{(x-x'')^2+y^2} \right) s(x'')dx''.$$
(5.105)

If we take the limit $y \to 0$ in these expressions, we obtain formulae for the boundary displacements in terms of the imposed tractions, which take the form

$$u_x(x, 0) = \frac{1-\nu}{\pi\mu} \int_{-\infty}^{\infty} \ln|x-x''| s(x'')dx'',$$

$$u_y(x, 0) = \frac{1-2\nu}{2\mu} \int_{x}^{\infty} s(x'')dx''.$$

These formulae together with the corresponding ones for normal tractions, which can be obtained from the results of the next section, were first obtained by Muskhelishvili.[21,22]

c Normal tractions and displacements in plane strain

Consider the mixed boundary-value problem for $y > 0$ with boundary conditions $u_y(x, 0+) = u(x)$, $\sigma_{xy}(x, 0+) = 0$. The corresponding problem for $y < 0$ with

$$u_y(x, 0-) = -u(x), \qquad \sigma_{xy}(x, 0-) = 0$$

leads to a displacement and stress-field defined for all values of y in which u_x, σ_{xx}, and σ_{yy} are even functions of y and u_y and σ_{xy} are odd functions. We can again regard the deformation as caused by a sheet of dislocations on $y = 0$, this time edge dislocations with their Burgers vectors in the y-direction (Fig. 5.9). Using the same sign convention as in §5.4, the density of edges is given by

$$bf(x) = 2du/dx. \tag{5.106}$$

The stress-field, using equation (3.16), is then

$$\sigma_{ij}(x, y) = -\frac{\mu}{2\pi(1-v)} \int_{-\infty}^{\infty} \frac{H_{ij}}{[(x-x')^2 + y^2]^2} bf(x')dx'$$

$$= -\frac{\mu}{\pi(1-v)} \int_{-\infty}^{\infty} \frac{H_{ij}u'(x')}{[(x-x')^2 + y^2]^2} dx', \tag{5.107}$$

where

$$H_{xx} = (x-x')[(x-x')^2 - y^2], \quad H_{yy} = (x-x')[(x-x')^2 + 3y^2],$$
$$H_{xy} = y[(x-x')^2 - y^2]. \tag{5.108}$$

If the normal traction is specified, so that $\sigma_{yy}(x, 0+) = p(x)$, the resulting density must satisfy

$$\sigma_{yy}(x, 0) = \frac{\mu b}{2\pi(1-v)} P \int_{-\infty}^{\infty} \frac{f(x')dx'}{x'-x} = p(x),$$

so that

$$bf(x) = -\frac{2(1-v)}{\pi\mu} P \int_{-\infty}^{\infty} \frac{p(x'')dx''}{x''-x}. \tag{5.109}$$

Then substituting into equation (5.107) and integrating over x' gives the stresses in the form

G

$$\sigma_{xx}(x, y) = \frac{2}{\pi} \int_{-\infty}^{\infty} \frac{(x-x'')^2 y}{[(x-x'')^2 + y^2]^2} \, p(x'') dx'',$$

$$\sigma_{yy}(x, y) = \frac{2}{\pi} \int_{-\infty}^{\infty} \frac{y^3}{[(x-x'')^2 + y^2]^2} \, p(x'') dx'', \qquad (5.110)$$

$$\sigma_{xy}(x, y) = \frac{2}{\pi} \int_{-\infty}^{\infty} \frac{(x-x'') y^2}{[(x-x'')^2 + y^2]^2} \, p(x'') dx''.$$

Finally the corresponding displacements are

$$u_x(x, y) = -\frac{1}{2\pi\mu} \int_{-\infty}^{\infty} \left((1-2\nu) \tan^{-1} \frac{y}{x-x''} + \frac{(x-x'')y}{(x-x'')^2 + y^2} \right) p(x'') dx'',$$

$$u_y(x, y) = \frac{1}{2\pi\mu} \int_{-\infty}^{\infty} \left((1-\nu) \ln[(x-x'')^2 + y^2] + \frac{(x-x'')^2}{(x-x'')^2 + y^2} \right) p(x'') dx''. \qquad (5.111)$$

D *Point force on a half-space*

As an example of the application of the above results we derive the stress-fields caused by the action of a point force located at the origin. We take as the definition of a point force that $p(x)$ and $s(x)$ are zero except in a small region $|x| < \varepsilon$ where they take large constant values, p and s, such that the products $p.2\varepsilon = P$ and $s.2\varepsilon = S$ are finite. Then for $x^2 + y^2 \gg \varepsilon^2$ the integration in equations (5.104) and (5.110) can be performed simply by setting $x'' = 0$, and we obtain, on combining the two solutions,

$$\sigma_{xx}(x, y) = x^2 Q(x, y), \quad \sigma_{yy}(x, y) = y^2 Q(x, y), \quad \sigma_{xy} = xy Q(x, y), \quad (5.112)$$

where

$$Q(x, y) = \frac{2}{\pi} \frac{Py + Sx}{(x^2 + y^2)^2}. \qquad (5.113)$$

The displacement components are similarly obtained from equations (5.105) and (5.111).

The stress components obtained here are singular of order r^{-1} as $r \to 0$ (where $r = (x^2 + y^2)^{1/2}$) and even the displacements are logarithmically singular. In a physical situation, for small values of r, we would have to take account of the fact that the force is not applied at a single point but is spread over a range $|x| < \varepsilon$, and this would remove the physically spurious singularity.

This leads us to consider the following problem more fully:

$$p(x) = \begin{cases} p & \text{for } |x| < a, \\ 0 & \text{for } |x| > a, \end{cases} \qquad (5.114)$$

and a similar problem for $s(x)$. For values of r of the order of a the solution of these problems will describe the stress distributions in the immediate vicinity of the 'point' force. From equation (5.110) we obtain

$$\sigma_{xx}(x, y) + \sigma_{yy}(x, y) = \frac{2p}{\pi}\left(\tan^{-1}\frac{y}{x-a} - \tan^{-1}\frac{y}{x+a}\right),$$

$$\sigma_{xx}(x, y) - \sigma_{yy}(x, y) = \frac{4p}{\pi} ay \frac{(x^2 - a^2) - y^2}{[(x+a)^2 + y^2][(x-a)^2 + y^2]}, \qquad (5.115)$$

$$\sigma_{xy}(x, y) = \frac{4p}{\pi} a \frac{xy^2}{[(x+a)^2 + y^2][(x-a)^2 + y^2]}.$$

For $r \gg a$ these approximate to equation (5.112) with $P = 2pa$, and for small values of r they remain bounded.

E *Contact problems*

Besides the traction boundary-value problems, certain mixed problems can also be solved by the dislocation method. We discuss here the indentation of a half-space in plane strain by a smooth, rigid punch. We suppose that the punch occupies the region $-a < x < a$ and that within this region the normal displacement $u_y(x, 0)$ of the half-space is determined by the profile of the punch (Fig. 5.12). For a smooth punch, σ_{xy} is zero in $|x| < a$, and the regions $|x| > a$ are taken to be traction-free. Thus the boundary conditions are

$u_y(x, 0) = u(x) \quad (|x| < a),$

$\sigma_{yy}(x, 0) = 0 \qquad (|x| > a),$

$\sigma_{xy}(x, 0) = 0 \qquad (\text{all } x).$

As in §C we replace the boundary $y = 0$ by a climb array of edge dis-

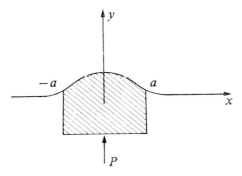

FIGURE 5.12

locations of density $f(x)$. Then for $|x| < a$, $bf(x) = 2du/dx$ (cf. eq. (5.106)), while for $|x| > a$, $f(x)$ remains unknown. However, using equations (3.16),

$$\sigma_{yy}(x, 0) = \frac{\mu b}{2\pi(1-\nu)} P \int_{-\infty}^{\infty} \frac{f(x')dx'}{x'-x}$$

$$= \frac{\mu}{2\pi(1-\nu)} \left(P \int_{|x'|>a} \frac{bf(x')dx'}{x'-x} + P \int_{-a}^{a} \frac{2u'(x')dx'}{x'-x} \right) \qquad (5.116)$$

and we must set this equal to zero in $|x| > a$. Consequently,

$$P \int_{|x'|>a} \frac{bf(x')dx'}{x'-x} = -2 \int_{-a}^{a} \frac{u'(x')dx'}{x'-x} \quad \text{for } |x| > a. \qquad (5.117)$$

Although in the Appendix we do not consider singular integral equations with this kind of region of integration, the same methods of solution apply and in fact the solution is still given by equation (5.A.17) with the range of integration being $|x'| > a$. In the case of equation (5.117), we have

$$bf(x) = \frac{1}{(x^2-a^2)^{1/2}} \left(D + \frac{2}{\pi^2} \int_{|x''|>a} \frac{(x''^2-a^2)^{1/2}}{x''-x} dx'' \int_{-a}^{a} \frac{u'(x')dx'}{x'-x''} \right)$$

$$= \frac{1}{(x^2-a^2)^{1/2}} \left(D + \frac{2}{\pi} \int_{-a}^{a} \frac{(a^2-x'^2)^{1/2}u'(x')dx'}{x'-x} \right). \qquad (5.118)$$

Knowing the equivalent dislocation density, it is a simple matter to calculate the complete stress-field. As a rule, the quantity of most interest is the pressure under the punch, $\sigma_{yy}(x, 0)$, and the total force on the punch,

$$P = -\int_{-a}^{a} \sigma_{yy}(x, 0)dx. \qquad (5.119)$$

The first of these is obtained by using equation (5.116) for $|x| < a$, with $f(x)$ given by equation (5.118). After performing one integration, we have

$$\frac{2\pi(1-\nu)}{\mu} \sigma_{yy}(x, 0)$$

$$= \frac{\pi D}{(a^2-x^2)^{1/2}} + \frac{2}{(a^2-x^2)^{1/2}} P \int_{-a}^{a} (a^2-x'^2)^{1/2} u'(x') \frac{dx'}{x'-x}. \qquad (5.120)$$

Then from equation (5.119) $P = -\pi\mu D/2(1-\nu)$.

Special cases
(a) Flat-ended punch; $u'(x) = 0$:

$$\sigma_{yy}(x, 0) = -\frac{P}{\pi(a^2-x^2)^{1/2}} \quad \text{for } |x| < a.$$

(b) Punch with inclined-plane base; $u'(x) = \alpha$, a constant:

$$\sigma_{yy}(x, 0) = -\frac{P}{\pi(a^2 - x^2)^{1/2}} - \frac{\alpha\mu}{(1-\nu)} \frac{x}{(a^2 - x^2)^{1/2}}.$$

In this case, unless P is sufficiently large, $\sigma_{yy}(x, 0)$ takes positive values over some range $-a < x < -b$. This means that insufficient pressure has been applied to the punch to attain penetration over the whole width $-a < x < a$, and the above solution breaks down. The critical value of P is reached when $\sigma_{yy}(x, 0)$ vanishes just at $x = -a$, and is given by $P_{crit} = \pi\alpha\mu a/(1-\nu)$ (see Prob. 5.19).

(c) Parabolic punch; $u(x) = $ const. $-x^2/2R$, where R is the radius of curvature at the base. From equation (5.120),

$$\sigma_{yy}(x, 0) = -\frac{1}{(a^2 - x^2)^{1/2}}\left(\frac{P}{\pi} + \frac{\mu}{2(1-\nu)}\frac{a^2 - 2x^2}{R}\right).$$

In this case, the size a of the zone of contact is determined by the condition that $\sigma_{yy}(x, 0)$ should be bounded at $x = \pm a$. This condition gives $a^2 = 2PR(1-\nu)/\pi\mu$.

For more details of contact problems, the reader is referred to the book by Galin.[23]

APPENDIX: THE SOLUTION OF CERTAIN SINGULAR INTEGRAL EQUATIONS[4-6, 24]

A *Infinite integration region*

Consider the equation, holding for all $-\infty < x < \infty$,

$$P\int_{-\infty}^{\infty} \frac{f(x')dx'}{x' - x} = F(x), \tag{5.A.1}$$

where $F(x)$ is a given function and we wish to find $f(x)$. Let z be a complex variable, and define the function $\phi(z)$ as

$$\phi(z) = \int_{-\infty}^{\infty} \frac{f(x')dx'}{x' - z}. \tag{5.A.2}$$

We shall assume that both the given function $F(x)$ and the unknown function $f(x)$ satisfy a Hölder condition[24] for all finite x. In addition, we shall assume that $f(x)$ tends to zero as $|x| \to \infty$ at least as fast as some negative power of $|x|$ in order to have convergence of the above integrals. (This latter condition can, in fact, be relaxed to requiring that $|f(x) - C| \sim$ (const.) $|x|^{-\alpha}$ as $|x| \to \infty$ for some constant C and some power $\alpha > 0$; the infinite integrals must then be interpreted in the principal-value sense.) With these conditions, $\phi(z)$ as defined in equation (5.A.2) is defined and analytic in each of the half-planes Im $z > 0$ and Im $z < 0$.

One of the basic results in the theory of singular integral equations is the formula of Plemelj, which gives an expression for the limiting value of the integral expression in equation (5.A.2) as z approaches the real axis. Calling the limits from above and below $\phi(x \pm i0)$, we have

$$\phi(x \pm i0) = \pm i\pi f(x) + P \int_{-\infty}^{\infty} \frac{f(x')dx'}{x'-x}, \tag{5.A.3}$$

provided $f(x)$ satisfies a Hölder condition. Comparing this with equation (5.A.1), we can rephrase the integral equation problem as requiring that we find a function $\phi(z)$, analytic in, say, the upper half-plane, whose limiting value on the real line has a real part equal to $F(x)$. The imaginary part of the limiting value is then $i\pi f(x)$:

$$\phi(x + i0) = F(x) + i\pi f(x). \tag{5.A.4}$$

By Cauchy's integral formula,

$$\phi(z) = \frac{1}{2\pi i} \int_\Gamma \frac{\phi(\zeta)d\zeta}{\zeta-z}, \tag{5.A.5}$$

where Γ is any closed contour in the upper half-plane and z is any point inside it. Taking Γ to be the real axis and large upper semicircle (Fig. 5.13), the semicircle contribution tends to zero as its radius tends to infinity (from eq. (5.A.2), $\phi(Re^{i\theta}) \to 0$ as $R \to \infty$ if $f(x) \sim |x|^{-\alpha}$ as $|x| \to \infty$) and we are left with

$$\phi(z) = \frac{1}{2\pi i} \int_{-\infty}^{\infty} \frac{\phi(x'+i0)dx'}{x'-z} = \frac{1}{2\pi i} \int_{-\infty}^{\infty} \frac{F(x')+i\pi f(x')}{x'-z}dx'.$$

Here we have made use of equation (5.A.4). Using the Plemelj formula again gives

$$\phi(x+i0) = \frac{1}{2\pi i}\left(i\pi[F(x)+i\pi f(x)] + P \int_{-\infty}^{\infty} \frac{F(x')+i\pi f(x')}{x'-x}dx' \right)$$

$$= F(x) + \frac{i\pi}{2}f(x) + \frac{1}{2\pi i}P \int_{-\infty}^{\infty} \frac{F(x')dx'}{x'-x}.$$

Comparing this with equation (5.A.4) then gives the required solution of the original integral equation:

$$f(x) = -\frac{1}{\pi^2}P \int_{-\infty}^{\infty} \frac{F(x')dx'}{x'-x}. \tag{5.A.6}$$

If $f(x) \to C$ at infinity, then $\phi(z) \to i\pi C$ as $z \to \infty$ in the upper half-plane,

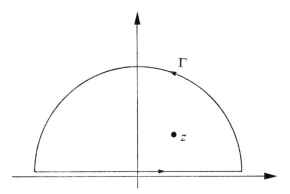

FIGURE 5.13

and there is a contribution from the upper semicircle in equation (5.A.5) equal to $i\pi C/2$. There is a corresponding term of $i\pi C/2$ in the limit $\phi(x+i0)$, which leads to an extra term C on the right-hand side of equation (5.A.6).

B *Finite integration region*

Next consider a similar equation but with the variables restricted to a finite region of x:

$$P \int_a^b \frac{f(x')dx'}{x'-x} = F(x). \tag{5.A.7}$$

As before, we define the function $\phi(z)$,

$$\phi(z) = \int_a^b \frac{f(x')dx'}{x'-z}, \tag{5.A.8}$$

which is analytic in the whole z-plane apart from the part of the real axis from a to b. For $a \leq x \leq b$, the Plemelj formula yields

$$\phi(x+i0)+\phi(x-i0) = 2F(x),$$
$$\phi(x+i0)-\phi(x-i0) = 2i\pi f(x). \tag{5.A.9}$$

Thus, in order to solve the integral equation we are required to find the analytic function $\phi(z)$, the sum of whose boundary values onto the real axis in (a, b) is $2F(x)$.

The trick in solving this problem is to solve first the homogeneous problem – that is to find a function $g(z)$, analytic in the whole z-plane except the part (a, b) of the real axis, whose boundary values satisfy

$$g(x+i0)+g(x-i0) = 0. \tag{5.A.10}$$

One of the solutions, which happens to be the most convenient for our present purposes, is

$$g(z) = [(z-a)(z-b)]^{1/2}, \qquad (5.\text{A}.11)$$

where the branch cut for $g(z)$ is taken to be from a to b on the real axis. If we now define $\psi(z) = g(z)\phi(z)$, then the new function ψ is an analytic function of z except in (a, b), and combining equations (5.A.9) and (5.A.10) gives the following formulae for the boundary values:

$$\psi(x+i0) - \psi(x-i0) = g(x+i0) \cdot 2F(x),$$
$$\psi(x+i0) + \psi(x-i0) = g(x+i0) \cdot 2i\pi f(x). \qquad (5.\text{A}.12)$$

We now apply Cauchy's integral formula (5.A.5) to the function $\psi(z)$ with the contour Γ consisting of a large circle Γ_R and an inner contour around (a, b) (Fig. 5.14). Using equation (5.A.12) gives

$$\psi(z) = \frac{1}{\pi i} \int_a^b \frac{g(x'+i0)F(x')}{x'-z}\,dx' + \psi_1(z), \qquad (5.\text{A}.13)$$

where

$$\psi_1(z) = \frac{1}{2\pi i} \oint_{\Gamma_R} \frac{\psi(\zeta)d\zeta}{\zeta-z}, \qquad (5.\text{A}.14)$$

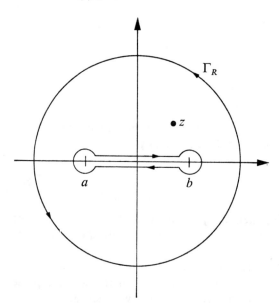

FIGURE 5.14

and we have, for the moment, ignored any contributions from the circuits around $\zeta = a$ and $\zeta = b$. Now, from its definition (5.A.8), $\phi(z)$ tends to zero as z^{-1} at infinity, so that $\psi(z) \to$ a constant, C, at infinity, since $g(z) \sim z$. Then from equation (5.A.14) $\psi_1(z)$ equals C so that the solution from equation (5.A.13) is

$$\psi(z) = g(z)\phi(z) = \frac{1}{\pi i} \int_a^b \frac{g(x'+i0)F(x')}{x'-z} \, dx' + C. \tag{5.A.15}$$

Using the Plemelj formula once again now gives

$$\psi(x+i0) = \frac{1}{\pi i}\left(\pi i g(x+i0)F(x) + P \int_a^b \frac{g(x'+i0)F(x')}{x'-x} \, dx' \right) + C$$

and comparing with equations (5.A.12),

$$g(x+i0)f(x) = -\frac{1}{\pi^2} P \int_a^b \frac{g(x'+i0)F(x')}{x'-x} \, dx' + \frac{C}{i\pi}. \tag{5.A.16}$$

Substituting from equation (5.A.11) for $g(x+i0)$ completes the solution of the integral equation. Note that an unknown constant has entered the solution and must be determined by an additional condition.

Let us return to the question of possible contributions to the solution from the circuits (γ_a and γ_b) around the end-points $\zeta = a$ and $\zeta = b$. If we assume that $(x'-a)^{1/2}F(x')$ is no more singular at $x' \to a$ than some power $(x'-a)^{-\alpha}(\alpha < 1)$, then the solution (5.A.15) satisfies $(\zeta-a)\psi(\zeta) \to 0$ as $\zeta \to a$ and so yields no contribution from γ_a. A similar result holds for the end $\zeta = b$. Thus equation (5.A.15) does always give a self-consistent solution of the problem.

However, let us suppose that the general solution has the form

$$\psi(z) = \frac{1}{\pi i} \int_a^b \frac{g(x'+i0)F(x')}{x'-z} \, dx' + C + \psi_2(z).$$

Then it is necessary that $\psi_2(z) \to 0$ at infinity since we know that $\psi(z) \to C$, and also, comparing with equation (5.A.12a), that $\psi_2(x+i0) = \psi_2(x-i0)$. Thus $\psi_2(z)$ has only isolated singularities at $z = a$ and $z = b$. Then utilizing its Laurent expansions about these two points and its known behaviour at infinity, it follows that $\psi_2(z)$ has the general form

$$\psi_2(x) = \sum_{n=1}^{\infty} \left(\frac{A_n}{(z-a)^n} + \frac{B_n}{(z-b)^n} \right).$$

It follows as before that we get similar extra terms on the right-hand side of equation (5.A.16). All of these contributions to $f(x)$ are non-integrable, so cannot be allowed in any solution of the integral equation (5.A.7). The most general solution is the one given already in equation (5.A.16).

c *Boundedness of the solution*

We can rewrite the solution (5.A.16) as

$$f(x) = \frac{1}{[(x-a)(x-b)]^{1/2}} \left(-\frac{1}{\pi^2} P \int_a^b \frac{[(x'-a)(x'-b)]^{1/2}}{x'-x} F(x')dx' + D \right).$$

$$(5.\text{A}.17)$$

It can be seen from this that $f(x)$ is in general unbounded at the two end-points a and b. An extra condition which often arises, and which enables the constant D to be determined, is that $f(x)$ is required to be bounded at one or the other of these points. Requiring for definiteness that $f(x)$ be bounded at $x = a$, we need to have

$$-\frac{1}{\pi^2} P \int_a^b \frac{[(x'-a)(x'-b)]^{1/2}}{x'-a} F(x')dx' + D = 0$$

and substituting back for D gives, after simplification,

$$f(x) = -\frac{1}{\pi^2} \left(\frac{x-a}{x-b} \right)^{1/2} P \int_a^b \left(\frac{x'-b}{x'-a} \right)^{1/2} \frac{F(x')dx'}{x'-x}. \qquad (5.\text{A}.18)$$

It is interesting to note that the solution is not only bounded as we required, but actually tends to zero as $x \to a$. Such a bounded solution is possible, however, only if $F(x')$ is not too singular at $x' = a$.

If instead we had required that $f(x)$ be bounded at $x = b$, the two square-root factors would simply be inverted.

Suppose now that $f(x)$ is required to be bounded at both a and b. Then the same arguments as before reduce the solution to equation (5.A.18), which is also required to be bounded at $x = b$. This requirement will be met if and only if

$$\int_a^b \frac{F(x')dx'}{[(x'-a)(x'-b)]^{1/2}} = 0. \qquad (5.\text{A}.19)$$

Then the solution can be rewritten as

$$f(x) = -\frac{1}{\pi^2} \left(\frac{x-a}{x-b} \right)^{1/2} \left[P \int_a^b \left(\frac{x'-b}{x'-a} \right)^{1/2} \frac{F(x')dx'}{x'-x} \right.$$

$$\left. - \int_a^b \frac{F(x')dx'}{[(x'-a)(x'-b)]^{1/2}} \right]$$

$$= -\frac{1}{\pi^2} [(x-a)(x-b)]^{1/2} P \int_a^b \frac{F(x')}{[(x'-a)(x'-b)]^{1/2}} \frac{dx'}{x'-x}. \qquad (5.\text{A}.20)$$

D *Solution for several disjoint intervals*

The above discussion can be generalized with very little change to the case when $f(x)$ is non-zero on several disjoint intervals:

$$P \int_L \frac{f(x')dx'}{x'-x} = F(x), \qquad (5.\text{A}.21)$$

where $L = (a_1, b_1) \cup (a_2, b_2) \cup \ldots \cup (a_p, b_p)$ is a set of intervals and the integral equation holds for all x on L. We can proceed exactly as before except that equation (5.A.11) must be changed to

$$g(z) = \left(\prod_{i=1}^p (z-a_i)(z-b_i) \right)^{1/2}. \qquad (5.\text{A}.22)$$

Then in equation (5.A.14) $\psi(\zeta) \sim$ a polynomial of degree $p-1$ as $\zeta \to \infty$, so that $\psi_1(z)$ equals this same polynomial. Proceeding to equation (5.A.16), the constant C is also replaced by this polynomial.

The general solution for p intervals thus contains p unknown constants which appear as coefficients of a polynomial in the solution. Again some or all of these constants can be removed by requiring that $f(x)$ be bounded at some of the end-points $\{a_i, b_i\}$. Suppose $f(x)$ is bounded at q of these points. Then for $q \leq p$ the q conditions leave the solution with $p-q$ unknown constants, and for $q > p$ all of the constants are eliminated, and there remain $q-p$ conditions of the type of equation (5.A.19) to be fulfilled in order that a solution should exist. The detailed results analogous to equations (5.A.17)–(5.A.20) can be summarized as follows. Let $f(x)$ be bounded at the q end-points $(c_1, c_2, \ldots, c_q)$ and unbounded at the remainder $(c_{q+1}, \ldots, c_{2p})$. Denote

$$R_1(x) = (x-c_1)(\ldots)(x-c_q),$$
$$R_2(x) = (x-c_{q+1})(\ldots)(x-c_{2p}). \qquad (5.\text{A}.23)$$

Then the general solution of equation (5.A.21) is

$$f(x) = -\frac{1}{\pi^2} \left(\frac{R_1(x)}{R_2(x)} \right)^{1/2} P \int_L \left(\frac{R_2(x')}{R_1(x')} \right)^{1/2} \frac{F(x')dx'}{x'-x} + \left(\frac{R_1(x)}{R_2(x)} \right)^{1/2} Q_{p-q-1}(x),$$
$$(5.\text{A}.24)$$

where (a) for $p > q$, $Q_{p-q-1}(x)$ is a polynomial of degree $(p-q-1)$; (b) for $p = q$, $Q_{p-q-1} \equiv 0$; (c) for $p < q$, $Q_{p-q-1} \equiv 0$ and a solution exists only if $F(x)$ satisfies the $q-p$ subsidiary conditions

$$\int_L \left(\frac{R_2(x)}{R_1(x)} \right)^{1/2} x^n F(x)dx = 0 \quad (n = 0, 1, \ldots, q-p-1). \qquad (5.\text{A}.25)$$

Finally, we observe in passing that there is no essential reason for tying the above arguments to the real axis. The integral equation (5.A.21) can be posed with L being a smooth arc, or series of smooth arcs, in the complex plane (x and x' now being complex variables on L), and the solution is again given by equation (5.A.24) et seq.

PROBLEMS

1 A straight dislocation is held fixed at the origin and n other dislocations occupy positions $x_1, x_2, \ldots, x_n$ on the plane $y = 0$ (as in Fig. 5.1). A uniform shear stress σ acts so as to provide a force $b\sigma$ in the negative x-direction on each of the dislocations. Show that the total force necessary to hold the leading dislocation fixed is $(n+1)b\sigma$. Generalize this result to the case when the dislocations have differing Burgers vectors and the shear stress is non-uniform.

2 In Problem 1 find the exact equilibrium positions of the dislocations for $n = 2$.

3 In Problem 1 replace the n dislocations by a continuous array, leaving the fixed dislocation at $x = 0$ as a discrete dislocation. Write down and solve the integral equation giving the equilibrium distribution for the array. Show that the density of dislocations is non-zero in a region $a_- < x < a_+$ where $a_\pm = (\mu b/4\pi\sigma)\{n+1 \pm [n(n+2)]^{1/2}\}$ (for screw dislocations). As $n \to \infty$ retrieve the solution of equation (5.8), and explain why this result is to be expected. For $n = 2$ compare graphically the density obtained here with the exact solution in Problem 2.

4 An array contains n positive screw dislocations and is held together by a shear stress $\sigma(x) = -Cx$. Find the equilibrium dislocation density and show that it is non-zero over a region $-a < x < a$ where $a^2 = \mu nb/\pi C$.[7]

5 n positive screw dislocations occupy a region $a < x < b$ and n negative dislocations occupy $-b < x < -a$. They are driven together by a uniform shear stress σ, as well as by their own mutual attractions, but are prevented from coalescing by blocks at $x = \pm a$. Show that the equilibrium density is $f(x) = (2\sigma/\mu b) [(b^2 - x^2)/(x^2 - a^2)]^{1/2}$ in $x > a$. Obtain an equation relating b to the number of dislocations.[7]

6 The complete stress-field of the double pile-up (example 4, §5.1B) is obtained in §5.2A. In particular, near the tip of the pile-up the stresses have a dominant behaviour given by equations (5.23)–(5.25) for edge dislocations and (5.36) for screws. Obtain the corresponding complete solutions for the single pile-up of example 3, and find the asymptotic behaviour of the stresses near $x = 0$. Check your results with equation (5.10).

7 An array of screw dislocations has zero total Burgers vector. Express the energy of the corresponding strain-field in terms of the density $f(x)$. Use both equations (3.26) and (3.39) to derive this energy, and verify that the two results so obtained are identical.

8 Use the methods of the Appendix to solve the integral equation

$$P \int_{-\infty}^{0} \frac{f(x')dx'}{x'-x} = F(x).$$

[Defining $\phi(z)$ by analogy with equation (5.A.8), show that if $f(x') \sim (-x')^{-\alpha}$ as $x' \to -\infty$, then $\phi(z) \sim R^{-\alpha}h(\theta)$ when $z = Re^{i\theta}$ with $R \to \infty$. This result must be used in evaluating the integral in equation (5.A.14). Consider separately the cases $\alpha = \frac{1}{2}$ and $\alpha > \frac{1}{2}$.] Obtain the solution under the additional requirement that it be bounded at $x = 0$, and any conditions that must be imposed in order to ensure the existence of such a solution.

9 An infinite body contains a semi-infinite crack occupying the half-plane $y = 0$, $x < 0$. Normal loads are applied to the two surfaces of the crack given as in conditions (5.85), by $\sigma_{yy}(x, 0) = p(x)$. Find the equivalent dislocation density which possesses the same stress-field as the crack. Show that the stress-intensity factor is

$$K_I = -\frac{1}{\pi} \int_{-\infty}^{0} \frac{p(x)dx}{(-x)^{1/2}}.$$

10 A point force P is applied normally to each of the faces of a crack of length $2a$ at a distance c from the right-hand crack-tip. Show that

$$K_I = \frac{P}{\pi}\left(\frac{2a-c}{2ac}\right)^{1/2}.$$

Verify that the result for a semi-infinite crack can be obtained as a suitable limit of this result.

11 The solutions obtained in §5.4 for tensile cracks remain valid only as long as the crack does not close. Obtain conditions (a) on the equivalent dislocation density and (b) on the loading $p(x)$ which suffice to guarantee that this does not happen.

12 A tensile crack carries loads on its internal surfaces as in equations (5.85). The loads are such that the crack is closed over a region $-a < x < d$ of its length, but is open over the remainder, $d < x < a$. Show how the method of solution in §5.4B must be modified in this case. Obtain the equivalent dislocation density and write down an equation which enables d to be found.

13 Consider the integral equation

$$P \int \frac{f(x')dx'}{x'-x} = F(x),$$

where the variables x and x' cover the two intervals $(-\infty, -a)$ and (a, ∞). Applying the methods in the Appendix obtain the general solution. Consider separately the cases in which (a) $f(x) \sim |x|^{-\alpha}$ for some $\alpha > 0$ as $|x| \to \infty$ and (b) $f(x) \sim$ const. as $|x| \to \infty$ (see eq. (5.117) et seq.).

14 Verify that equations (5.60) and (5.64) give correctly the energies for a BCS crack.

15 In Figure 5.6, rather than representing the plastic regions at the crack-tips by continuous linear dislocation arrays, suppose that we represent them by a single dislocation of Burgers vector nb situated at $x = d$ and one of Burgers vector $-nb$ at $x = -d$ (with $d > c$ of course). Write down the integral equation which holds in the crack region $-c < x < c$ and the equations describing the equilibrium of the two big dislocations. Find the solutions. Show that

$$nb = \frac{\pi\sigma}{\mu}(d^2-c^2)^{1/2} \quad \text{and} \quad \frac{3d^2-c^2}{4d(d^2-c^2)^{1/2}} = \frac{\sigma_1}{\sigma}.$$

Compare the plastic displacement nb and plastic zone size $d-c$ with the corresponding quantities Φ and $a-c$ in the BCS model. Hence assess the usefulness and limitations of the big dislocation model.

16 Write down the integral equations of equilibrium for the dislocation array of Figure 5.11 (their solution is not known).

17 A half-space $y > 0$ is deformed in antiplane strain by the application to the
face $y = 0$, at the points $x = \pm a$, of two point forces of equal magnitude P
but opposite sign. Find the solution as in §5.5D and obtain the limiting
solution when $P \to \infty$, $a \to 0$ in such a way that $aP \to M$.

18 Verify that the stress-field $\sigma_{ij}{}'$ which satisfies the boundary conditions in
equations (3.51) and (3.52) is given by equations (3.53)–(3.56).

19 In §5.5E, example (b), we examined the stresses produced by a punch
occupying the region $-a < x < a$ of the surface $y = 0$ of a half-space, and
such that the base of the punch is flat and inclined at some angle $\tan^{-1} \alpha$ to
the surface. If the total force on the punch is less than a certain critical
value, contact does not occur over the whole base of the punch. Obtain the
solution in this case.

20 A punch has a semiparabolic end, as shown in Figure 5.15. It has contact
with the half-space $y > 0$ over a region $0 < x < a$, in which it produces a
displacement $u(x) = $ const. $- x^2/2R$. Find the size a of the zone of contact
in terms of the total pressure P on the punch, and the distribution of
tractions on the end of the punch.

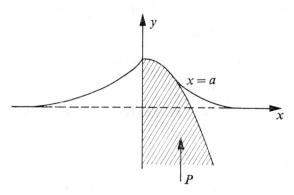

FIGURE 5.15

21 Apply the methods of §5.5A–C to express any traction boundary-value
problem in plane or antiplane strain in terms of a layer of dislocations
around the boundary of the region concerned.[20] Show that any such problem
can be reduced to an integral equation in the antiplane case or to a pair of
integral equations in the plane case. Consider a circular cylinder of radius a
deformed in antiplane strain by tractions $t(\beta)$ at the points $(a \cos \beta, a \sin \beta)$
on the surface, acting in the z-direction. Show that the deformation is the
same as that produced by a layer of screw dislocations around the surface
of the cylinder with Burgers vector $bf(\alpha)d\alpha$ between angles α and $\alpha + d\alpha$,
where

$$t(\beta) = \frac{\mu b}{4\pi} P \int_0^{2\pi} \cot \frac{\alpha - \beta}{2} f(\alpha) d\alpha.$$

(For the solution of this and of the corresponding plane strain problem, see
§31 of reference 24 and reference 20.)

REFERENCES

1 Eshelby, J.D., Frank, F.C., and Nabarro, F.R.N., Phil. Mag. **42** (1951), 351
2 Leibfried, G., Z. Phys. **130** (1951), 214
3 Mitchell, T.E., Hecker, S.S., and Smialek, R.L., Phys. Stat. Solidi **11** (1965), 585
4 Muskhelishvili, N.I., *Singular integral equations* (Amsterdam: Noordhoff, 1953)
5 Tricomi, F.G., *Integral equations* (New York: Interscience, 1957)
6 Mikhlin, S.G., *Integral equations and their applications* (Oxford, Pergamon, 1964)
7 Head, A.K., and Louat, N., Australian J. Phys. **8** (1955), 1
8 Sneddon, I.N., and Elliott, H.A., Quart. Appl. Math. **4** (1946), 262; Sneddon, I.N., *Fourier transforms* (New York: McGraw-Hill, 1951); Sneddon, I.N., and Lowengrub, M., *Crack problems in the classical theory of elasticity* New York: Wiley, 1969)
9 Williams, M.L., J. Appl. Mechs. **24** (1957), 109
10 Savin, G.N., *Concentration of stress around holes* (Oxford: Pergamon, 1963)
11 Hult, J.A.H., and McClintock, F.A., 9th Int. Congress on Applied Mechs., Brussels **8** (1956), 51
12 Rice, J.R., and Rosengren, G.F., J. Mech. Phys. Solids **16** (1968), 1
13 Hutchinson, J.W., J. Mech. Phys. Solids **16** (1968), 13
14 Bilby, B.A., Cottrell, A.H., and Swinden, K., Proc. Roy. Soc. **A272** (1963), 304
15 Bilby, B.A., Cottrell, A.H., Smith, E., and Swinden, K., Proc. Roy. Soc. **A279** (1964), 1
16 Dugdale, D.S., J. Mech. Phys. Solids **8** (1960), 100
17 Goodier, J.N., and Field, F.A., in *Fracture of solids*, ed. D. C. Drucker and J.J. Gilman (New York: Wiley, 1963), p. 103
18 Hahn, G.T., and Rosenfield, A.R., *Local yielding and extension of a crack under plane stress*, Batelle Memorial Institute Rept. ssc-165 (1964)
19 Bilby, B.A., and Swinden, K., Proc. Roy. Soc. **A285** (1965), 22
20 Lardner, R.W., Quart, J. Mech. Appl. Math. **25** (1972), 45
21 Muskhelishvili, N.I., *Some basic problems in the theory of elasticity*, 4th ed., trans. J.R.M. Radock (Amsterdam: Noordhoff, 1963)
22 Muskhelishvili, N.I., Doklady Akad. Nauk **8** (1935), 51
23 Galin, L.A., *Contact problems in the theory of elasticity*, trans. Mrs H. Moss (Raleigh: Applied Research Group, North Carolina State College, 1961)
24 Gakhov, F.D., *Boundary value problems* (London: Pergamon, 1966)

6
FRACTURE

MACROSCOPIC FEATURES

A *Basic fracture types*

In this section we shall describe some basic observations concerning the fracture process. For simplicity we shall consider a cylindrical specimen loaded in uniaxial tension (or compression) which is increased monotonically from zero to the fracture level. The situation in which fracture occurs under fluctuating loads (fatigue) will be discussed later (§6.4).

The simplest type of fracture, which is observed in some single crystals, is illustrated in Figure 6.1. It corresponds to a continuation of the planar glide process of Figure 1.2 within a single slip-band until eventually complete sliding off occurs. This type of fracture is favoured by crystal types such as hexagonal in which usually only one set of slip-planes can easily be activated, and occurs with very little strain-hardening, so that the fracture stress is almost the same as the yield stress. The fracture stresses in tension and compression are roughly equal.

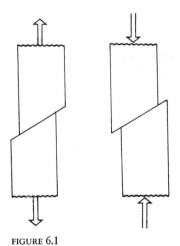

FIGURE 6.1

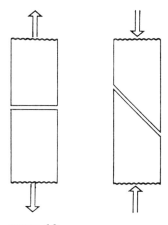

FIGURE 6.2

In polycrystalline metals, sliding off is prevented as a rule by the difficulty of propagating the planar slip from one grain to another across the specimen, and two quite distinct types of fracture are observed, called *brittle* and *ductile* fracture. Brittle fracture is characterized by the lack of any plastic flow on a macroscopic scale preceding fracture, and results typically in the sort of fracture illustrated in Figure 6.2. Under tension the fracture faces are plane and smooth and are normal to the tensile axis, and under compression the separation occurs on a plane inclined at an angle θ to the axis, with θ often being about 45° but always less than this value. Under both tension and compression the fracture process occurs very quickly once it has begun. The stress needed to produce fracture in compression is much larger (by a

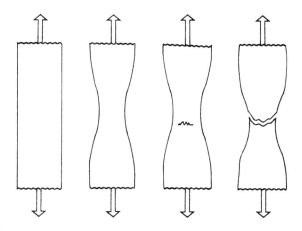

FIGURE 6.3

factor of 10 or 20) than the tensile fracture stress. Examples of brittle materials are glass, chalk, glassy polymers, cast iron, and, in fact, many polycrystalline metals at a sufficiently low temperature.

The features of ductile fracture under simple tension are quite different, the most common sequence being illustrated in Figure 6.3. Fracture is preceded by a certain amount, often a large amount, of plastic strain which culminates in an instability, necking. Then a crack develops in the necked region with an average direction normal to the imposed tension but with irregular faces,[2] and spreads relatively slowly across the specimen. Finally, near the sides, sliding off occurs on one or other of the 45° planes of maximum shear stress, leaving the familiar 'cup and cone' fracture faces.

B *Macroscopic fracture theories*

The earliest attempts to devise a criterion, or set of criteria, for the occurrence of fracture, dating back to the last century, are based on the hypothesis that fracture occurs when a certain function of the stresses attains a critical value. The three most plausible criteria are as follows. Let t_n and t_s denote the normal traction and the magnitude of the shear traction across any plane in the body. Then criterion 1 states that fracture occurs when the maximum value of t_n across all planes reaches a critical value. Criterion 2 similarly uses t_s, and in the third the critical quantity is the distortional strain energy. If $\sigma_1 \geq \sigma_2 \geq \sigma_3$ are the three principal stresses, then the critical quantities in the three cases are respectively σ_1, $\frac{1}{2}(\sigma_1 - \sigma_3)$, and $[(\sigma_1 - \sigma_2)^2 + (\sigma_2 - \sigma_3)^2 + (\sigma_3 - \sigma_1)^2]/2\mu$. The last two correspond to the Tresca and von Mises criteria for yield. Since the first of these three criteria predicts that fracture would never occur in simple compression whereas the last two predict equal strengths in compression and tension, it is clear that none of them is adequate, even as a rough model.

An improvement, due to Navier, is to maintain the view that the shear stress across a plane is the critical quantity involved in fracture on that plane, but to suppose that the limiting value that t_s can attain is reduced in proportion to the normal stress t_n. Then the fracture criterion takes the form $t_s = t_0 - kt_n$, where t_0 and k are certain constants which give respectively the basic shear strength and the rate at which this strength is reduced by normal traction. It can be shown[3] that fracture occurs across a plane which makes an angle $\frac{1}{2}\tan^{-1}(k^{-1})$ with the σ_1 principal direction and (see Prob. 6.1) certain other relations can also be deduced.

Although representing a considerable improvement over the shear stress condition, this approach to fracture is still much too crude to explain all the details. For example, the fact that fracture in simple tension occurs normal to the σ_1 direction would require an infinite value for k, whereas for compression $\theta \neq 0$, implying a finite value for k. The observed values of the tensile and compressive strengths also indicate a finite (but usually different) value for k. Further refinements can be made by taking a more general functional relationship $t_s = f(t_n)$ between the shear and normal tractions, rather than Navier's linear one, as the criterion for failure. The curve $t_s = f(t_n)$ in the t_s–t_n plane is called the Mohr envelope, and bounds the region $\{t_s < f(t_n)\}$ of safe stresses.

All of these macroscopic criteria are unsatisfactory, not only because they fail to fit more than approximately all of the facts, but also because they oversimplify what is a rather complex phenomenon. In the following sections we shall describe some of the more recent work on fracture which considers more deeply the details of what is occurring during this process, on both a macroscopic and a microscopic level.

c Brittle fracture

The fracture of a brittle material occurs through the growth of a crack across the specimen, the rate of growth being very high, of the same order of magnitude as the speed of propagation of surface waves. The first question to be raised is how such cracks start, whether they start from imperfections or microcracks present in the original unstressed material, or whether they are formed during the application of stress. The answer here depends on the material considered. For glass, for example, which was the material considered in the pioneer work of Griffith,[4, 5] it is currently believed that small cracks are present in the surface layers of a specimen, and it is the catastrophic growth of the longest and most suitably oriented of these which results in failure. The origin of such surface cracks would in many cases arise during manufacture, perhaps through imperfect polishing, or during handling; in his original work[4] Griffith observed how, by simply allowing it to come into contact with another solid, one could greatly reduce the fracture stress of a high-strength glass specimen. But it is clear that there must be other ways in which the surface flaws can form, and these are at present only partially understood. It is possible to make very-high-strength glass specimens by careful drawing of fibres from the melt, and such fibres can have a tensile fracture stress approaching $E/10$ ($E =$ Young's modulus), which is the ideal strength as calculated from interatomic force considerations.[6, 7] But if such specimens are left to age for a few hours, their strength falls to the commonly observed value.[4] Thus the aging process itself gives rise to the development of flaws.

If tractions are applied to a material containing microcracks, the stresses in the neighbourhoods of the crack-tips will be very high because of the stress concentration effect. The longer the crack, the greater will these stresses be. If the stress at any point reaches the cohesive strength of the material ($\sim E/10$), then fracture will ensue. It appears then that from our knowledge of the stresses around cracks we can calculate the level of applied tractions at which fracture will occur for a given body and a given crack-length (see §6.2) and this leads to the so-called Griffith criterion for crack instability. An application of these considerations to glass can explain the observed reduction in strength from the ideal value provided that micro-cracks of lengths of about 10^{-4} cm are present.

One interesting result of the natural occurrence of flaws in materials is a dependence of the fracture stress on the size of the specimen, for the larger the specimen the more likely is it to contain large flaws. Developments of this idea lead to a statistical theory of strength[8,9] in which the probability distribution of the fracture stress is related to the distribution of the sizes of flaws over the surface or throughout the volume, whichever is appropriate in any particular case.

Crystalline materials which exhibit brittle fracture, such as some body-centred cubic metals and in particular iron and steel under suitable circum-stances, can fracture without the existence of microcracks in the unstressed state. The question arises as to how such cracks can be found. There is a good deal of experimental evidence, summarized by Smith,[10] indicating that the nucleation of microcracks occurs within or near regions in which plastic slip has occurred on a microscopic scale, and especially where the propaga-tion of such slip has been blocked. Direct observations have been made of cracks nucleated by slip-bands in silicon iron [11,12] and ionic crystals[13] and of cracks formed at the tips of blocked twins,[14] besides other indirect observations.[10] We shall discuss this nucleation process more fully in §6.3.

The main aim in constructing a theory of fracture is to obtain a criterion for its occurrence, and in order to do this for brittle crystalline materials, in which there are two stages to the fracture process – namely the nucleation of a microcrack and its unstable growth – it is necessary to decide which stage is the critical one. If nucleation is the critical stage, then immediately a microcrack is formed it would grow across the specimen, and the fracture stress could be identified with the nucleation stress. On the other hand, if growth is critical, the microcrack when formed would exist with a certain equilibrium length until the stress level is sufficiently increased that the microcrack becomes unstable. In this case the fracture stress is given by the Griffith criterion for cracks of length equal to the stable length of the micro-cracks.

It is clear, however, from the following considerations, that there is likely to be no unique answer to this dilemma, but rather, for a given material, whichever of the two events is critical must depend on the detail of the stresses. For suppose that for a certain set of stress components nucleation is critical. As we have remarked earlier, nucleation is caused by microscopic amounts of slip and hence is dependent on the level of maximum shear stress, $(\sigma_1 - \sigma_3)/2$. Now we can superimpose a hydrostatic pressure, p, on the given stresses without affecting the shear stress, so that nucleation occurs exactly when it did before, but if we choose p sufficiently large we can reasonably expect to prevent the microcrack from spreading catastrophically as soon as it forms. Thus for high enough values of p, a stable microcrack is formed and growth becomes critical. Now suppose that for principal stresses $(\sigma_1, \sigma_2, \sigma_3)$ microcracks are nucleated with a certain stable length, and do not cause failure until a higher stress level. If we superimpose an isotropic tension on the nucleation stresses, again we do not affect the shear stress, and by choosing the level of the tension sufficiently high we can make the microcrack immediately unstable. Later in this chapter, we shall discuss in more detail the nucleation of microcracks in the presence of general stress states, and we shall show more fully how the stability or instability of the microcrack is influenced by the presence of tensile or compressive stresses.

It may, of course, happen that the isotropic tension or compression of the preceding conceptual experiments is much higher than the stress levels attainable in a physical experiment. It is then possible that within the physical range of stresses one or other of the two events, nucleation or growth, is always critical. If nucleation were critical in any region of stress-space, the fracture criterion would be a condition on the maximum shear stress, $\frac{1}{2}(\sigma_1 - \sigma_3)$, and this fact offers a relatively easy test for the hypothesis that nucleation is critical. On this basis, a number of observations on iron and steels indicate conclusively[10,15,16] that nucleation is not the controlling event, but that at least for some, if not all, stress systems stable microcracks can be formed.

There are several factors that can favour the occurrence of stable microcracks, apart from hydrostatic pressure. It is possible for the microcracks to be blunted by plastic flow, thus reducing the stress concentration at their tips. Such blunting on a microscopic scale would require the operation of Frank-Read sources near the tip of the growing microcrack, and would result in an equilibrium length of the order of the average spacing between these sources. Also, growing microcracks can be stopped at a grain boundary: in crystalline materials, separation can occur most easily on certain crystal planes, called *cleavage planes*, and at a grain boundary a crack has to change to a new cleavage plane which may not be one across which the stresses are high enough to cause growth to continue. Finally, a possibility which seems particularly applicable to mild steel,[10] the microcracks can form initially in second-phase particles which may exist as precipitates in the parent material (e.g. carbide particles in mild steel). These particles may be unable to deform plastically in order to accommodate the plastic strains in the parent matrix, so that they crack, giving microcracks of lengths equal to the particle diameters. These cracks might then be unable to propagate out of the brittle precipitate and into the matrix until the stress is raised.

D *Ductile fracture*

In both brittle and ductile crystalline materials the early stages of deformation are similar, being essentially elastic with small amounts of plastic flow within individual grains. This microplasticity is prevented by the grain boundaries, or sometimes by obstacles within the grains, from propagating across the specimen and it is the stress concentration caused by the blocked slip that commonly results in the formation of microcracks in a brittle medium. If the material is ductile, however, there will be many easily activated Frank-Read sources within the neighbourhood of the stress concentration, and these will emit dislocations in sufficient numbers to neutralize the original concentration and to propagate the slip into the next grain (cf. Fig. 1.25). Thus microcracks form much less readily in ductile materials, and furthermore those that do form tend to be blunted by plastic flow at their tips, so that they become rounded pores rather than sharp cracks. The development of pores within the necked region of a ductile specimen is a common observation in the earlier stages of ductile fracture,[2,7,15] although in many cases they originate at precipitates or surface inclusions rather than at blocked slip-bands.

Once formed, the pores are enlarged by further plastic strain, and their growth and coalescence leads to the observed irregular-faced crack. The continued growth of this dominant crack proceeds through linking with new pores, whose formation and enlargement is enhanced by the large plastic strains near the crack-tips. This process is commonly terminated in the tensile case by sliding off on one of the planes of maximum shear, producing a cup and cone, or less commonly by sliding on both of these planes simultaneously, which produces necking in the outer ring of material and leads to a double cup fracture. Suggestions that have been made to account for the preference for the activation of one rather than both of these slip-planes are as follows. Large amounts of slip localized on one plane lead to a heating and hence softening of the material near that plane, thus making it easier for slip to continue there. Alternatively, such softening could occur without heating, for example by the localized slip breaking down some of the obstacles to continued slip (called *strain-softening*). A similar sliding off probably occurs in the bridges between the enlarged pores immediately before their coalescence into a crack.

In view of all this, it becomes important to study the ductile growth of holes up to such a size that they can link together. For plane specimens, this coalescence marks the critical event and occurs at or near the point of maximum load. Working from this point of view, McClintock et al.[17,18] have obtained a criterion for ductile fracture.

6.2
CRACK PROPAGATION

A *Growth modes*

Both brittle and ductile fracture involve at some stage the growth of a single crack across a large proportion of the specimen. In the general case the crack will have an irregular shape, and the stress distribution will be complex, so that a fracture criterion is difficult to obtain. In order to simplify the problem to its essentials it is customary to consider three basic configurations. In each of these the crack occupies the infinite strip $y = 0$, $-c < x < c$, $-\infty < z < \infty$ in an infinite body. Mode I (tensile mode, or crack-opening mode) arises when $\sigma_{yy} \to \sigma$ at infinity and the other stresses tend to zero. This particular boundary-value problem was solved in §5.4A for a purely elastic body, and we showed there that it gives rise to large tensile stresses in the neighbourhood of the crack-tip: $\sigma_{yy}(x, 0) \sim K_{I}/(x-c)^{1/2}$ for $x \gtrsim c$. Mode II (plane shear mode) corresponds to shear stresses $\sigma_{xy} \to \sigma$ at infinity (see §5.2A) and in the purely elastic case, $\sigma_{xy}(x, 0) \sim K_{II}/(x-c)^{1/2}$ near $x = c$. Mode III (tearing mode) is a state of antiplane strain given by $\sigma_{yz} \sim \sigma$ at infinity. Here (cf. §5.2C) it is σ_{yz} that has the singularity near the crack-tips: $\sigma_{yz}(x, 0) \sim K_{III}/(x-c)^{1/2}$. For these three simple geometries, the stress-intensity factors have the same value: $K_{I, II, III} = \sigma(c/2)^{1/2}$.

B *Griffith theory of brittle crack growth*

In considering the growth of cracks in brittle materials, Griffith[4, 5] made the assumption that the deformation throughout the material at any stage was purely elastic. He considered elliptical holes rather than sharp cracks, to allow for the radius of curvature of the crack-tip, although it is not necessary to do this. For the crack deformed in mode I, the energy of the body and its surroundings (cf. eq. (5.82)) is, per unit length in the z-direction,

$$U^{\text{tot}} = -\frac{\pi(1-\nu)}{2\mu} \sigma^2 c^2 + 4\gamma c, \tag{6.1}$$

where the last term represents the surface energy of the crack, γ being the energy per unit area of a free surface. If U^{tot} is plotted against c (Fig. 6.4) for a fixed stress σ, it has a maximum when

$$\sigma^2 c = 4\gamma\mu/\pi(1-\nu). \tag{6.2}$$

For crack lengths smaller than the value determined by this equation, a small increase in length involves an increase in U^{tot}, but for larger crack lengths an increase in length leads to a decrease in energy. Griffith argued

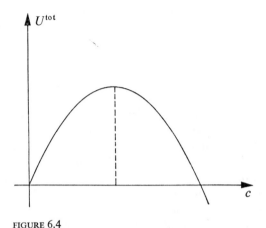

FIGURE 6.4

that this constitutes a suitable criterion for crack growth, and concluded that a crack will grow if

$$\sigma^2 c > 4\gamma\mu/\pi(1-\nu). \tag{6.3}$$

A similar argument leads to an identical criterion for mode II, and for mode III the same holds but with $1-\nu$ replaced by 1 (eqs. (5.30) and (5.38)).

It should be mentioned that a similar derivation has been given by Sack[19] for a penny-shaped crack rather than our infinite slit. The fracture criterion (6.3) is replaced by $\sigma^2 c > \pi\gamma\mu/(1-\nu)$, which is not very different from equation (6.3). It is expected that most cracks occurring in practice will be more nearly circular than infinite slits, and it is encouraging that there is so little difference between the criteria for the two cases, since calculations based on the plane geometry are usually much easier.

An alternative way of looking at crack propagation dispenses with the energy approach and looks instead at the stress distribution ahead of the crack.[5,20] For a mode I crack, for $x \gtrsim c$, $\sigma_{yy}(x, 0) \sim K_I/(x-c)^{1/2}$ and crack growth can be expected if the value of this stress on the atoms nearest the crack-tip exceeds the ideal cohesive strength of the material ($\sim E/10$).[6,7] Taking $x - c = b$, the interatomic spacing in the crack direction, we obtain a crack-growth criterion

$$K_I > E\sqrt{\bar{b}}/10. \tag{6.4}$$

Many materials have a surface energy which is given approximately by $\gamma \sim 0.06\mu b$, so that the two criteria (6.3) and (6.4) are roughly identical.

Griffith verified the criterion (6.3) for glass by introducing cracks of known length and measuring the resulting tensile fracture stress. He found quite good agreement with the theoretical prediction, particularly with the $\sigma^2 c$ interrelation, but also with the constant on the right-hand side. In order to explain the low strength of the usual 'uncracked' glass specimens on the basis of this theory, it is necessary that small cracks of lengths about 10^{-4} cm be present (in the surface). Although a number of attempts to observe such flaws directly have not given conclusive results, there is a great deal of indirect evidence in support of their existence.[7]

For brittle crack growth in crystalline metals, although it is found that a good fracture criterion is obtained by setting $\sigma^2 c$ equal to some critical value, this value is much larger than would be predicted by equation (6.3). An example is the result of Felbeck and Orowan[21] on a low-carbon steel, who found the critical value of $\sigma^2 c$ to be two thousand times larger than would be given by equation (6.3). The explanation for this, provided by Orowan (ref. 7, p. 214), is that in a crystalline material, even a brittle one, there will be small amounts of plastic flow near the crack-tip in its successive instantaneous positions,[21,22] and this will introduce a sizable dissipation of energy which must be overcome before the crack is free to grow. Thus the energy criterion for growth to occur should be $dU^{tot}/dc + D < 0$, where D is the rate of dissipation per unit crack extension and U^{tot} is given by equation (6.1). This leads, in place of equation (6.3), to the growth criterion in mode I,

$$\sigma^2 c > \frac{\mu}{\pi(1-\nu)}(4\gamma + D).$$ (6.5)

D is often referred to as an effective surface energy.

The simplest assumption we can make is that D is a constant, independent of σ and c. Then in most cases D will be much greater than 4γ which can be dropped from equation (6.5). There is a further check on this condition, since D can be estimated from an examination of the amount of plastic distortion on the faces of the cracks as well as from the critical value of $\sigma^2 c$. In this way Felbeck and Orowan[21] obtained reasonable agreement between the two measurements. However, it is in actual fact unlikely that D is independent of σ and c since the greater these quantities are, the more plastic flow is expected at the crack-tip. In fact, we would expect at low stresses, when the plastic zone is contained within the region dominated by the elastic singularity, that D would depend on the combination $\sigma^2 c$, that is on the stress-intensity factor.

A further complication in the use of the energy method arises if there is full plastic relaxation of the stress concentration at the crack-tip. In this case it might be anticipated that the dissipation D exactly balances the rate of release of elastic energy. This has, in fact, been shown to be the case for a growing BCS crack[23,24] (see §5.3B). In this case it becomes inappropriate to use an energy condition for fracture, since energy automatically balances.

In any case, whether the plastic relaxation is complete, partial, or non-existent, it is to be expected that the growth of a crack will depend on the stresses near its tip, which in turn, at least for low stresses, are determined by the elastic stress-intensity factor. Thus, on these grounds alone, it seems reasonable to suppose that a suitable fracture criterion must take the form $K_I = K_{Ic}$ for mode I, where K_{Ic} is the critical stress-intensity factor. Similar conditions occur for modes II and III. The three critical intensities, K_{Ic}, K_{IIc}, K_{IIIc}, would usually be taken from experiment, although there are theoretical relationships between them (see next section). The advantage of this type of criterion is that it can be applied to any complex geometry and state of stress once the elastic stress-intensity factors are known.[25] Thus, for example, equation (5.76) gives the fracture criterion for an antiplane shear crack in a finite plate of width $2h$ as

$$c\left(\frac{h}{\pi}\tan\frac{\pi c}{2h}\right)^{1/2} = K_{IIIc}.$$

As a second example of this approach, consider a crack in an infinite medium under arbitrary normal loads $p(x)$ on its two faces. The stress-intensity factor near the tip $x = +c$ is given by equation (5.91) with a replaced by c and near $x = -c$ by a similar expression with the square-root factor inverted. Thus the fracture criteria are

$$-\frac{1}{\pi\sqrt{(2c)}}\int_{-c}^{c}\left(\frac{c+x'}{c-x'}\right)^{1/2}p(x')dx' = K_{Ic} \quad \text{for growth at } +c,$$

$$-\frac{1}{\pi\sqrt{(2c)}}\int_{-c}^{c}\left(\frac{c-x'}{c+x'}\right)^{1/2}p(x')dx' = K_{Ic} \quad \text{for growth at } -c.$$

(6.6)

In the particular case when the two faces are wedged apart by point forces of magnitude P at the point $x = d > 0$, then fracture occurs preferentially at the end $x = +c$, when

$$\frac{P}{\pi\sqrt{(2c)}}\left(\frac{c+d}{c-d}\right)^{1/2} = K_{Ic}.$$

This result is obtained from equation (6.6) by taking $p(x) = -P\delta(x-d)$.

c Brittle crack growth under combined stresses

We have seen in the last section that one approach to a criterion for crack growth concerns itself with some critical feature of the stress-field directly ahead of the crack, typically the maximum tensile stress at a distance of about one lattice spacing from the crack-tip. With such a view of the fracture process, it is not difficult to extend the considerations to a crack under combined stresses, i.e. under a mixture of modes I and II, simply by super-imposing the stress-fields for the two modes. We shall consider the geometry

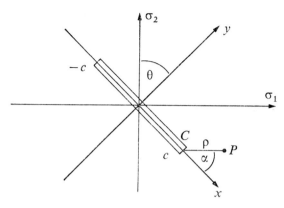

FIGURE 6.5

of Figure 6.5, with principal stresses σ_1 and σ_2 at infinity ($\sigma_1 > \sigma_2$) and a crack of length $2c$ making an angle θ with the σ_1-principal direction. Taking x, y-axes as shown, the asymptotic stresses are

$$\sigma_{xx} \to p + \sigma \cos 2\theta, \qquad \sigma_{yy} \to p - \sigma \cos 2\theta, \qquad \sigma_{xy} \to \sigma \sin 2\theta, \qquad (6.7)$$

where $p = \frac{1}{2}(\sigma_1 + \sigma_2)$ and $\sigma = \frac{1}{2}(\sigma_1 - \sigma_2)$.

The first of these stresses merely leads to a uniform σ_{xx} throughout the body, and the last two give the solutions obtained in §5.2A and §5.4A. At a point P near the crack-tip $x = c$, the dominant terms in the stresses are obtained by adding the stresses in equations (5.23)–(5.25) and (5.84), with σ there set equal to $\sigma \sin 2\theta$ and $p - \sigma \cos 2\theta$ respectively. Now we are interested in the stress component tending to cause the small element CP of material near the crack-tip to tear apart. This is given by

$$\sigma_{\alpha\alpha} = \sigma_{xx} \sin^2 \alpha + \sigma_{yy} \cos^2 \alpha - 2\sigma_{xy} \sin \alpha \cos \alpha, \qquad (6.8)$$

$$\approx \left(\frac{c}{2\rho}\right)^{1/2} \cos^2 \frac{\alpha}{2} \left((p - \sigma \cos 2\theta) \cos \frac{\alpha}{2} - 3\sigma \sin 2\theta \sin \frac{\alpha}{2}\right). \qquad (6.9)$$

For given values of p, σ, and θ, if we now take the maximum value of this stress over all values of α and equate the result to the cohesive strength of the material ($\sim E/10$) we obtain the required growth criterion. It reduces to equation (6.4), of course, when $\theta = 0$.

The brittle fracture of non-crystalline materials such as glass and rocks is attributed to the catastrophic growth under stress of small cracks present within the original unstressed material. A whole range of crack-lengths and crack orientations will be present, and in order to calculate the state of biaxial stress at which fracture occurs, it is necessary to consider the peripheral stress $\sigma_{\alpha\alpha}$ in equation (6.9) for the most dangerous of all the cracks. That is, we must maximize $\sigma_{\alpha\alpha}$ over c and θ as well as over α, and then set the result equal to the cohesive strength. Assuming that there are many cracks present, we can simply give c the value of the greatest crack length, and if the cracks are randomly oriented (which may not be correct in particular cases) we can maximize over all values of θ, to give

$$\sigma_{\alpha\alpha}^{max} = \left(\frac{c}{2\rho}\right)^{1/2} \cos^2\frac{\alpha}{2}\left[p\cos\frac{\alpha}{2} + \sigma\left(1+8\sin^2\frac{\alpha}{2}\right)^{1/2}\right]. \tag{6.10}$$

Maximizing over α and setting the result equal to a certain material constant $[(c/2\rho)^{1/2}]\,\sigma_0$ for fracture, we get finally the following criteria:

for $p > 2\sigma$, fracture occurs when $\sigma + p = \sigma_0$; $\tag{6.11}$

for $p < 2\sigma$, fracture occurs when

$$\left[p\cos\frac{\alpha}{2} + \sigma\left(1+8\sin^2\frac{\alpha}{2}\right)^{1/2}\right]\cos^2\frac{\alpha}{2} = \sigma_0, \tag{6.12}$$

where

$$\sin^2\frac{\alpha}{2} = \frac{1}{16(p^2+8\sigma^2)}\left[7p^2 + 32\sigma^2 - p(9p^2+64\sigma^2)^{1/2}\right]. \tag{6.13}$$

The first of these corresponds to a normal fracture under the maximum tensile stress, at $\sigma_1 = \sigma_0$, and with $\theta = \pi/2$ and $\alpha = 0$. For $\sigma_2 > \sigma_1/3$ the occurrence of fracture is not influenced by the value of the second principal stress. For lower values of σ_2, however, there is some effect of the shear. For example, in simple tension ($\sigma_2 = 0$), equations (6.12) and (6.13) lead to fracture at $\sigma_1 \approx 0.97\sigma_0$, and the angles are $\theta = 112°$ and $\alpha \approx 36°$. In pure shear ($p = 0$), fracture occurs at $\sigma \approx 0.77\sigma_0$ at orientations $\theta = 2\pi/3$ and $\alpha = \pi/3$. The two parts of the fracture locus corresponding to conditions (6.11) and (6.12) are shown in Figure 6.6, as segments AB and BC respectively.

An approach of this kind to brittle fracture under biaxial stress was first considered by Griffith.[5] In his work, an elliptical hole was considered rather than the sharp crack of the present discussion, and it was supposed that growth of the hole occurs when the maximum peripheral tensile stress at some point around it attains a critical value. The results obtained for the elliptical hole are not very different from those of Figure 6.6 for $\sigma_2 > 0$, but depart somewhat for negative values of σ_2. In fact, the fracture criterion $\sigma_1 = \sigma_0$ is found to hold for an elliptical hole for $p > -2\sigma$. For negative values of p, there is even greater divergence between the results for the two types of crack.

When σ_2 becomes negative, however, some care must be taken since for certain values of the orientation θ the normal stress ($p - \sigma\cos 2\theta$) becomes negative. Such compressive stresses can be transmitted across the crack

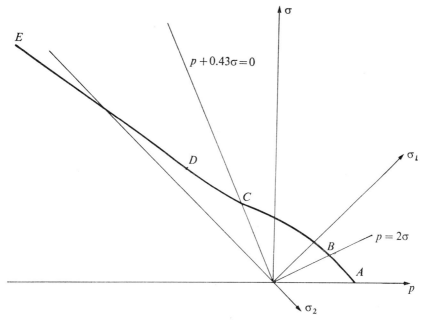

FIGURE 6.6

without producing a stress concentration, and hence require some modification of the results, both for the sharp crack and for Griffith's elliptical hole. It can be shown that for the value of θ giving the maximum in equation (6.9), $p - \sigma \cos 2\theta$ is positive as long as $p > -0.43\sigma$, corresponding to the point C in Figure 6.6.

For smaller values of p, the maximum occurs at orientations at which the normal stress across the crack is negative, and in the expression (6.9) for the local stresses at the crack-tip, the first term is absent. However, as remarked by McClintock and Walsh,[26] the compressive stresses across the crack will induce some frictional resistance to relative sliding of the two crack faces under shear. We can account for this by reducing the effective shear stress from $\sigma \sin 2\theta$ to $\sigma \sin 2\theta + \mu(p - \sigma \cos 2\theta)$, where μ is the coefficient of friction. Then equation (6.9) becomes

$$\sigma_{\alpha\alpha} - -3 \left(\frac{c}{2\rho} \right)^{1/2} \cos^2 \frac{\alpha}{2} \sin \frac{\alpha}{2} [\sigma \sin 2\theta + \mu(p - \sigma \cos 2\theta)]. \tag{6.14}$$

The maximum value occurs at $\tan 2\theta = -1/\mu$ and $\sin \alpha/2 = -1/\sqrt{3}$ and gives the fracture criterion

$$\mu p + \sigma(1 + \mu^2)^{1/2} = \sqrt{3}\sigma_0/2. \tag{6.15}$$

Taking $\mu = 1$, this corresponds to the segment DE in Figure 6.6. The curve cannot be continued below D since at that point $p - \sigma \cos 2\theta$ ceases to be negative at the fracture orientation. The gap CD can be completed by interpolation.

The surfaces of microcracks are expected to be clean and, since they are of the same material, the coefficient of friction should be high, a value of unity not being unrealistic. With this value, McClintock and Walsh found that they could fit very well the experimental fracture strengths of a large number of rocks under biaxial compression (σ_1 and $\sigma_2 < 0$). They choose σ_0 to fit the observed strength under simple compression ($\sigma_1 = 0$), in which case the present theory gives a much better fit to fracture strengths in the region $\sigma_1 < 0$ than does the original Griffith theory, which ignores frictional binding of the crack faces. The results for a sharp crack are, however, worse than for an elliptical crack in that they predict too low a ratio for the strengths under simple compression and under simple tension (the ratio is about 4 for $\mu = 1$ rather than 8 as for Griffith's theory).

The analysis of compressive strengths based on equations (6.14) and (6.15) is mathematically identical with Navier's macroscopic theory outlined in §6.1B. In the Navier theory, the Mohr envelope consists of the line ED only. There is a difference between the two approaches, however, in that the crack model of fracture makes more detailed statements about the growth process. For example, for $\mu = 1$, the critical values of θ and α are $67\frac{1}{2}°$ and $-70\frac{1}{2}°$ respectively, so that during growth the critical cracks tend to become more nearly parallel to the compressive σ_2-axis. This is probably the cause of the almost longitudinal fractures observed in some brittle materials under simple compression. Considerations of this kind also suggest that equations (6.13) and (6.15) probably underestimate the fracture strengths in the region $BCDE$ of Figure 6.6 relative to the strengths in the part AB, because in the first region the critical value of α is non-zero, and any increment of crack growth will not be parallel to the original crack. Thus, after some growth the average orientation of the crack may change to a safe value. Then fracture will eventually occur at some higher stress, perhaps due to cracks whose initial orientation made them less dangerous, but whose growth does not diminish their effectiveness as stress raisers. This could account for the discrepancy between the predicted ratio of compressive to tensile strengths and the observed ratio.

D *Barenblatt's theory*

The Griffith approach to crack propagation is based on a use of linear elasticity theory to solve for the stress-fields. As we have seen, this results in singularities in the stress and strain components at the crack-tips, and the application of linear elasticity is invalid in the neighbourhood of these points, since this theory is restricted not only to finite strains but to strains much smaller than unity. It is necessary, at the very least, to include a region of non-linear elastic behaviour at each crack-tip. Barenblatt[27] does this for a tensile crack by supposing that there are large cohesive forces acting across the plane of the crack in the neighbourhood of the crack tips and that these forces are related in some non-linear way to the separation of the atomic layers in these neighbourhoods.

We consider a state of plane strain in an infinite medium with a crack occupying the region $-c < x < c$, $y = 0$, and for definiteness we consider the case of normal internal loading: $\sigma_{yy}(x, 0) = p(x)$ $(-c < x < c)$ (where $p(x)$ is negative) and we take $p(x)$ to be symmetric about $x = 0$. Then we suppose that the medium can be treated by linear elasticity theory except for two edge regions $c < |x| < a$, $y = 0$, in which large (non-linear) cohesive forces act across the plane of the crack. These forces are equivalent to certain tractions acting on the material above and below the crack-plane, and so give conditions of the type $\sigma_{yy}(x, 0) = G(x)$ $(c < |x| < a)$. The configuration of forces and the displacement of the upper crack face are shown in Figure 6.7. It is to be expected that the edge region $c < |x| < a$ will be quite small, but that the cohesive stress $G(x)$ will be large, approaching the ideal cohesive strength ($\sim E/10$) as conditions for crack growth are attained. For the general state, there will be a relationship

$$G(x) = g[\Delta u(x)] \quad (c < |x| < a) \tag{6.16}$$

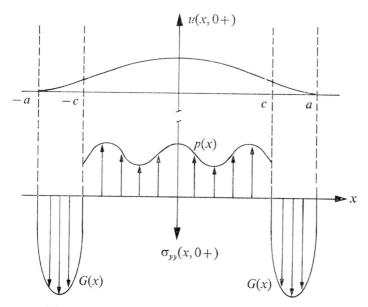

FIGURE 6.7

between the traction and the separation $\Delta u(x) = u(x, 0+) - u(x, 0-)$ of the atomic layers at x. The detailed form of this equation will depend on the nature of the interatomic forces acting across the edge region.

The distribution of stress caused by the tractions $p(x)$ and $G(x)$ can be obtained by the methods of §5.4B. Comparing the boundary conditions of the present case (Fig. 6.7) with conditions (5.85), we see that the two are identical if we define $p(x) \equiv G(x)$ in $c < |x| < a$. Then the equivalent density of dislocations, $f(x)$, satisfies the integral equation (5.86) whose solution is equation (5.87), the stress-field is given by equation (5.88), and the displacement across the crack (cf. eq. (5.81)) satisfies

$$d\Delta u(x)/dx = bf(x). \tag{6.17}$$

At this stage we introduce the fundamental hypothesis that for any equilibrium crack the cohesive forces, $G(x)$, and the width $d = a - c$ of the edge region will adjust themselves so that the crack profile is cusped at $x = \pm a$. In other words, we require $f(x) \to 0$ as $x \to \pm a$. This situation should be compared with that in the linear theory in which the crack-tips $x = \pm c$ are completely rounded; for example, under uniform loading $(p(x) \equiv -\sigma)$ the crack profile is elliptical. The conditions under which the solutions of singular integral equations are bounded at the end-points of the integration range were investigated in the Appendix to Chapter 5(§C), and for the equation (5.A.7) the condition is (5.A.19), with equation (5.A.20) giving the solution. In the present case these are, respectively,

$$\int_{-c}^{c} \frac{p(x)dx}{(a^2 - x^2)^{1/2}} + \int_{c < |x| < a} \frac{G(x)dx}{(a^2 - x^2)^{1/2}} = 0, \tag{6.18}$$

$$bf(x) = -\frac{2(1-v)}{\pi\mu} \frac{1}{(a^2 - x^2)^{1/2}} \left(\int_{-c}^{c} (a^2 - x'^2)^{1/2} \frac{p(x')dx'}{x' - x} \right.$$
$$\left. + \int_{c < |x'| < a} (a^2 - x'^2)^{1/2} \frac{G(x')dx'}{x' - x} \right) \tag{6.19}$$

$$= -\frac{2(1-v)}{\pi\mu} (a^2 - x^2)^{1/2} \left(\int_{-c}^{c} \frac{1}{(a^2 - x'^2)^{1/2}} \frac{p(x')dx'}{x' - x} \right.$$
$$\left. + \int_{c < |x'| < a} \frac{1}{(a^2 - x'^2)^{1/2}} \frac{G(x')dx'}{x' - x} \right). \tag{6.20}$$

Comparing the first term in equation (6.18) with the expression (5.91) for the stress-intensity factor, K_I, and remembering that $p(x)$ is an even function, we have the condition

$$\int_{c}^{a} \frac{G(x)dx}{(a^2 - x^2)^{1/2}} = \frac{\pi K_I}{(2a)^{1/2}}. \tag{6.21}$$

We can further simplify this result on the assumption that $d = a - c$ is small compared with a. Using the variable $\xi = a - x$ and writing $G(x) \equiv G(\xi)$, we have, to lowest order in d/a,

$$K_{\mathrm{I}} = \frac{1}{\pi} \int_0^d \frac{G(\xi)}{\sqrt{\xi}} \, d\xi. \tag{6.22}$$

Barenblatt now makes the further hypothesis that at the point of fracture the size d of the edge region and the detailed form $G(\xi)$ of the cohesive forces always have the same values, say d_m and $G_m(\xi)$, regardless of the form of $p(x)$ and the crack-length. Then it follows from equation (6.22) that fracture occurs when $K_{\mathrm{I}} = K_{\mathrm{Ic}}$, where K_{Ic} is the material constant

$$K_{\mathrm{Ic}} = \frac{1}{\pi} \int_0^{d_m} \frac{G_m(\xi)}{\sqrt{\xi}} \, d\xi. \tag{6.23}$$

In this way, the critical stress-intensity factor, K_{Ic}, is related to the distribution of cohesive forces at the crack-tip; indeed, for this reason it is sometimes termed the 'modulus of cohesion.'

This final hypothesis of the Barenblatt theory can be made plausible following an argument by Willis.[28] Let $p(x)$ be of order of magnitude σ and $G(x)$ of order G; then, assuming all functions to be sufficiently smooth, equation (6.18) implies that the two quantities $\sigma\sqrt{a}$ and $G\sqrt{d}$ are of the same order. If we now look at the two terms in the bracket in equation (6.20), taking x to lie in the interval (c, a), the first is of order σ/a and in the second the integral over $(-a, -c)$ is of order $G\sqrt{d}/a^{3/2}$ and the integral over (c, a) is of order $G/(ad)^{1/2}$. Clearly this last is dominant, and we can write, for points x in the edge region,

$$\begin{aligned} bf(x) &= -\frac{2(1-\nu)}{\pi\mu}(a^2 - x^2)^{1/2} \int_c^a \frac{1}{(a^2 - x'^2)^{1/2}} \frac{G(x')dx'}{x' - x}\left[1 + O\left(\frac{d}{a}\right)\right] \\ &= -\frac{2(1-\nu)}{\pi\mu} \int_0^d \left(\frac{\xi}{\xi'}\right)^{1/2} \frac{G(\xi')d\xi'}{\xi - \xi'}\left[1 + O\left(\frac{d}{a}\right)\right]. \end{aligned} \tag{6.24}$$

Writing $\Delta u(x) = \Delta u(a - \xi) = v(\xi)$ and using equations (6.16) and (6.17), we can write to leading order

$$\frac{dv}{d\xi} = \frac{2(1-\nu)}{\pi\mu} \int_0^d \left(\frac{\xi}{\xi'}\right)^{1/2} \frac{g[v(\xi')]}{\xi - \xi'} \, d\xi'. \tag{6.25}$$

For a given atomic-force function $g(v)$, this equation is an integral equation for $v(\xi)$, which is the separation of the atomic layers in the edge region. It does not involve either the crack length or the loading function $p(x)$, the connection with these quantities being maintained through equation (6.22). Thus the solution $v(\xi)$ and the edge size d, and hence $G(\xi)$ also, depend on c

H

and $p(x)$ only through K_I. At the point of fracture, when K_I has its critical value K_{Ic}, $v(\xi)$, c, and $G(\xi)$ will always have the same values, and will not be influenced by any other features of c and $p(x)$ than that they should lead to the critical stress-intensity factor.

E Connection between the Barenblatt and Griffith theories

Both the theories of Barenblatt and Griffith lead to a criterion of fracture based on a critical stress-intensity factor, and in fact equations (6.2) and (6.23) become identical if

$$K_{Ic}^2 = 2\gamma\mu/\pi(1-\nu). \tag{6.26}$$

A discussion of the meaning of 'surface energy' within the Barenblatt theory has been given by Willis,[28] and it can be concluded that such a quantity can be defined and that it has the correct value γ implied by equation (6.26). The two theories are thus seen not merely to lead to the same fracture criterion but to be equivalent at the physical level also.

In the Barenblatt theory a traction $g(v)$ operates across a surface with separation v of the atomic layers, and the work done in causing complete separation is

$$\int_0^{v_m} g(v)dv$$

per unit area, where v_m is the maximum separation at which the cohesive forces act. This creates two surfaces, so we define a surface energy

$$\gamma = \tfrac{1}{2} \int_0^{v_m} g(v)dv. \tag{6.27}$$

Considering now a crack just about to extend, $v(\xi)$ is zero at $\xi = 0$ $(x = a)$ and it is reasonable to assume that $v(\xi)$ reaches the maximum value v_m at the crack-tip $\xi = d_m (x = c)$, where growth is just about to occur. Thus we can write

$$\gamma = \frac{1}{2} \int_0^{d_m} g[v(\xi)] \frac{dv}{d\xi} d\xi. \tag{6.28}$$

At the growth point, $g[v(\xi)]$ has the critical form $G_m(\xi)$ and we can use equation (6.25) to give

$$\gamma = \frac{1-\nu}{\pi\mu} \int_0^{d_m} G_m(\xi)d\xi \int_0^{d_m} \left(\frac{\xi}{\xi'}\right)^{1/2} \frac{G_m(\xi')}{\xi-\xi'} d\xi'. \tag{6.29}$$

If in this expression we interchange ξ and ξ', and then form the average of the resulting expression and the original, we obtain

$$\gamma = \frac{1-v}{2\pi\mu} \int_0^{dm} \frac{G_m(\xi)d\xi}{\sqrt{\xi}} \int_0^{dm} \frac{G_m(\xi')d\xi'}{\sqrt{\xi'}} \tag{6.30}$$

$$= \frac{\pi(1-v)}{2\mu} [K_{1c}]^2 \tag{6.31}$$

after using equation (6.23). This is the required result. It is worth noting that the definition (6.28) of the material constant γ does not correspond precisely to the energy of separation of new surfaces starting from a state of zero strain. Rather it corresponds to the extra energy needed to cause separation when the surface element in question has already been subjected to a strain equal to that at the outer limit ($x = a$) of the non-linear region.

An alternative derivation of the equivalence between cohesive-force models of cracks and the Griffith theory has been given by Rice[29] on the basis of a certain path-independent integral. This approach is more general than the one described here, and makes no assumption about the shape of the cohesive zone, although this zone is still assumed to be small.

6.3
MECHANISMS OF CRACK NUCLEATION

A Dislocation pile-up

We have seen in §5.1 that when the continuance of slip on an active slip-plane is blocked the dislocations pile up behind the barrier and so cause large concentrations of stress there. It was suggested by Zener[30] that these stress concentrations could be sufficiently high to break the cohesive bonds between the atoms in the vicinity of the pile-up, and a theory of this process was later provided by Stroh.[31-33] From this point of view, a typical crack nucleation event is illustrated in Figure 6.8. In this case the barriers are grain boundaries, and the dislocations produced by the source S are piled up at the two sides A and B of the grain in which they are produced. If the pile-up is sufficiently severe, the stress at A or B can cause microcracks to form as shown.

The pile-ups will be longest when S is at the middle of the grain, and we shall examine that case. The problem of finding the equilibrium density of dislocations was solved in the fourth example of §5.1, the solution being given in equation (5.11). We wish, however, to generalize that example by including in the problem a frictional resistance $b\sigma_0$ per unit length of dislocation line, arising for example from the Peierls-Nabarro force or from the frictional drag of impurity atoms. The effect of this change is simply to change σ to $\sigma - \sigma_0$ in the previous results, where σ is the external shear stress on the array. We have earlier used the equality of the stress-fields of the pile-up and the shear crack, and we can use it again to claim that the stresses in the neighbourhood of the point B are given by equation (5.23) with σ replaced by $\sigma - \sigma_0$, where $2a$ is the length AB and (ρ, α) are polar coordinates relative to B of the point P in question (Fig. 6.8).

The normal tension acting across the radial element BP at P, $\sigma_{\alpha\alpha}$, is obtained from equations (6.8) and (5.23) as

$$\sigma_{\alpha\alpha} \sim -3(\sigma - \sigma_0)\left(\frac{a}{2\rho}\right)^{1/2} \sin\frac{\alpha}{2}\cos^2\frac{\alpha}{2}. \tag{6.32}$$

As a function of α this is maximum at $\cos\alpha = \frac{1}{3}$, $\alpha \approx -70\frac{1}{2}°$. Thus the most favourable orientation for a microcrack to start is at an angle of $70\frac{1}{2}°$ to the slip-plane direction. The tension across this plane is

$$\sigma_{\alpha\alpha}^{max} = \frac{2}{\sqrt{3}}(\sigma - \sigma_0)\left(\frac{a}{2\rho}\right)^{1/2}. \tag{6.33}$$

The simplest method of estimating the criterion for a crack to nucleate is to require that at a distance $\rho = b$, the interatomic spacing, from the pile-up the stress should attain a value comparable with the cohesive strength $E/10$ as calculated from atomic-force models. Thus we obtain, for nucleation,

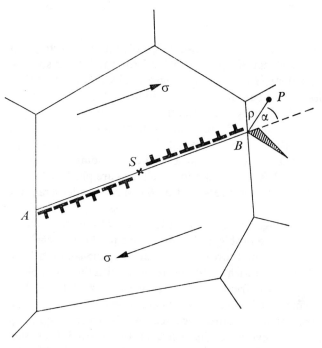

FIGURE 6.8

$$\sigma > \sigma_0 + \frac{1}{(2a)^{1/2}} \frac{E(3b)^{1/2}}{10}. \tag{6.34}$$

An alternative estimate, following Stroh,[31] is obtained using energy arguments of the Griffith type. A crack can form from zero length if, by doing so, it attains a negative total energy (cf. eq. (6.1) and Fig. 6.4). Taking the microcrack length to be 2ρ and the average stress across it to be $\sigma_{\alpha\alpha}^{max}$, the criterion of formation is

$$4\gamma\rho - \frac{\pi(1-\nu)}{2\mu} (\sigma_{\alpha\alpha}^{max})^2 \rho^2 < 0,$$

i.e.

$$\sigma > \sigma_0 + \frac{1}{(2a)^{1/2}} \left(\frac{24\gamma\mu}{\pi(1-\nu)}\right)^{1/2}. \tag{6.35}$$

Since we have the approximate result, for many materials, that $\gamma \sim 0.06\,\mu b$, these two estimates are not very different.

The number n of dislocations on each side of a double pile-up was found in §5.1; for the present problem $n = 2(\sigma-\sigma_0)a(1-\nu)/\mu b$. Thus, using the second of the above estimates, we have at fracture

$$n(\sigma-\sigma_0) \simeq 24\gamma/\pi b \approx 0.5\mu, \tag{6.36}$$

if we take $\nu = \frac{1}{3}$ and the above approximation for γ. For a typical case σ/μ may be about 10^{-3} at fracture, so that there should be about 500 dislocations in the pile-up. Because of the crudity of the estimates we have made, this figure turns out to be too large, and a more detailed investigation along lines we shall outline later shows that a number of dislocations of the order of two or three hundred may be sufficient to initiate fracture.

B *Nucleation mechanisms in general*

Direct observations of the pile-up mechanism of crack formation have been made in some bcc and ionic crystals.[11-13,34] In addition there is indirect evidence for its occurrence in the observed sizes of piled-up arrays in brittle materials[33] which are found to be of the size necessary to initiate fracture, and in observations centred around the Hall-Petch relation (see next section). Nevertheless, the simple dislocation pile-up is not the only mechanism; others have been suggested and in some cases observed also.

Many metals contain second-phase particles existing as precipitates within the parent crystals, often introduced deliberately in order to inhibit dislocation motion and thus raise the yield stress. When pile-ups occur next to one of these particles, either within a grain or at a grain boundary, a microcrack can form within the precipitate if it is sufficiently brittle. Such

cracks have been observed, and there is reason to believe[10] that they con-
stitute the dominant fracture mechanism in mild steel. In such a case it can
be expected that macroscopic fracture occurs at a stress level sufficient to
cause one of these precipitate-sized microcracks to grow unstably.

Other mechanisms that have been suggested for bcc metals are the
collision of two slip systems which are thus able to block one another,
and the collision of two deformation twins. The first of these, suggested by
Cottrell,[35] hinges on the fact that two edge dislocations with Burgers
vectors $\frac{1}{2}[1, -1, 1]$ and $\frac{1}{2}[-1, 1, 1]$ (components relative to the cube axes
with the lattice spacing taken as unit of length), if made to combine, form a
dislocation of Burgers vector [0, 0, 1] which can be immobile. The resulting
pile-ups behind this immobile dislocation (Fig. 6.9) could result in a stress
sufficient to nucleate a crack. Such a fracture mechanism has in fact been
observed,[36] although there are reasons for believing that it is not very
common.[37] The occurrence of fractures at the tips of blocked twins has also
been observed,[14, 38] and this is probably the dominant cause of crack
nucleation at very low temperatures.

In any crystal structure there are certain planes called *cleavage planes*
across which brittle cracking can occur more easily than for any other
planes. Thus, for example, for the simple pile-up crack formation of Figure
6.8, the microcracks will probably form not at the ideal $70\frac{1}{2}°$ angle to the
slip-plane but on the cleavage plane nearest to this direction. In general, for
a polycrystal, this will make little difference, since there will usually be
grains with a suitable relative orientation to enable the microcracks to form.
However, for single crystals, even if there exist obstacles strong enough to
cause the pile-up, the geometrical relationships between the slip-planes and

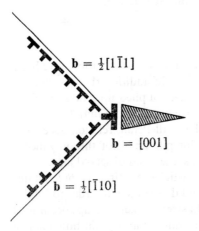

$$\mathbf{b} = \tfrac{1}{2}[1\bar{1}1]$$

$$\mathbf{b} = [001]$$

$$\mathbf{b} = \tfrac{1}{2}[\bar{1}10]$$

FIGURE 6.9

cleavage planes play an important part. This is particularly true for hexagonal crystals in which the only easily activated slip-planes are the basal
planes, which are also the cleavage planes. Since a glide pile-up produces no
normal stress directly in front of itself (cf. eq. (6.32) with $\alpha = 0$), it can be
seen that microcracks in hexagonal single crystals are unlikely to be formed
in this way.

An alternative mechanism has been proposed by Stroh[39] to account for
crack nucleation in such cases, where the glide and cleavage planes coincide.
We have seen in §3.3c that parallel edge dislocations on different glide-planes
have a position of minimum energy in which one is directly above the other.
Thus, with several active parallel slip-planes, the dislocations on the different
planes will tend to line up to form an array called a *low-angle boundary*
perpendicular to the slip-planes (Fig. 6.10a). Under a shear stress σ_s this
will tend to move through the crystal. If h is the average separation of the
dislocations, the angle of rotation of the slip-planes produced by the extra
half-planes of the dislocations in the boundary is $\theta = b/h$.

Now suppose that half of the boundary is blocked, and the other half
moves away and reaches an equilibrium position at a distance x from the
blocked half, at which the force on it due to the external stress just balances
the attraction of the blocked array. Following Stroh, we can estimate the
value of x. There are of the order of x/h dislocations in each array which
exert sizable interactions on one another, so that there are $(x/h)^2$ interacting
pairs, each producing a force of the order $\mu b^2/2\pi x$. Thus the total force on
the mobile half is $(\mu b^2/2\pi x)\,(x/h)^2 = (\mu \theta^2 x/2\pi)$. This must balance the
force $Nb(\sigma_s - \sigma_0)$, where N is the number of dislocations in that half and σ_0 is
the frictional resistance to dislocation motion, so that

$$x = 2\pi Nb(\sigma_s - \sigma_0)/\mu \theta^2.$$

There will exist large tensile forces near the broken ends of the two
boundaries which can cause cleavage on the slip-plane as in Figure 6.10(b).
The effect of the two boundaries is to wedge one end of the crack open by a
displacement of magnitude θx. Now we shall show later on (§E) that a
wedged crack attains an equilibrium length $2c$, where in this case (cf. eq.
(6.51))

$$2c = \mu(\theta x)^2/8\pi\gamma(1-v) = \pi(\sigma_s - \sigma_0)^2(Nh)^2/2\gamma\mu(1-v).$$

Such a crack would then be unstable under a normal stress σ_n if $\sigma_n^2 c$ exceeds
the critical value given by Griffith (eq. (6.3)), and thus we obtain the criterion
for complete fracture

$$\sigma_n(\sigma_s - \sigma_0) = 4\gamma\mu/\pi(Nh). \tag{6.37}$$

The approximations made in deriving this result can be improved, and a
detailed calculation by Stroh[39] gives essentially the same result. Stroh is

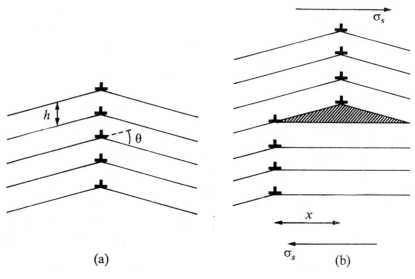

FIGURE 6.10

also able to allow for the anistropy of the stress-strain relations. A comparison of the above criterion with the known fracture strength of zinc crystals under simple tension with various orientations of the basal planes relative to the tensile axis showed good agreement with the predicted linear relation between σ_s and σ_n^{-1} and reasonable agreement for the slope of that relation provided Nh is taken as the crystal diameter.

c The Hall-Petch relations

The stress in equation (6.35) at which microcracks are nucleated can be written in the form

$$\sigma = \sigma_0 + Kd^{-1/2}, \tag{6.38}$$

where $d = 2a$ is the length of the slip-line, or in other words, for the most effective slip-lines, d is the grain diameter. If nucleation were the critical event in causing fracture, equation (6.38) would give the macroscopic fracture stress. That the fracture stress depends on the grain size in the way indicated in this equation was first reported by Petch[40] for some irons and steels at liquid nitrogen temperatures, at which they show brittle fracture behaviour, and since then the same connection has been observed in other metals. A comparison of the observed values of K and values calculated from equation (6.35), or rather from the more accurate version of that equation which we shall derive later on, shows fairly good agreement,[33] and the observed dependence of σ_0 on the impurity content of the steel is consistent with its interpretation in the pile-up model as a friction stress. Thus it seems plausible that, at least at these very low temperatures, crack nucleation becomes the controlling event in fracture.

A similar dependence on grain size of the macroscopic yield stress σ_y has been found by Hall[41] and Petch[40]:

$$\sigma_y = \sigma_0 + K^* d^{-1/2}. \tag{6.39}$$

The stress σ_y is the stress at which plastic flow first takes place on a macroscopic scale. It is in most cases the stress required to propagate slip out of the individual grains into neighbouring grains, that is to break down the barriers presented by the grain boundaries. A similar relation has since been found for materials in which the principal barriers to slip are other than the grain boundaries, for example precipitates or barriers formed by dislocation networks, with d being the average spacing of the barriers in all cases.

It is possible to explain this Hall-Petch relation for the yield stress in a way similar to the explanation of equation (6.38) for the fracture stress. The early stages of plastic flow in a polycrystal take place within individual grains, and result in dislocation pile-ups at the grain boundaries. In a ductile material, however, before such pile-ups have time to become severe enough to form microcracks, they will activate Frank-Read sources in the neighbouring grain. These sources emit dislocation loops, which on the one hand neutralize the stress concentration of the original pile-up and on the other enable the slip to continue across the second grain (see Fig. 1.25). Now, in order to activate a source of length l a shear stress of magnitude $\max[\sigma_0'$, $2T/bl]$ is needed. Here σ_0' is the shear stress on the line segment of the source necessary to enable it to break free from the impurities which pin it down and $2T/bl$ is the stress necessary to bend the segment to its critical radius of curvature (see §1.3D). T is the line tension of the source segment, and is of the order of magnitude μb^2. Such a source at distance r from the pile-up will then be activated if

$$(\sigma - \sigma_0)\beta\left(\frac{d}{r}\right)^{1/2} = \max\left[\sigma_0', \frac{2T}{bl}\right],$$

where β is a factor of order unity depending on the orientations involved. A source of length l cannot be on average closer than l to the pile-up, so that the optimum conditions for operating sources of this length are obtained by putting $r = l$:

$$(\sigma - \sigma_0)\beta d^{1/2} = \max[\sigma_0' l^{1/2}, (2T/b) l^{-1/2}].$$

The minimum value of σ occurs when $l = 2T/b\sigma_0'$, and is

$$\sigma = \sigma_0 + K^* d^{-1/2},$$

where $K^* = (2T\sigma_0'/\beta^2 b)^{1/2}$. Thus we have the Hall-Petch relation for the minimum shear stress needed to propagate slip from one grain to another.

Besides the form of the dependence of yield stress on d, the predicted value of K^* agrees as well as can be expected with the observed values and σ_0 both agrees with the value from fracture data and has the correct dependence on impurity content expected of a friction stress.[33,42]

With this picture, then, the essential difference between brittle and ductile behaviour is that in the latter the pile-ups are able to activate dislocation sources before they become severe enough to create a crack. Now the activation of sources becomes the easier the higher the temperature, since the greater thermal fluctuations enable the line segment to free itself more readily from the pinning impurities. Thus we are able to explain why metals which are brittle at low temperatures show a relatively sharp transition to ductile behaviour as the temperature is raised. Furthermore, it seems reasonable to expect that the ductile-brittle transition temperature will depend on the applied strain-rate, as indeed it does. For, at high strain rates, the sources will have less time to operate, and may be unable to do so, even with the aid of thermal fluctuations, so that brittle behaviour becomes more likely. Quantitative expressions of these ideas have been obtained by Stroh.[33]

D *Microcrack formation*

In the preceding sections we have given several rough estimates for various quantities associated with microcrack formation, and we shall now improve them. Considering first crack initiation from a pile-up, we examine the situation of Figure 6.11 in which a crack of length $2c$ has formed at the

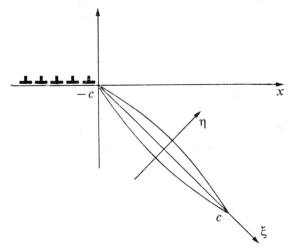

FIGURE 6.11

optimum $70\frac{1}{2}°$ angle to the slip-plane of the pile-up. For the moment we ignore any change that this might make on the pile-up itself, so that the effect of the crack is simply to relax the stress-field of the pile-up on the two faces $\eta = 0$, $-c < \xi < c$. Thus, from equation (6.33), we get the change in the stress-field equal to the solution of the elastic boundary-value problem with

$$\sigma_{\eta\eta} = -(\sigma - \sigma_0)\left(\frac{2a}{3}\right)^{1/2}(\xi + c)^{-1/2}, \quad \sigma_{\xi\eta} = 0 \quad (\eta = 0, -c < \xi < c).$$

(6.40)

The vanishing shear stress condition is obtained from equation (5.23) for $\alpha = -70\frac{1}{2}°$.

According to the note after equation (5.91) the reduction in total energy caused by the opening up of a crack under tension can be obtained from equation (5.46). Carrying out the two integrations, we obtain for the present case an energy change (see Probs. 8 and 9)

$$\Delta U = -\frac{16(1-\nu)}{3\pi\mu}(\sigma - \sigma_0)^2 ac + 4\gamma c$$

(6.41)

including the surface energy. Thus the criterion $\Delta U < 0$ for microcrack nucleation becomes

$$(\sigma - \sigma_0)^2 a > 3\pi\gamma\mu/4(1-\nu).$$

(6.42)

Introducing the number of dislocations n in the pile-up from §5.1, this becomes

$$n(\sigma - \sigma_0) > 3\pi\gamma/2b,$$

(6.43)

which is about half the estimate in equation (6.36). Thus, in fact, the number of dislocations necessary to form the microcrack is about 250, much less than our earlier estimate.

E Wedge-crack formation

In the calculation of the preceding section, the length c of the microcrack drops out of the equations, indicating that, once formed, the crack will continue to grow with a reduction in energy at each stage. This conclusion would be unsound, however, since when the microcrack becomes long the stress-field of the pile-up can no longer be approximated by the square-root singularity term. But even more important is the effect of the microcrack on the dislocations of the pile-up. Stroh has examined the consequences of a second stage in the cracking process in which the n dislocations run into the microcrack, thus wedging it open at one end by a displacement (nb) (see Fig. 6.12).

It turns out that such a crack has an equilibrium length, in the absence of external stress. In applying this wedge-crack model we are making the assumptions that the source of the slip dislocations has been blocked from producing more than the original n dislocations and that all n of these dislocations run into the crack. These assumptions are, of course, an oversimplification, but the more complete calculation becomes prohibitively difficult. It has been given by Smith[43] for a microcrack which is coplanar with the original microcrack, with results that will be mentioned later.

The n dislocations can be regarded as one big dislocation of Burgers vector nb, and give rise to the stress-field of equation (3.15). On the plane of the microcrack, these components are

$$\sigma_{\eta\eta}{}^{(d)} = +\hat{\sigma}_{\alpha\alpha} = -D \sin \alpha/r = -D \sin \alpha/(\xi+c),$$
$$\sigma_{\xi\eta}{}^{(d)} = \hat{\sigma}_{r\alpha} = D \cos \alpha/r = D \cos \alpha/(\xi+c),$$

(6.44)

where $D = \mu nb/2\pi(1-\nu)$ and $\alpha = -70\frac{1}{2}°$. To find the effect of the crack we must superimpose the solution of the boundary-value problem with the negative of the above stress components acting on the surface of the crack. Denoting by $\mathbf{\sigma}^{(e)}$ the extra stress-field, we have the boundary conditions on $\eta = 0, |\xi| < c$:

$$\sigma_{\eta\eta}{}^{(e)} = D \sin \alpha/(\xi+c), \qquad \sigma_{\xi\eta}{}^{(e)} = -D \cos \alpha/(\xi+c). \tag{6.45}$$

Now the stress-fields of arbitrarily loaded cracks were obtained in Chapter 5, being given in full by equations (5.48)–(5.50) for shear loading and by equation (5.89) and the preceding sentence for normal loading. We need here only the stress components on the half-plane $\eta = 0, \xi < -c$, where $\theta = \pi/2$ and $R = (\xi^2 - c^2)^{1/2}$ (cf. eq. (5.19)) and so

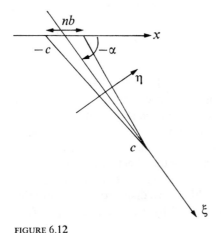

FIGURE 6.12

$$\sigma_{\eta\eta}^{(e)}(\xi, 0) = \frac{D \sin \alpha}{\pi(\xi^2 - c^2)^{1/2}} \int_{-c}^{c} \frac{(c^2 - \xi'^2)^{1/2} d\xi'}{(\xi - \xi')(\xi' + c)}$$

$$= D \sin \alpha \left(\frac{1}{\xi + c} + \frac{1}{(\xi^2 - c^2)^{1/2}} \right), \tag{6.46}$$

$$\sigma_{\xi\eta}^{(e)}(\xi, 0) = -\frac{D \cos \alpha}{\pi(\xi^2 - c^2)^{1/2}} \int_{-c}^{c} \frac{(c^2 - \xi'^2)^{1/2} d\xi'}{(\xi - \xi')(\xi' + c)}$$

$$= -D \cos \alpha \left(\frac{1}{\xi + c} + \frac{1}{(\xi^2 - c^2)^{1/2}} \right). \tag{6.47}$$

Adding on the stresses of the big dislocation, we obtain the total stress-field of the wedge-crack on this half-plane:

$$\sigma_{\eta\eta}(\xi, 0) = \frac{D \sin \alpha}{\pi(\xi^2 - c^2)^{1/2}}, \qquad \sigma_{\xi\eta}(\xi, 0) = -\frac{D \cos \alpha}{(\xi^2 - c^2)^{1/2}}. \tag{6.48}$$

In order to calculate the elastic energy stored in a medium with a wedge crack, we can proceed as for dislocations themselves (cf. §3.3A). The crack can be formed by making a cut in the material on the half-plane $\eta = 0$, $\xi < -c$ and displacing the two faces relative to one another by an amount nb, or ($nb \cos \alpha$, $-nb \sin \alpha$) (note that α is negative) (Fig. 6.13). Since the resulting tractions are given in equation (6.48), the work done per unit length in the z-direction is

$$U = \frac{1}{2} \int_{-R}^{-c} [-\sigma_{\xi\eta}(\xi, 0) \, nb \cos \alpha + (-\sigma_{\eta\eta}(\xi, 0)) (-nb \sin \alpha)] d\xi \tag{6.49}$$

$$= nb \, D \int_{-R}^{-c} \frac{d\xi}{(\xi^2 - c^2)^{1/2}}$$

$$= \frac{\mu(nb)^2}{4\pi(1 - \nu)} \cosh^{-1} \frac{R}{c} \approx \frac{\mu(nb)^2}{4\pi(1 - \nu)} \ln \frac{2R}{c}. \tag{6.50}$$

Here R is the outer radius of the specimen, which must be introduced for the same reasons as for single dislocations.

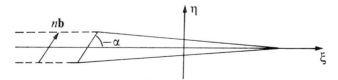

FIGURE 6.13

Adding the surface energy $4\gamma c$, we find that the total energy has a minimum when

$$2c = \frac{\mu n^2 b^2}{8\pi(1-\nu)\gamma}. \tag{6.51}$$

Therefore wedge-cracks can exist with a stable length, and do not necessarily lead to large-scale fracture as soon as they form. The equilibrium length obtained from this formula when $n = 200$ turns out to be $2c \sim 10^{-3}$ cm, a not unreasonable value.

It will be of interest later to calculate the relative displacements of the wedge-crack faces. The two densities of dislocations corresponding to the tensile and shear parts are obtained from equation (5.87) as

$$f(\xi) = -\frac{2(1-\nu)}{\pi\mu b} D \sin \alpha \frac{1}{(c^2-\xi^2)^{1/2}} \int_{-c}^{c} (c^2-\xi'^2)^{1/2} \frac{d\xi'}{(\xi'+c)(\xi'-\xi)}, \tag{6.52}$$

and a similar result with $-\sin \alpha$ replaced by $\cos \alpha$. Carrying out the integration and substituting for D then gives

$$f(\xi) = \frac{n \sin \alpha}{\pi} \frac{1}{(c^2-\xi^2)^{1/2}}. \tag{6.53}$$

The relative displacement from equation (5.81) is then

$$\Delta u_\eta(\xi, 0) = -nb \sin \alpha \frac{1}{\pi} \cos^{-1} \frac{\xi}{c}, \tag{6.54}$$

and a similar derivation gives for the shear displacement

$$\Delta u_\zeta(\xi, 0) = nb \cos \alpha \frac{1}{\pi} \cos^{-1} \frac{\xi}{c}. \tag{6.55}$$

F *Influence of external stress on the stability of microcracks*

In the previous section we ignored the effect of any external stress-field on the stability of the microcrack. Clearly the presence of a sufficiently large tensile component $\sigma_{\eta\eta}$ will cause microcracks of the length in equation (6.51) to propagate immediately to failure, so that crack nucleation becomes critical in such a case. More generally, by examining the formation of wedge cracks in the presence of external stress, we shall be able to determine in any particular case which of the two stages is the critical event.[44]

We shall use the notation of §3.3A in which u_i and σ_{ij} denote the final displacement and stress fields of the wedge crack when the external boundaries are subjected to the conditions in equations (3.22). The cut C for the wedge crack of Figure 6.13 consists of the half-plane $\eta = 0$, $\xi < c$. It is convenient to split C into two parts, C_1 corresponding to the crack itself $(-c < \xi < c)$ and C_2 corresponding to the dislocation cut $(\xi < -c)$. On C_1 the traction t_i^c is specified (it is, in fact, zero), and on C_2 it is the relative displacement $\Delta\mathbf{u}$ which is given, equal to $n\mathbf{b}$.

The displacement field u_i^0 in conditions (3.23) is the result of applying the same external boundary conditions to the body when both the crack and the dislocations are absent. The change in energy upon forming the wedge crack is given by equation (3.30), which we can rewrite as

$$\Delta U^{\text{tot}} = \tfrac{1}{2}\int_{C_{1^+}} \Delta u_i t_i^0 dS + \tfrac{1}{2}\int_{C_{2^+}} nb_i t_i^c dS + \tfrac{1}{2}\int_{C_{2^+}} nb_i t_i^0 dS. \tag{6.56}$$

Here t_i^c denotes the value of $\sigma_{ij}n_j$ on C_2^+. Then, assuming that the stresses σ_{ij}^0 are known, all we need in order to evaluate this energy are the values of the displacement discontinuity Δu_i across the crack and of $\sigma_{ij}n_j$ on the cut C_2.

We shall be interested primarily in the change in ΔU^{tot} as the crack length changes. Since crack growth occurs only from the end $\xi = +c$, the cut C_2 keeps a constant length and the last term in equation (6.56) remains fixed. Hence for our purposes we can ignore it.

As with all dislocation problems, the energy can only be maintained finite by introducing an outer cut-off radius. We suppose that the body is a cylinder centered at the wedging dislocations $(\xi = -c)$ and of radius R. The surface C_2 then occupies the region $-R-c < \xi < -c$.

We shall consider the state σ_{ij}^0 to be one of uniform stress for which $\sigma_{\eta\eta}^0 = \sigma_n$ and $\sigma_{\xi\eta}^0 = \sigma_s$. Let us denote by u_i^D and σ_{ij}^D the deformation fields for the wedge crack alone which we found in the last subsection. Then defining $u_i' = u_i - u_i^D$ and $\sigma_{ij}' = \sigma_{ij} - \sigma_{ij}^D$, we have the following boundary conditions:

$$\sigma_{ij}^D n_j = 0 \text{ on } C_1, \quad \Delta u_i^D = nb_i \text{ on } C_2, \quad \sigma_{ij}^D \to 0 \text{ at infinity};$$

$$\sigma_{ij}' n_j = 0 \text{ on } C_1, \quad \Delta u_i' = 0 \text{ on } C_2, \quad \sigma_{ij}' \to \sigma_{ij}^0 \text{ at infinity}. \tag{6.57}$$

Consequently σ_{ij}' is the field due to just a crack $|\xi| < c$ in a body subject to uniform stresses σ_{ij} at infinity. This problem was solved in detail in Chapter 5, the stress-fields being given in equations (5.18), (5.21), (5.22), and in (5.83) for the shear and tensile parts respectively. In the region $\xi < -c$, $\eta = 0$, we therefore obtain

$$\sigma_{\eta\eta}'(\xi, 0) = -\frac{\sigma_n\xi}{(\xi^2 - c^2)^{1/2}}, \quad \sigma_{\xi\eta}'(\xi, 0) = -\frac{\sigma_s\xi}{(\xi^2 - c^2)^{1/2}} \quad (\xi < -c). \tag{6.58}$$

The terms in ΔU^{tot} involving σ_{ij}' then reduce to

$$\frac{1}{2}\int_{C_1+}\Delta u_i' t_i^0 dS + \frac{1}{2}\int_{C_2+} nb_i t_i' dS$$

$$= -\frac{\pi(1-\nu)}{2\mu}(\sigma_n^2 + \sigma_s^2)c^2 + \frac{1}{2}nb(\sigma_s\cos\alpha - \sigma_n\sin\alpha)\int_{-R-c}^{-c}\frac{\xi d\xi}{(\xi^2 - c^2)^{1/2}}$$

$$= -\frac{\pi(1-\nu)}{2\mu}(\sigma_n^2 + \sigma_s^2)c^2 - \frac{1}{2}nbc(\sigma_s\cos\alpha - \sigma_n\sin\alpha) + \text{const.} \qquad (6.59)$$

Here we have used the fact, which is clear from equations (5.33) and (5.82), that the first integral is the energy of a crack under uniform tractions. The terms labelled const. are independent of c apart from small terms of order c/R.

The remaining terms in ΔU^{tot} are

$$\tfrac{1}{2}\int_{C_1+}\Delta u_i^D t_i^0 dS + \tfrac{1}{2}\int_{C_2+} nb_i t_i^D dS.$$

The second of these terms is just the energy of the wedge crack evaluated in equation (6.49), and the first can be evaluated using the displacements in equations (6.54) and (6.55). We obtain

$$\frac{\mu(nb)^2}{4\pi(1-\nu)}\ln\frac{2R}{c} + \frac{1}{2}\int_{-c}^{c}(\sigma_n nb\sin\alpha - \sigma_s nb\cos\alpha)\frac{1}{\pi}\cos^{-1}\frac{\xi}{c}d\xi.$$

Thus altogether, and including the surface energy of the microcrack,

$$\Delta U^{\text{tot}} = \frac{\mu(nb)^2}{4\pi(1-\nu)}\ln\frac{2R}{c} + 4\gamma c$$

$$-\frac{\pi(1-\nu)}{2\mu}(\sigma_n^2 + \sigma_s^2)c^2 - nbc(\sigma_s\cos\alpha - \sigma_n\sin\alpha), \quad (6.60)$$

ignoring small terms and a constant.

The question of whether stable microcracks can form can now be answered by examining whether or not ΔU^{tot} has a minimum at a positive value of c. The equation $d\Delta U^{\text{tot}}/dc = 0$ is simply a quadratic in c, and has two real positive roots, the smaller of which is an energy minimum, provided

$$[4\gamma - nb(\sigma_s\cos\alpha - \sigma_n\sin\alpha)]^2 > n^2b^2(\sigma_n^2 + \sigma_s^2). \qquad (6.61)$$

If this condition is not met, U^{tot} is a monotonically decreasing function of c, and all microcracks are unstable. Now for wedge cracks formed from a dislocation pile-up, α is $-70\tfrac{1}{2}°$ and the critical number of wedging dislocations is obtained in terms of the shear stress σ on the pile-up from equation (6.43). The condition for stable microcracks is then

$$(\sigma_n^2 + \sigma_s^2)^{1/2} + (\sigma_s + 2\sqrt{2}\sigma_n)/3 < 8(\sigma - \sigma_0)/3\pi. \qquad (6.62)$$

To illustrate the implications of this result, we examine the case of a cylindrical specimen under simple axial tension T with superimposed hydrostatic pressure p. The slip-planes most likely to be activated are those whose planes coincide with the 45° directions of maximum shear stress. Then the formation of microcracks will occur in the way illustrated in Figure 6.14. We have

$$\sigma = \tfrac{1}{2}T, \qquad \sigma_n = \sigma_{\eta\eta} = T\cos^2\beta - p, \qquad \sigma_s = \sigma_{\xi\eta} = -T\sin\beta\cos\beta,$$

where $\beta = -\alpha - \pi/4$, so that $\cos\beta = (2\sqrt{2}+1)/3\sqrt{2}$ and $\sin\beta = (2\sqrt{2}-1)/3\sqrt{2}$. If we now ignore the friction stress σ_0, the criterion (6.62) becomes

$$\left(\frac{9+4\sqrt{2}}{2}T(T-2p)+9p^2\right)^{1/2} + \frac{1+2\sqrt{2}}{2}T-2\sqrt{2}p < \frac{4}{\pi}T. \tag{6.63}$$

This inequality is not satisfied for pure tension ($p = 0$) or for pure shear ($p = T/2$) so that in both of these cases microcrack nucleation causes fracture. Stable microcracks are, in fact, possible only for $p > 10.36T$. Thus in most cases of practical importance microcrack initiation is the critical event according to this theory.

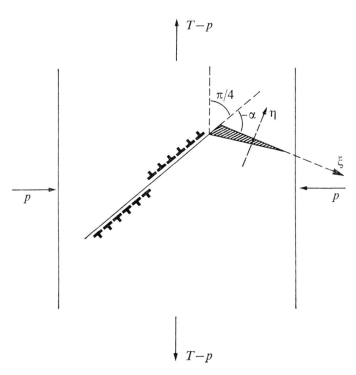

FIGURE 6.14

A related calculation by Smith,[43] in which microcracks are considered parallel to the slip-plane, leads to the conclusion that wedged microcracks are always unstable, provided they can be formed at all, no matter how high the hydrostatic pressure. Although this analysis is idealized in that the microcrack plane will in general be different from the slip-plane, it improves the present calculation in that allowance is made for the continued operation of the dislocation source after the wedge crack has been formed. It is, in fact, the continued production of wedging dislocations that is essentially responsible for the complete instability of the cracks in this case. Whether such indefinitely continued operation of the source is physically realistic is an open question. However, both theories agree in predicting that it is the initiation of microcracks which is critical, at least in all but the high-pressure region.

A factor which can favour the occurrence of stable microcracks is some ductility in the material which can blunt the microcrack before it can become too long. As shown in equation (6.5), this is equivalent to having an extra surface-energy term in the growth criterion, arising from the plastic dissipation of energy by the growing crack. Since the criterion (6.61) for wedge-crack stability applies to a growth situation, again the surface energy γ must there be regarded as including the extra plastic term. That is, γ should be replaced by $\gamma + D/4$ as in equation (6.5). On the other hand, the surface energy appearing in the initiation criterion (6.43) should be the true surface energy γ, since crack initiation occurs on such a small scale (of atomic dimensions) that there can be no plastic dissipation. Therefore the right-hand side of equation (6.62) contains a factor $1 + D/4\gamma$, which can be much greater than unity, and hence will promote stability of the microcracks. Such blunted microcracks are, in fact, a common feature in the fracture of ductile materials.[2]

In materials containing large brittle precipitate particles, it is possible for the microcracks to be formed in the precipitate particles, and there is substantial evidence in favour of this occurring in mild steel.[10, 43] Then the surface energy γ entering equation (6.61), which expresses the growth criterion for the cracks, corresponds to the surrounding material, and the criterion for crack initiation concerns the properties of the precipitate particle. Thus equation (6.43) cannot be used as the initiation criterion in such a case, and our discussion following equation (6.62) would not apply.

6.4

CRACK GROWTH UNDER CYCLIC STRESSING

A *Survey of the fatigue process*[45]

When the applied stress fluctuates in time, fracture can occur at levels of stress far lower than the monotonic fracture strength, an occurrence termed metal fatigue. The simplest case to consider is repeated tension-compression in which the stress varies regularly between constant maximum and minimum levels, σ_{max} and σ_{min}, and particularly the completely reversed case in which $\sigma_{min} = -\sigma_{max}$. Data from this kind of test can be summarized in a graph of the number N of cycles to failure against the stress level σ_{max}, the so-called N–S curve, a typical example of which is shown in Figure 6.15. The occurrence of a fairly sharp knee in the NS curve at around $N \sim 10^5$ cycles is a common observation, and separates what, from many points of view, appear to be two different processes.

In the low-cycle range ($N \lesssim 10^5$), the stresses are usually sufficiently high to cause an amount ε_p of macroscopic plastic strain on each cycle. For a fatigue test between fixed stress levels, ε_p changes during the test,[46] usually decreasing, a phenomenon called cyclic hardening. (Occasionally softening occurs, depending on the treatment, such as prestraining, which has been given to the specimen before the test.) For this reason low-cycle fatigue tests are often made with cycling taking place between fixed levels of plastic strain. In fact, for such tests, a very simple phenomenological relation is found, called the Manson-Coffin law, which states that for any material[47, 48]

$N\varepsilon_p^2 = \text{const.}$

At lower stress levels, the plastic strain is limited to dislocation motion within individual grains, and the NS relation takes on a different character.

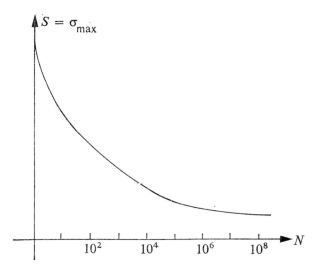

FIGURE 6.15

Here it is possible to subdivide the fatigue process into a number of approximately distinct stages. First of all, after perhaps a few hundred cycles, there appear on the surface of the material what are termed persistent slip-bands, somewhat like the slip-bands which occur during monotonic loading at higher stress levels. All of the microplasticity seems to be concentrated in these slip-bands, through dislocations moving backwards and forwards during the two halves of the stress cycle. As the number of cycles increases, the appearance of the slip-bands becomes roughened, and extrusions and intrusions appear.[49] Eventually an intrusion becomes deep enough to be called a crack, that is deep enough to give rise to a stress-concentration effect at its tip. When this stage is reached, fatigue proceeds through growth of this dominant crack by a small but quite regular increment for each cycle until it attains a sufficient length to fail catastrophically when the stress next reaches its maximum value. This last, failure, stage would fall into the category of phenomena discussed in §6.2.

The mechanism by which fatigue cracks are initiated on the surface of a specimen is still not fully understood, although the following simplest explanation still seems the most likely. With the slip taking place only on slip-planes within the relatively narrow persistent slip-bands, an arrangement of surface slip steps such as the one shown in Figure 6.16 can arise. Here the forward slip is concentrated in the lower half of the band and the reverse slip in the upper half. Clearly such a situation will eventually lead to a deep intrusion and then a crack. It could occur simply as a result of a

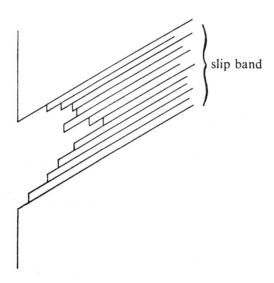

slip band

FIGURE 6.16

random distribution of forward and backward slip-steps within the slip-bands, although mechanisms have been suggested by Mott[50] and Cottrell and Hull[51] by means of which such intrusions could develop systematically. For example, Mott considers the slip-steps to be produced by screw dislocation lines moving into and out of the plane of the diagram in Figure 6.16. If towards the end of their forward motion on, say, one of the lower slip-planes, they encounter an obstacle which forces them to cross-slip on to one of the upper slip-planes, their reverse motion will then occur on this second plane. If again there is a second obstacle which forces them to cross-slip down again at the end of the reverse motion, then repeated cycling will drive the screws round and round a parallelogram-like path, thus eventually producing an intrusion. Although such a coincidence of obstacles might seem rather improbable, and undoubtedly most of the dislocations do simply see-saw backwards and forwards on the same slip-plane, it is equally true that only a few cases of cross-slip are necessary in order to produce large effects. For example, if only 10 dislocations are driven around such a path, in only 10^4 cycles, they will produce an intrusion of depth equal to a typical grain diameter.

The growth of fatigue cracks can itself be divided into two stages, referred to as I and II.[49] In stage I the crack continues to grow in a slip-plane, like the intrusion that caused it, the growth being essentially shear-controlled. Often stage I ceases once the first grain boundary is reached, but it can last much longer and sometimes stage II does not occur. In stage II, the tensile stress becomes a controlling factor, and the plane of crack-growth turns to become normal to the tensile axis. The faces of the crack become rippled with striations,[52] one for each stress cycle, in contrast with stage I in which the crack faces are relatively smooth. The striations caused by stage II growth are a characteristic feature used to diagnose fatigue failure.

There is available a great deal of experimental data on the rates of fatigue crack-growth[53] as well as a number of theoretical models of the process.[54] It is well established that, for plane geometries, the rate of growth depends on the applied stress and crack length only through the stress-intensity factor K_I (for tension-compression), at least as long as the stress is not high enough to produce large plastic zones at the crack-tip. It is also argued on dimensional grounds[55,56] that for a crack of length $2c$ in an infinite specimen the rate of growth per cycle dc/dn must be proportional to c since there is no other length in the problem. This would mean that $dc/dn \propto K_I^2$. This argument is questionable, however, since there may be lengths associated with the microstructure of the material which enter the phenomenon. In fact the balance of experimental evidence indicates that in stage II the rate of growth is proportional to some higher power of K_I, most commonly K_I^3 or K_I^4.

B *Shear mode of fatigue crack growth*

In §5.3 we described the BCS model[57] of an elastoplastic shear crack in which the crack is replaced by a linear array of freely slipping dislocations and the plastic zones by coplanar arrays moving against a frictional resistance (Fig. 5.6). The solution for the dislocation density, sketched in Figure 5.7, was obtained from the equilibrium equation (5.52). This solution can be extended[58] to allow for unloading from some maximum shear stress σ to a lower shear stress $(\sigma - \Delta\sigma)$, such as occurs with a fatigue crack, and the result can be used to explain some features of fatigue crack growth.

When σ is reduced, a reverse plastic flow occurs immediately near the crack-tip. The zone $c < |x| < d$ over which this takes place does not extend completely out to the limits $x = \pm a$ of the original plastic zone, so that the new dislocation density $g(x)$ has the form indicated in Figure 6.17. In the regions of reverse flow, the friction stress σ_1 is reversed in direction, so that the equilibrium equation satisfied by $g(x)$, analogous to equation (5.52), is

$$\frac{\mu b}{2\pi(1-\nu)} P \int_{-a}^{a} \frac{g(x')dx'}{x-x'} = \begin{cases} -(\sigma - \Delta\sigma) & \text{for } -c < x < c, \\ -(\sigma - \Delta\sigma) - \sigma_1 & \text{for } c < |x| < d. \end{cases} \quad (6.64)$$

The value of the left-hand side is not known for $d < |x| < a$; however, we do know that $f(x) = g(x)$ in that region. Therefore, if we subtract equation (6.64) from equation (5.52), we obtain

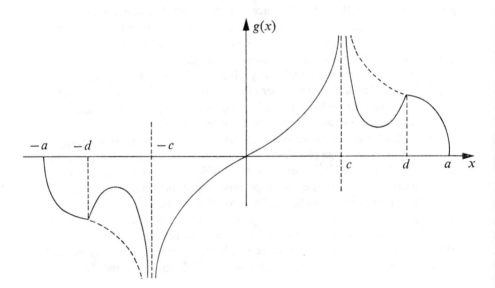

FIGURE 6.17

$$\frac{\mu b}{2\pi(1-\nu)} P \int_{-d}^{d} \frac{f(x')-g(x')}{x-x'} dx' = \begin{cases} -\Delta\sigma & \text{for } -c < x < c, \\ -(\Delta\sigma-2\sigma_1) & \text{for } c < |x| < d. \end{cases} \quad (6.65)$$

This equation is identical with the original BCS equation (5.52) provided we make the changes $f(x') \to f(x')-g(x')$, $\sigma \to \Delta\sigma$, $\sigma_1 \to 2\sigma_1$, and $a \to d$, and therefore we can take over immediately the results of §5.3. In particular

$$c/d = \cos(\pi\Delta\sigma/4\sigma_1),$$

or at low stresses,

$$(d-c)/c \approx \pi^2\Delta\sigma^2/32\sigma_1^2. \quad (6.66)$$

The reverse plastic displacement is defined as

$$\Delta\Phi = b \int_c^d [f(x')-g(x')] \, dx' = \frac{8(1-\nu)\sigma_1 c}{\pi\mu} \ln \sec \frac{\pi\Delta\sigma}{4\sigma_1}, \quad (6.67)$$

and at low stresses

$$\Delta\Phi \approx \frac{\pi(1-\nu)}{4\mu\sigma_1} \Delta\sigma^2 c. \quad (6.68)$$

For repeated cycling between σ and $\sigma-\Delta\sigma$ we associate $\Delta\Phi$ with the amount of crack growth per cycle as follows. On the initial stress rise to σ an amount Φ of new crack face appears at $x = c$ on the bottom crack face and a similar amount at $x = -c$ on the top face. On the stress fall to $\sigma-\Delta\sigma$ the dislocation motion causes a length $\Delta\Phi$ of new crack face at $x = c$ on the top face and at $x = -c$ on the bottom face. Then on the next stress rise new lengths $\Delta\Phi$ appear on the bottom face at $x = c$ and the top face at $x = -c$, and so on, with the result that $dc/dn = \Delta\Phi$. In making this correspondence we are assuming that new crack face, once formed, remains as surface and cannot be rewelded into the material on the following half-cycle of stress. Such rewelding might occur with fatigue cracks in high vacuum, and could be the explanation of the lower rates of crack-growth observed there.[59,60]

The above discussion applies only to cracks growing under plane shear. However, in stage I growth under tension-compression, growth occurs on a single slip-plane which will lie approximately on a plane of maximum shear stress at roughly 45° to the tensile axis. Then the range of shear stress on the plane of the crack is $\Delta\sigma = \Delta T/2$, where ΔT is the range of tensile stress. The resulting expression

$$\frac{dc}{dn} = \frac{\pi(1-\nu)}{16\mu\sigma_1} \Delta T^2 c \quad (6.69)$$

is then found to give reasonable agreement with the experimental observations over a wide range of materials.[58] It becomes violated at high growth rates, where sizable plastic zones occur at the crack-tip, and the growth rate is found to be proportional to K_I^4 rather than K_I^2 as indicated by equation (6.69). It is also violated at very low growth rates in some materials, such as some aluminium alloys, in which the fatigue cracks can even become non-propagating at sufficiently low stresses, a phenomenon which is still not satisfactorily explained.

c *Damage mode of fatigue crack growth*

At higher growth rates, the dependence of dc/dn on K_I is not K_I^2 as follows from the model of the previous section, but usually some higher power of K_I. In this connection, the results of Pelloux[61] are interesting. In experiments on an aluminium alloy, the rate of crack growth was determined for a number of stress-intensity factors both by measuring the spacings of the stage II striations on the crack faces and by measuring in the usual way the macroscopic growth rates. These were different, and furthermore, while the first increased as K_I^2, the second increased on average as K_I^3. The interpretation of these results, which has been given by McClintock,[62] is that there are two essentially distinct types of crack growth occurring. One is the ductile type of growth described in the last section, with some changes to allow for the fact that the crack is growing normal to the tensile axis, which, in view for example of the results described in §5.4c, probably introduces no very great differences. In addition, there is a second growth mechanism, which at higher stress levels occurs side by side with the first, each in turn operating at different parts of the crack front, so that areas of crack face with striations are interspersed with areas in which the second mechanism has operated. As K_I is increased, the second mechanism assumes a more and more dominant role, so that the apparent K_I dependence increases from K_I^2 to the dependence appropriate to this second type.

The second mechanism proposed by McClintock[62,63] is based on the occurrence of microscopic damage ahead of the growing crack. When the stress-intensity factor is large, the plastic zone at the crack-tip will also be large, and it is to be expected that a number of microcracks will form ahead of the main crack. These will occur at grain boundaries or other obstacles to slip, and particularly at brittle precipitates, just as in the case of fracture under monotonic loading. Then it becomes possible for rapid crack growth to occur through linking of the main crack with the microcracks in front of it. A quantitative theory of this type of growth has been given by McClintock, using results for plastic strains obtained earlier for the antiplane shear crack and an estimate for the rate of accumulation of damage in the form of microcracks ahead of the crack based on the Manson-Coffin law. The resulting growth rate turns out to be proportional to K_I^4, and to be in quite good agreement numerically with Pelloux's observations.

As described earlier (§6.1D), ductile fracture under monotonic loading is regarded as taking place via this same mechanism of crack growth through linking with pores in the plastic zones ahead of the crack. Thus we can anticipate some connection between the two processes of crack growth under monotonic and cyclic loadings. In McClintock's analysis, for example, the crack growth rate is found to depend on the fracture ductility (i.e. plastic strain at fracture) under monotonic loading. The BCS crack model[57] has also been used by Cottrell[64,65] to discuss the growth of cracks in semi-ductile materials through microcrack formation, and this theory has been extended to fatigue cracks by Weertman.[66] Although this theory bears a strong resemblance to McClintock's in its basic approach, it differs in certain important respects which are not obviously reconcilable.

For monotonic fracture, Cottrell supposes that crack growth occurs when the plastic displacement at the crack-tip (Φ in §5.3) attains a critical value, say Φ_c. Such a criterion may be rationalized by supposing that, as Φ measures the number of dislocations in the plastic zone at the crack-tip, it also gives a measure of the number of microcracks which are formed by the dislocations. For cyclic growth, considering a point at x ahead of the crack whose current length is c, the plastic displacement at x during any half-cycle of stress is

$$\Delta\Phi(x, c) = b \int_x^d [f(x') - g(x')] \, dx', \tag{6.70}$$

using the notation of the last section. If we now denote by $c_0 = x \cos(\pi\Delta\sigma/4\sigma_1)$ the crack length at which plastic cycling first starts at x and by c_1, $c_2, \ldots, c_n$ succeeding crack lengths as the crack-tip approaches x, then the total cyclic plastic displacement at x is

$$2 \sum_{i=0}^n \Delta\Phi(x, c_i) \approx 2 \int_{c_0}^{c_n} \Delta\Phi(x, c) \, \frac{dc}{dc/dn}.$$

The factor 2 includes both halves of the cycle. We have also used the fact that, as a rule, many cycles are needed to cause the crack to grow from c_0 to x, but that the rate of growth remains roughly constant over this range. We now extend Cottrell's criterion and require that this total cyclic displacement should exactly equal Φ_c when the crack-tip reaches x. Thus

$$\frac{dc}{dn} = \frac{2}{\Phi_c} \int_{c_0}^x \Delta\Phi(x, c) \, dc. \tag{6.71}$$

In reaching this result we have ignored the plastic displacement which occurs before cycling starts, that is for x between d and a on Figure 6.17, and also the displacement due to the increments of crack growth. At high rates of growth these will both become important corrections.

Integration by parts and use of the fact that $b[f(x') - g(x')]$ is a function of x'/c only gives, after a slight change of notation,

$$\frac{dc}{dn} = c\Phi_c \int_c^d \frac{x^2 - c^2}{x^2} b[f(x) - g(x)]\, dx. \tag{6.72}$$

Substituting the dislocation density derived in the last section now gives

$$\frac{dc}{dn} = \frac{8(1-v)\sigma_1}{\pi\mu\Phi_c} c^2 \left(2\ln\frac{d}{c} - \frac{(d^2 - c^2)^{1/2}}{d} \ln\frac{(d+c)^{1/2} + (d-c)^{1/2}}{(d+c)^{1/2} - (d-c)^{1/2}} \right) \tag{6.73}$$

$$\approx \frac{\pi^3(1-v)}{192\mu\sigma_1{}^3\Phi_c} \Delta\sigma^4 c^2, \tag{6.74}$$

when $\Delta\sigma \ll \sigma_1$. Weertman's calculation varies from the one given above and the two results differ by a numerical factor, with the present one being very close to the experimental results quoted by Weertman. Here Φ_c is taken from experiments on monotonic loading. Apart from the numerical agreement, it is interesting to note the high-power ($K_1{}^4$) dependence on the stress-infinity factor.

Finally, we note that, although these results based on the BCS model apply strictly only to shear cracks, we can expect to be able to use them, at least in some cases, for tensile cracks, just as the Dugdale model (§5.3) can sometimes be applied. It must be observed, however, that if in a cyclic tension experiment the applied stress becomes compressive ($\Delta\sigma > \sigma_{max}$), then the formulae should be applied with $\Delta\sigma$ set equal to σ_{max}, since the compressive part of the cycle produces no stress concentration.

PROBLEMS

1 With the notation of §6.1B, show that the quantity $t_s + kt_n$ assumes its maximum value for a plane which contains the σ_2 principal direction and whose normal makes an angle $\theta = \frac{1}{2}\tan^{-1}(k^{-1})$ with the σ_1 direction. Let σ_T and σ_C denote the fracture stresses in simple tension and compression respectively. Show that $|\sigma_C|/\sigma_T = [(k^2+1)^{1/2}+k]^2$ when Navier's fracture criterion is used. Hence show that Navier's criterion is equivalent to the relation

$$\frac{\sigma_1}{\sigma_T} - \frac{\sigma_3}{|\sigma_C|} = 1.$$

2 Suppose that the criterion for fracture across a plane takes the form $t_s{}^2 = \sigma_0{}^2 - \beta t_n$ relating the shear and normal stress across that plane. Show that fracture occurs normal to the σ_1 principal direction as long as $\sigma_1 - \sigma_3 < \beta$, but that if this condition does not hold fracture occurs across a plane whose normal makes an angle θ with the σ_1 direction, where $\cos 2\theta = \beta/(\sigma_1 - \sigma_3)$. Show that if $\sigma_0 < \beta$ the fracture stresses under simple tension and simple compression are respectively $\sigma_T = \sigma_0{}^2/\beta$ and $-\sigma_C = 2\sigma_0 + \beta$. Sketch the form of the fracture criterion as a curve in the σ_1, σ_3 plane (as in Fig. 6.6). Show that for a material in which $|\sigma_C| = 15\sigma_T$ the plane of fracture under simple compression makes an angle of about $26\frac{1}{2}°$ with the compressive axis.

3 An infinite body contains the usual strip crack on $y = 0$, $-c < x < c$. A stress $\sigma_{yy} \to \sigma$ is imposed at infinity, and in addition the crack is wedged apart by normal point forces of magnitude P at the points $x = d$, $y = 0\pm$ on its two faces. Using a stress-intensity factor criterion for fracture as in equation (6.6), find the conditions on P and σ which will cause the crack to grow.

4 A crack as in Problem 3 is wedged apart by normal point forces at $x = d$ which are of such magnitude that the relative displacement of the crack faces at $x = d$ equals Δ. How big must Δ be in order to cause crack growth? What is likely to happen subsequent to crack growth occurring in this situation? To what kind of experimental situation would this problem apply?

5 With the geometry of Figure 6.5, suppose that the crack makes an angle of $45°$ with the σ_1 direction. Show that crack growth can be expected at an angle α such that $\sigma(3 \cos \alpha - 1) + p \sin \alpha = 0$. Discuss the cases $\sigma = 0$, $p = 0, p < 0$.

6 Derive the results (6.11)–(6.13). Show that the normal component of applied stress across the crack is tensile under the conditions leading to these fracture criteria provided that $p > -0.43\sigma$.

7 What is the connection between Dugdale's model of the crack-tip mentioned in §5.4c and the Barenblatt model?

8 Use the general result (5.91) to show that the energy reduction given in equation (6.41) corresponds to the stresses (6.40) on the crack surfaces.

9 In equation (6.41) we have claimed that the change in total elastic energy when the crack in Figure 6.11 is formed is just the negative of the strain energy that the crack alone would have. Verify, using Betti's theorem, that this property takes account of the interaction energy between the stresses of the crack and those of the dislocation pile-up.

10 Derive the equivalent formula to equation (6.43) for a microcrack formed at the tip of a single pile-up (see example 3 in §5.1b).

11 An infinite medium contains a crack in the form of an infinite plane slit of initial width $2c_0$. The crack is filled with a gas whose pressure is p_0. Assuming that p_0 is high enough to start the crack growing, find the length of the crack at which the growth stops. (Assume that the growth is sufficiently slow that dynamical effects may be ignored and the gas expands isothermally. Also ignore any effect the gas might have on the surface energy of the medium.)

12 Suppose that the microcrack in §6.3f contains a gas at pressure p. Modify equation (6.56) to take account of this, and evaluate the energy corresponding to equation (6.60). [This problem is related to the effect of dissolved hydrogen in steel as an embrittling agent.[44]]

13 Use the results of §6.3e to discuss whether a single edge dislocation should reasonably be taken to have a wedge crack attached to it, on the opposite side of the slip-plane from its extra half-plane.

14 Two opposite groups containing n and m edge dislocations and gliding on parallel planes are both blocked in a configuration in which the line joining the two pile-ups makes an angle $-\alpha$ with the slip-planes. The two groups run together to form a double wedge-crack, as shown in Figure 6.18. Use the methods of §5.3e to find the energy of such a crack. Find a limit on the separation, d, of the two slip-planes such that the crack has a lower energy than the total energy of two composite dislocations with Burgers vectors nb and mb in the same positions as the wedging dislocations in Figure 6.18.

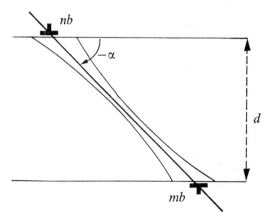

FIGURE 6.18

REFERENCES

1 General references on fracture: *Fracture*, Swampscott Conference, ed. B.L. Averbach, D.K. Felbeck, G.T. Hahn, and D.A. Thomas (New York: Wiley, 1959); *Fracture of solids*, Seattle Conference, ed. D.C. Drucker and J.J. Gilman (New York: Wiley, 1963); *Fracture*, ed. H. Liebowitz (New York: Academic Press, 1968); *Mechanical behaviour of materials*, by F.A. McClintock and A.S. Argon (Reading, Mass: Addison-Wesley, 1966); *Physical basis of yield and fracture*, Oxford conference, ed. A.C. Strickland (London: Institute of Physics and Physical Society, 1966)
2 Puttick, K.E., Phil. Mag. **4** (1959), 964
3 Jaeger, J.C., *Elasticity, fracture and flow* (London: Methuen, 1956)
4 Griffith, A.A., Phil. Trans. **A221** (1920), 163
5 Griffith, A.A., Proc. First. Int. Cong. Appl. Mechs., Delft (1924), p. 55
6 Zwicky, F., Z. Phys. **24** (1923), 131
7 Orowan, E., Repts. Progr. Phys. **12** (1949), 185
8 Weibull, W., Ingen. Vetenskaps Akad., Handlingar, nos. 151 and 153 (1939)
9 Fisher, J.C., and Holloman, J.H., Trans. A.I.M.E. **171** (1947), 546
10 Smith, E., Int. J. Fracture Mechs. **4** (1968), 131
11 Hull, D., Proc. Roy. Soc. **A285** (1965), 148
12 Honda, R., J. Phys. Soc. Japan **16** (1961), 1309
13 Johnston, T.L., and Parker, E.R., *Fracture of solids* (ref. 1), p. 267 (1963)
14 Hull, D., *Fracture of solids* (ref. 1), p. 417 (1963)
15 Cottrell, A.H., *Fracture* (ref. 1), p. 20 (1959)
16 Petch, N.J., *Fracture* (ref. 1), p. 54 (1959)
17 McClintock, F.A., Kaplan, S.M., and Berg, C.A., Int. J. Fracture Mechs. **2** (1966), 614
18 McClintock, F.A., Int. Conf. on Fracture, Kiruna, 1967
19 Sack, R.A., Proc. Phys. Soc. **58** (1946), 729
20 Orowan, E., Z. Kristall. **89** (1934), 327
21 Felbeck, D.K., and Orowan, E., Welding J. **34** (Suppl., 1955), 570
22 Gell, M., and Robertson, W.D., Acta Metall. **14** (1966), 481
23 Bilby, B.A., and Eshelby, J.D., *Fracture* (ref. 1), vol. 1, p. 99 (1968)

24 Lardner, R.W., Proc. Roy. Soc. **A317** (1970), 199
25 Irwin, G.R., *Handbuch der Physik* VI, p. 551 (Berlin: Springer, 1958)
26 McClintock, F.A., and Walsh, J.B., Proc. 4th U.S. Nat. Cong. Appl. Mech. **2** (1962), 1015
27 Barenblatt, G.I., J. Appl. Math. Mechs. **23** (1959), 622 and 1009; Adv. Appl. Mech. **7** (1962), 55
28 Willis, J.R., J. Mech. Phys. Solids **15** (1967), 151
29 Rice, J.R., J. Appl. Mech. **35** (1968), 379
30 Zener, C., *Fracturing of metals* (New York: Amer. Soc. Metals, 1948), p. 3
31 Stroh, A.N., Proc. Roy. Soc. **A223** (1954), 404
32 Stroh, A.N., Proc. Roy. Soc. **A232** (1955), 548
33 Stroh, A.N., Adv. Phys. **6** (1957), 418
34 Stokes, R.J., Johnston, T.L., and Li, C.H., Phil. Mag. **3** (1958), 718
35 Cottrell, A.H., Trans. A.I.M.E. **212** (1958), 192
36 Washburn, J., Gorum, A.E., and Parker, E.R., Trans. A.I.M.E. **215** (1959), 230
37 Stroh, A.N., *Fracture* (ref. 1), p. 117 (1959)
38 Hull, D., Acta Metall. **8** (1960), 11
39 Stroh, A.N., Phil. Mag. **3** (1958), 597
40 Petch, N.J., J. Iron and Steel Inst. **174** (1953), 25
41 Hall, E.O., Proc. Phys. Soc. **B64** (1951), 747
42 Cracknell, A., and Petch, N.J., Acta Metall. **3** (1955), 186
43 Smith, E., *The physical basis of yield and fracture* (ref. 1), p. 36 (1966).
44 Bilby, B.A., and Hewitt, J., Acta Metall. **10** (1962), 587
45 General references on fatigue: Kennedy, A.J., *Processes of creep and fatigue in metals* (Edinburgh: Oliver and Boyd, 1962); Freudenthal, A.M., *Handbuch der Physik* VI, p. 591 (Berlin: Springer, 1958); Freudenthal, A.M. (ed.) *Fatigue in aircraft structures* (New York: Academic Press, 1956); Thompson, N., and Wadsworth, N.J., Adv. Phys. **7** (1958), 72; Proceedings of Crack Propagation Symposium (Cranfield, 1961); Proceedings of Int. Conf. on Fatigue of Metals (New York: A.S.M.E. and I.M.E. 1956)
46 Wood, W.A., and Segall, R.L., Proc. Roy. Soc. **A242** (1957), 180
47 Coffin, L.F., Trans. A.S.M.E. **76** (1954), 931
48 Manson, S.S., NACA Tech. Note 2933 (1953); Experimental Mechs. **5** (1963), 193
49 Forsyth, P.J.E., *Crack propagation symposium* (ref. 45), p. 76 (1961)
50 Mott, N.F., Acta Metall. **6** (1958), 195
51 Cottrell, A.H., and Hull, D., Proc. Roy. Soc. **A242** (1957), 211
52 McMillan, J.C., and Pelloux, R.M.N., Boeing S.R.L. Rept. no. D1-82-0553 (1966)
53 Donaldson, D.R., and Anderson, W.E., *Crack propagation symposium* (ref. 45), p. 375 (1961)
54 Paris, P., and Erdogan, F., J. Basic Engng. **85** (1963), 528
55 Frost, N.E., and Dugdale, D.S., J. Mech. Phys. Solids **6** (1958), 92
56 Liu, H.W., J. Basic Engng. **83** (1961), 23
57 Bilby, B.A., Cottrell, A.H., and Swinden, K.H., Proc. Roy. Soc. **A272** (1963), 304
58 Lardner, R.W., Phil. Mag. **17** (1968), 71
59 Wadsworth, N.J., *Internal stresses and fatigue in metals* (Amsterdam: Elsevier, 1958), p. 382
60 Bradshaw, F.J., and Wheeler, C., Appl. Mater. Res. **5** (1966), 112
61 Pelloux, R.M.N., Trans. Am. Soc. Metals **57** (1964), 511

62 McClintock, F.A., M.I.T. Research Memo. no. 108 (July 1966)
63 McClintock, F.A., *Fracture of solids* (ref. 1), p. 65 (1963)
64 Cottrell, A.H., Proc. Roy. Soc. **A285** (1965), 10
65 Knott, J.F., and Cottrell, A.H., J. Iron and Steel Inst. **201** (1963), 249
66 Weertman, J., Northwestern Univ. Materials Science Tech. Rept. no. 1
    (Aug. 1965)

7

CONTINUOUS DISTRIBUTIONS
OF DISLOCATIONS

7.1

DISLOCATION DENSITY TENSOR

A *Physical basis*

So far we have been concerned with the properties of single dislocations. But when we wish to consider large-scale plastic phenomena, for example plastic flow above the yield stress, it is known that very large numbers of dislocations are involved, of the order of 10^{11} to 10^{12} lines/cm^2. In such a situation a reasonable point of view is to treat the aggregate of dislocations and to ignore the individual, just as for instance in electromagnetic theory one generally ignores the discrete nature of electric charges and deals with properties of the aggregate, such as the charge density and electric current. Such a view leads to the theory of continuous distributions of dislocations.

This theory was originated independently by Kondo[1] and Bilby, Bullough, and Smith,[2] following the pioneer work of Nye,[3] and has been developed principally by these and their co-workers[4, 5] and by Kröner[6,7] and Seeger.[8] All of this work is based on the concept of an atomistic and crystalline structure of the dislocated materials. More recently Noll and Truesdell[9, 10] have given an approach to the problem which is based entirely on continuum mechanical notions, and as such is rationally self-consistent as well as being more widely applicable – in fact to internal stresses in any elastic continuum, crystalline or not. In §7.3 we shall describe Noll and Truesdell's theory.

In order to construct the dislocation density tensor from a physical point of view we classify the possible elementary dislocations into a number of types. The ith type is specified by its Burgers vector, $\mathbf{b}^{(i)}$, and its line direction, $\boldsymbol{\xi}^{(i)}$, and we suppose that in the neighbourhood of the point in question there are $N^{(i)}$ dislocations of this type crossing a unit area normal to the $\boldsymbol{\xi}^{(i)}$ direction. Then we define a dislocation density $a_{kl}^{(i)}$ for each type by

$$a_{kl}^{(i)} = N^{(i)} b_k^{(i)} \xi_l^{(i)} \tag{7.1}$$

and the total dislocation density by

$$a_{kl} = \sum_{(i)} a_{kl}^{(i)}. \tag{7.2}$$

Suppose that we have a small planar circuit $\mathscr{C}$ in the body, of area dS and normal $\mathbf{n}$. Then the number of dislocation lines of type i threading this circuit is $N^{(i)}\boldsymbol{\xi}^{(i)} \cdot \mathbf{n} \, dS$. Let the circuit now be mapped onto a perfect reference crystal by mapping lattice step to lattice step as in Chapter 1 (Fig. 1.10). The mapped circuit does not close in general, and the vector needed to close it is the sum of the true Burgers vectors of all the dislocation lines threading $\mathscr{C}$. This closure failure can now be mapped back onto the real crystal to give the local Burgers vector for the circuit, $\mathbf{b}(\mathscr{C})$, which is given by

$$\mathbf{b}(\mathscr{C}) = \sum_{(i)} (N^{(i)}\boldsymbol{\xi}^{(i)} \cdot \mathbf{n} \, dS) \, \mathbf{b}^{(i)},$$

i.e.

$$b_k(\mathscr{C}) = a_{kl} n_l \, dS. \tag{7.3}$$

Screw dislocations have their Burgers vector parallel to their line direction, and hence provide a diagonal term in the dislocation density tensor when one of the coordinate axes lies along the dislocation line. With a similar choice of coordinates, however, edge dislocations provide off-diagonal components. In simple cubic crystals we can achieve complete simplicity by choosing axes in the cube directions, in which case the three diagonal components and the six off-diagonal components represent individually the nine possible dislocation types. With other crystal structures this simplicity is not in general attainable.

We have been led to equation (7.3) by a consideration of sets of discrete dislocation lines in a crystal. However, we can now dispense with the physical motivation and use equation (7.3) as the definition of the dislocation density tensor in terms of the closure failure of Burgers circuits. In this way we can separate dislocations from the notions of crystallinity and atomicity, and can define a dislocation density tensor for any elastic continuum.

B *Field equations for the linear theory*

In this section we shall derive the differential equations which describe the kinematical relationships for a continuously dislocated medium. We shall restrict our attention to the small strain case, so that classical elasticity may be used. Our procedure will be to obtain the equations for a continuous distribution of dislocations as a limiting case of those for a distribution of discrete dislocation lines, in much the same way as equation (7.3) was obtained. Later, in §§7.3 and 7.4, we shall reconsider this question within the non-linear theory of elasticity, giving there a more general and more rigorous derivation of the equations. Meanwhile the present section, together with the examples in §7.2, should form a sufficient basis for readers who wish to do so to proceed directly to Chapters 8 and 9.

An arrangement of discrete dislocation lines can be formed by making various cuts throughout the body and causing appropriate displacement discontinuities across them. The resulting deformation can be described within the linear theory by the displacement field $\mathbf{u}(\mathbf{x})$, which will be defined at each point in the cut body. Corresponding to this there will exist a strain tensor e_{ij} defined as in equation (2.1) as the symmetric part of the displacement gradient and also a rotation tensor ω_{ij} which is the antisymmetric part of $\partial_i u_j$. Thus we can write

$$\partial_i u_j = e_{ij} + \omega_{ij}, \qquad e_{ij} = e_{ji}, \qquad \omega_{ij} = -\omega_{ji}. \tag{7.4}$$

It then follows that the differentials of the displacement components are

$$du_j = (e_{ij} + \omega_{ij})dx_i. \tag{7.5}$$

Let us now suppose that the cuts in the body are all glued together again. The displacement field $\mathbf{u}$, which was well-defined in the cut body, is no longer single-valued in the healed body. We shall make the assumption that e_{ij} and ω_{ij} do exist as single-valued fields after the cuts are healed. (As we shall show in §7.4, this assumption eliminates the possibility of rotation dislocations.) Then in equation (7.5) the right-hand side is a well-defined differential form, although it may not be integrated, except in the cut body.

Let $\mathscr{C}$ be a circuit in the cut body which starts and finishes at opposite points on one of the cuts. Then $\oint_{\mathscr{C}} du_j$ gives the difference in the displacement component between the final and initial points on $\mathscr{C}$. According to the definition in §1.1A, this equals the Burgers vector b_j of the dislocation corresponding to the cut in question, provided that $\mathscr{C}$ is right-handed with respect to the line direction. Thus we have

$$\oint_{\mathscr{C}} (e_{ij} + \omega_{ij})dx_i = b_j. \tag{7.6}$$

This equation remains meaningful after the cuts are healed, and moreover there is no need then to specify a particular starting point for $\mathscr{C}$.

Now let $\mathscr{C}$ be a circuit enclosing several dislocation lines. Since $\mathscr{C}$ is equivalent to a number of circuits, one around each of the lines, the right-hand side of equation (7.6) is replaced by the sum of the Burgers vectors of all the dislocation lines threading $\mathscr{C}$. Combining this with equation (7.3), when $\mathscr{C}$ is sufficiently small, we have

$$\oint_{\mathscr{C}} (e_{ij} + \omega_{ij})dx_i = a_{jl}n_l dS.$$

Applying Stokes' theorem to reduce the line integral to a surface integral, and comparing the coefficients of $n_l dS$ on the two sides of the equation, then gives

I

$$a_{jl} = \varepsilon_{lip}\partial_i(e_{pj}+\omega_{pj}).$$ (7.7)

Thus we have been able to express the dislocation density, for small strain and rotation, in terms of the gradients of the strain and rotation tensors.

It follows immediately from equation (7.7) that

$$\partial_l a_{jl} = 0.$$ (7.8)

This result is a conservation law for Burgers vectors, representing the fact that the Burgers vector remains constant along a dislocation line or in particular that dislocation lines cannot end within the interior of a body (see Prob. 3).

Let us apply the operator $\varepsilon_{mkj}\partial_k$ to both sides of equation (7.7). Taking the symmetric and antisymmetric parts of the result leads to

$$\varepsilon_{lip}\varepsilon_{mkj}\partial_i\partial_k e_{pj} = (\varepsilon_{mkj}\partial_k a_{jl})^{(S)},$$ (7.9)

$$\varepsilon_{lip}\varepsilon_{mkj}\partial_i\partial_k \omega_{pj} = (\varepsilon_{mkj}\partial_k a_{jl})^{(A)}.$$ (7.10)

Here (S) and (A) indicate that the expressions inside the brackets are to be symmetrized and antisymmetrized respectively with respect to the two free indices, l and m. The first of these two equations reduces, when $a_{jl} = 0$, to the usual compatibility equations of linear elasticity, which are well known to be necessary and sufficient conditions for the existence of a single-valued elastic displacement field (a proof will be given in §8.1). Thus, to the extent that dislocations are present, there is no such displacement associated with the strain field (e_{kl}); this brings us back to our original starting point, which was the non-integrability of equation (7.5).

c *Lattice vectors*

Since many of the applications of the theory will concern dislocations in crystalline materials it is helpful to preserve the connection between the present continuum mechanical basis and the corresponding quantities which appear for crystals. In the unstrained state of the body, before the dislocations are formed, there will be defined for a crystal a triad of lattice vectors, which are constant throughout the body. We shall choose to represent this crystalline structure by an orthonormal triad of vectors $\{u_1, u_2, u_3\}$. Except for a simple cubic crystal, this triad will not be identical with the lattice vectors themselves, but will be some fixed linear combination of them. Nevertheless we shall refer to $\{u_k\}$ as the reference triad of lattice vectors. We shall also, for convenience, use $\{u_k\}$ as the basis for our Cartesian coordinate system.

In a distorted state of the crystal, whether dislocated or not, the triad of lattice vectors will be changed, and we shall denote them by $\{e_k(x): k = 1, 2, 3\}$ at the point x. We can write

$$e_k(\mathbf{x}) = \mathbf{u}_k + \alpha_{kl}(\mathbf{x})\mathbf{u}_l, \tag{7.11}$$

where α_{kl} is often referred to as the *lattice distortion*. If the strains and rotations are everywhere small, then the α_{kl} will be small, since in such a case the distorted lattice vectors will not differ very much from those in the unstrained state.

Consider a small element $\mathbf{dx} = dx_k\mathbf{u}_k$ of material located at the point $\mathbf{x}$. Under the deformation, this element moves with the lattice, and becomes an element $dx_k e_k(\mathbf{x})$ which equals $\mathbf{dx} + \alpha_{kl}dx_k\mathbf{u}_l$. Now from a continuum point of view, in terms of the displacement field, $\mathbf{dx}$ changes into $\mathbf{dx} + \mathbf{u}(\mathbf{x} + \mathbf{dx}) - \mathbf{u}(\mathbf{x})$, which equals $\mathbf{dx} + \partial_k u_l dx_k\mathbf{u}_l$. Thus we can identify the lattice distortion α_{kl} with the displacement gradient $\partial_k u_l$. In particular, from equation (7.4) we have

$$e_{kl} = \tfrac{1}{2}(\alpha_{kl} + \alpha_{lk}), \qquad \omega_{kl} = \tfrac{1}{2}(\alpha_{kl} - \alpha_{lk}). \tag{7.12}$$

From equation (7.7) the dislocation density has a simple form in terms of the lattice distortion tensor:

$$a_{jl} = \varepsilon_{lip}\partial_i \alpha_{pj}. \tag{7.13}$$

We define a *lattice curve* in a dislocated crystal as a curve which everywhere is tangent to one of the three families of lattice vectors. Consider an e_1 curve for example, and let it have a parametric equation $x_k = x_k(t)$ in terms of some parametrization t. The tangent vector dx_k/dt is then parallel to e_1 at each point on the curve. By appropriate choice of t, we can arrange for these two tangent vectors to be equal, which gives us the equations

$$dx_k/dt = e_1 \cdot \mathbf{u}_k.$$

Comparing with equation (7.11), for a lattice curve of the e_l family, we have the equation

$$dx_k/dt = \delta_{kl} + \alpha_{lk}. \tag{7.14}$$

7.2
SOME SPECIAL CASES

A Antiplane strain

In classical elasticity, antiplane strain corresponds to the deformation field $u_1 = u_2 = 0$, $u_3 = u_3(x_1, x_2)$. The corresponding state in a dislocated crystal will then have a lattice distortion of the form

$$e_1 = \mathbf{u}_1 + \alpha_1\mathbf{u}_3, \qquad e_2 = \mathbf{u}_2 + \alpha_2\mathbf{u}_3, \qquad e_3 = \mathbf{u}_3, \tag{7.15}$$

where α_1 and α_2 are functions of x_1 and x_2. Comparing with equations

(7.11) and (7.12), the corresponding non-zero components of strain and rotation are

$$e_{13} = \alpha_1/2, \qquad e_{23} = \alpha_2/2, \qquad \omega_{13} = \alpha_1/2, \qquad \omega_{23} = \alpha_2/2.$$

From equation (7.13), the only non-zero component of dislocation density is a_{33}, which is given by

$$a_{33} = \partial_1 \alpha_2 - \partial_2 \alpha_1. \tag{7.16}$$

The lattice curves, from equation (7.14), are solutions of the following equations:

for e_1-curves: $\quad x_2 = \text{const.}, \quad dx_3/dx_1 = \alpha_1(x_1, x_2),$

for e_2-curves: $\quad x_1 = \text{const.}, \quad dx_3/dx_2 = \alpha_2(x_1, x_2).$ $\tag{7.17}$

From equation (7.16), if $a_{33} = 0$, there exists a single-valued displacement field $w(x_1, x_2)$ such that $\alpha_i = \partial_i w$, just as in the classical case. An expression for w is

$$w(\mathbf{x}) = w(\mathbf{x}_0) + \int_{\mathbf{x}_0}^{\mathbf{x}} [\alpha_1(\boldsymbol{\xi})d\xi_1 + \alpha_2(\boldsymbol{\xi})d\xi_2]. \tag{7.18}$$

This result also holds in the general case in any simply connected region of the $x_1 x_2$ plane in which the dislocation density vanishes. For a multiply connected region, $w(\mathbf{x})$ as defined in equation (7.18) is single-valued if and only if the integral there vanishes if taken around any closed curve. Using Stokes' theorem and equation (7.16), this is equivalent to the requirement that

$$\int_S a_{33}(\boldsymbol{\xi})d\xi_1 d\xi_2 = 0 \tag{7.19}$$

for all areas S bounded by curves lying in the given region. Thus, in particular, if the dislocations are confined to certain bounded and simply connected regions, S_1, S_2, ... and if the total Burgers vector in each of these regions is zero, then there exists a single-valued displacement field in the part of the $x_1 x_2$-plane lying outside these regions. This was the situation encountered for example in the BCS model of a crack (§5.2), where the dislocation density is non-zero only in the 'region' $x_2 = 0$, $|x_1| < a$, and outside this a displacement exists.

B Crossed screws

Consider the generalization of the previous section which corresponds to the distortion

$$\mathbf{e}_1 = \mathbf{u}_1 + \alpha_{13}\mathbf{u}_3, \qquad \mathbf{e}_2 = \mathbf{u}_2 + \alpha_{21}\mathbf{u}_1 + \alpha_{23}\mathbf{u}_3, \qquad \mathbf{e}_3 = \mathbf{u}_3 + \alpha_{31}\mathbf{u}_1 \tag{7.20}$$

with α_{ij} being small. The strain and rotation components are now

$$e_{12} = \tfrac{1}{2}\alpha_{21}, \qquad e_{13} = \tfrac{1}{2}(\alpha_{13}+\alpha_{31}), \qquad e_{23} = \tfrac{1}{2}\alpha_{23},$$

$$\omega_{12} = -\tfrac{1}{2}\alpha_{21}, \qquad \omega_{13} = \tfrac{1}{2}(\alpha_{13}-\alpha_{31}), \qquad \omega_{23} = \tfrac{1}{2}\alpha_{23},$$

with the other components, apart from the symmetrically related ones, being zero.

From equation (7.13), the non-zero dislocation density components are

$$a_{11} = -\partial_3\alpha_{21}+\partial_2\alpha_{31}, \qquad a_{12} = -\partial_1\alpha_{31}, \qquad a_{13} = \partial_1\alpha_{21},$$

$$a_{31} = -\partial_3\alpha_{23}, \qquad a_{32} = \partial_3\alpha_{13}, \qquad a_{33} = -\partial_2\alpha_{13}+\partial_1\alpha_{23}. \qquad (7.21)$$

With this distortion, it is possible to achieve a strain-free state, by setting $\alpha_{21} = \alpha_{23} = 0$, $\alpha_{13} = -\alpha_{31}$. Then taking $\tfrac{1}{2}(\alpha_{13}-\alpha_{31}) = \omega$ gives

$$a_{11} = a_{33} = -\partial_2\omega, \qquad a_{12} = \partial_1\omega, \qquad a_{32} = -\partial_3\omega, \qquad (7.22)$$

with the other components zero. The case when ω depends only on x_2, which corresponds to a pure torsion of the lattice about the x_2-direction, is realized by the grid of crossed screws, $a_{11} = a_{33} = -\partial_2\omega$. On planes $x_2 = $ const., the lattice lines are simply rotated through an angle $\omega(x_2)$, the e_1-lines becoming lines $x_3 = \omega(x_2)x_1+$const. and the e_3-lines $x_1 = -\omega(x_2)x_3+$const.

The interest in strain-free states arises in the consideration of crystals which have been subjected to plastic flow which has produced large lattice rotations. For these the average dislocation density necessary to account for the observed rotations can be calculated approximately by ignoring any residual internal strain. The present results, for example, would give us the dislocation density in a crystal which had been twisted about the $\mathbf{u}_2$-direction, the angle of plastic twisting being a general function of the x_2-coordinate.

c Plane strain

A state of plane strain in classical elasticity has non-zero displacement components u_1 and u_2 which are functions of x_1 and x_2 only. Consequently we are led to consider the lattice distortion

$$\mathbf{e}_1 = \mathbf{u}_1+\alpha_{11}\mathbf{u}_1+\alpha_{12}\mathbf{u}_2, \qquad \mathbf{e}_2 = \mathbf{u}_2+\alpha_{21}\mathbf{u}_1+\alpha_{22}\mathbf{u}_2, \qquad \mathbf{e}_3 = \mathbf{u}_3, \qquad (7.23)$$

where the α_{kl} are (small) functions of x_1 and x_2. The non-zero strains and rotations are

$$e_{11} = \alpha_{11}, \qquad e_{22} = \alpha_{22}, \qquad e_{12} = \tfrac{1}{2}(\alpha_{12}+\alpha_{21}),$$

$$\omega_{12} = \tfrac{1}{2}(\alpha_{12}-\alpha_{21}). \qquad (7.24)$$

From equation (7.13) we obtain the only non-zero components of dislocation density to be a_{13} and a_{23}, which are given by

$$a_{13} = \partial_1\alpha_{21} - \partial_2\alpha_{11}, \qquad a_{23} = \partial_1\alpha_{22} - \partial_2\alpha_{12}. \tag{7.25}$$

Equivalently, we may use equation (7.7) to express these densities in terms of the strain and rotation as

$$a_{13} = -\partial_1\theta + \partial_1 e_{21} - \partial_2 e_{11}, \qquad a_{23} = -\partial_2\theta + \partial_1 e_{22} - \partial_2 e_{12}, \tag{7.26}$$

where we use θ to denote ω_{12}.

Such a state may arise through the operation of two systems of edge dislocations, both with lines along the $\mathbf{u}_3$-direction. One system has glide-planes normal to $\mathbf{u}_2$, the other normal to $\mathbf{u}_1$ (Fig. 7.1). The Burgers vectors $\mathbf{b}^{(1)}$ and $\mathbf{b}^{(2)}$ of typical members of two systems are shown for line directions $\boldsymbol{\xi}$ along the positive $\mathbf{u}_3$-direction (cf. Fig. 1.10). Thus $\mathbf{b}^{(1)}$ contributes positively to a_{13} and $\mathbf{b}^{(2)}$ contributes negatively to a_{23}.

A strain-free state of plane strain is possible, by choosing $\alpha_{11} = \alpha_{22} = 0$ and $\alpha_{21} = -\alpha_{12}$. Then the dislocation density is

$$a_{13} = -\partial_1\theta, \qquad a_{23} = -\partial_2\theta. \tag{7.27}$$

Conversely, a given state of dislocation density is strain-free in the absence of external tractions if and only if $\partial_2 a_{13} - \partial_1 a_{23} = 0$ (see Prob. 6). The simplest example of a strain-free state of this kind involves only one type, say $\mathbf{b}^{(2)}$, of dislocation with the density a_{23} being independent of x_1. The lattice is then rotated through an angle independent of x_1. An illustration of this is the tilt boundary in Figure 4.15, where the dislocations are concentrated on the single plane $x_2 = 0$, thus providing a sudden jump in θ rather than the smooth change in equation (7.27). The general state described by equation (7.27) contains a continuous distribution of tilt boundaries in both the x_1 and x_2 directions.

The $\mathbf{e}_1$ and $\mathbf{e}_2$ families of lattice lines for the general state of plane strain can be seen from equation (7.14) to have the respective equations

$$dx_2/dx_1 = \alpha_{12} \quad \text{and} \quad dx_1/dx_2 = \alpha_{21}$$

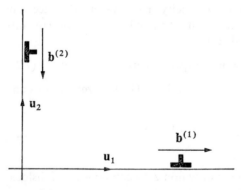

FIGURE 7.1

(with x_3 being constant on each family). In the strain-free case they become two families of orthogonal trajectories.

D *General single glide*

A more general example is obtained by allowing all four types of dislocation which can glide on a single set of planes.[11] For example, taking the glide-planes to be the ones containing u_1 and u_3, we can have dislocation density components a_{11}, a_{33}, a_{13}, and a_{31}. The first two correspond to two types of screw dislocation, and the last two represent edges. For the general case the relationship between the lattice distortion and these density components is contained in equation (7.13). Considering the strain-free case, we can set $\alpha_{kl} = \omega_{kl}$, the antisymmetric rotation tensor. Furthermore we can write $\omega_{kl} = \varepsilon_{klm}\omega_m$, where ω_m is the corresponding rotation vector, and substituting into equation (7.13) gives

$$a_{jl} = \varepsilon_{lip}\varepsilon_{pjm}\partial_i\omega_m = \delta_{jl}\partial_m\omega_m - \partial_j\omega_l. \tag{7.28}$$

This equation is so far quite generally applicable to any strain-free state. For the particular case of strain-free single glide, a number of the components of the left-hand side are zero, which enables us to derive some simpler forms for the rotations ω_k and the non-zero a_{kl}. Thus

$$a_{32} = -\partial_3\omega_2 = 0 \quad \text{and} \quad a_{12} = -\partial_1\omega_2 = 0$$

imply that $\omega_2 = f(x_2)$ only, and

$$a_{22} = \partial_1\omega_1 + \partial_3\omega_3 = 0$$

implies that there exists a function g such that

$$\omega_1 = \partial_3 g, \qquad \omega_3 = -\partial_1 g.$$

Then $a_{23} = -\partial_2\omega_3 = \partial_1\partial_2 g = 0$ and $a_{21} = -\partial_2\omega_1 = -\partial_2\partial_3 g = 0$ together imply that g has the most general form

$$g = \phi(x_2) + \psi(x_1, x_3).$$

Thus we have the general solutions for strain-free single glide:

$$a_{13} = \partial_{11}\psi, \qquad a_{11} = f'(x_2) - \partial_{13}\psi, \tag{7.29}$$

$$a_{31} = -\partial_{33}\psi, \qquad a_{33} = f'(x_2) + \partial_{13}\psi,$$

$$\omega_2 = f(x_2), \qquad \omega_1 = \partial_3\psi, \qquad \omega_3 = -\partial_1\psi, \tag{7.30}$$

where $f(x_2)$ and $\psi(x_1, x_3)$ are arbitrary functions.

Again we can write down differential equations for the three sets of lattice curves, using equation (7.14). They are:

$\mathbf{e}_1$-lines: $dx_2/dx_1 = \omega_3$, $dx_3/dx_1 = -\omega_2$,

$\mathbf{e}_2$-lines: $dx_1/dx_2 = -\omega_3$, $dx_3/dx_2 = \omega_1$,

$\mathbf{e}_3$-lines: $dx_1/dx_3 = \omega_2$, $dx_2/dx_3 = -\omega_1$.

It then follows readily that (to first order) these three lines through any one point are mutually orthogonal. The equations of the glide surfaces can also be derived. These surfaces contain the $\mathbf{e}_1$ and $\mathbf{e}_3$ directions at any point, and hence are orthogonal to $\mathbf{e}_2$. Thus, for any differential dx_k in the surface, $\mathbf{dx} \cdot \mathbf{e}_2 = 0$, i.e. $dx_2 - \omega_3 dx_1 + \omega_1 dx_3 = 0$. Substituting from equation (7.30) and integrating gives the glide surface equations as

$$x_2 + \psi(x_1, x_3) = \text{const.} \tag{7.31}$$

To zero order, they are of course simply $x_2 = \text{const.}$; the function $\psi(x_1, x_3)$ determines the first-order deviation of the glide surfaces from x_2-planes.

Knowledge of the configurations of the glide surfaces enables the densities of edge dislocations to be determined completely from equation (7.29). The screw densities still contain unknown contributions $f'(x_2)$, which account for the possible presence of cross-grids of screws, producing rotation of the glide surfaces about the x_2-direction (cf. §B).

7.3
NON-LINEAR THEORY OF DISTRIBUTIONS OF DISLOCATIONS

A *Dislocations as inhomogeneities*[9, 10, 12]

In this subsection we shall show how a dislocation density arises naturally in considering any elastic continuum, linear or non-linear, in which internal stresses are present. The theory will be more generally applicable than to dislocations alone. Examples of other internal stress problems which can fit into it are thermal stress-fields produced by non-uniform temperature distributions and the internal stresses produced by point defects (see Prob. 13).

In §2.7 we discussed various properties of the strain energy density $W(E_{\alpha\beta}, \mathbf{X})$ of a continuous elastic material, which in general is a function of the strain tensor $E_{\alpha\beta}$ at the point under consideration and also of the particle involved, as described by its position $\mathbf{X}$ in a reference configuration. At that time we said that for a homogeneous material we can find a uniform reference configuration with the property that W is independent of $\mathbf{X}$ when the strains are measured from this uniform reference. We then further specialized to take the natural state as reference configuration, so that the stresses and strains vanish together. Now it is clear that if the material contains dislocations, or indeed any source of internal stress, this procedure breaks down, and there is no natural state for the whole body in which the stresses are zero at every point.

There are two quite different reasons why W can depend on X, and we need to distinguish between them. The most direct way is for there to be a variation in material properties from one point of the body to another, due, say, to a variation in the material constitution. In the cases we wish to consider, however, the material is uniform throughout, and the particle-dependence of W arises from the presence of internal stress. Such a body is termed *materially uniform* but *inhomogeneous*.

Although in such a situation there is no global natural state, such a state does exist for any sufficiently small element of the body. If such an element were to be cut out from the body, its stresses would relax, and it would automatically assume a natural state. And furthermore, because of the assumed material uniformity, an element when cut out and allowed to relax is indistinguishable from any other element treated in the same way, in the sense that if strains are measured from this relaxed state, the energy density function would be the same for both elements. Thus we are led to consider a second reference configuration which consists in fact of a collection of configurations, one for each neighbourhood of the body, which cannot in general be joined continuously into a global reference configuration for the whole body.

Let $\mathcal{N}(X^0)$ be a sufficiently small neighbourhood of the particle X^0; then we denote by $\xi = \xi(X)$ the position of the particle $X \in \mathcal{N}(X^0)$ in the local reference configuration in which $\mathcal{N}(X^0)$ is in its natural state. The deformation gradients measured from this local reference state to the global reference state (X) and to the current configuration (x) are respectively $\partial X_\alpha/\partial \xi_k$ and $\partial x_i/\partial \xi_k$, where we have the chain-rule relation

$$\frac{\partial x_i}{\partial \xi_k} = \frac{\partial x_i}{\partial X_\alpha} \frac{\partial X_\alpha}{\partial \xi_k} \tag{7.32}$$

between the components of the deformation gradients. The three configurations are illustrated in Figure 7.2. The assumption of material uniformity can now be expressed by requiring that when W is expressed as a function of the gradients $\partial x_i/\partial \xi_k$ rather than $\partial x_i/\partial X_\alpha$, then there is no explicit dependence on X. The previous particle dependence enters through the factors $\partial X_\alpha/\partial \xi_k$ when the second of these gradients is used.

It has been assumed implicitly in this discussion that the local reference configurations are all oriented parallel to one another – that is, for example, for a material with cubic symmetry the three cube axes for each neighbourhood in its local reference state are taken to point in the same three directions, say along the coordinate axes. This requirement is, however, not sufficient to determine the local reference configurations uniquely, and in fact makes no restriction whatsoever on their orientations for an isotropic

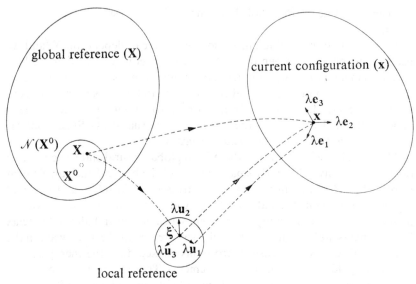

global reference (**X**)

current configuration (**x**)

$\lambda\mathbf{e}_3$

$\mathbf{x}$ $\lambda\mathbf{e}_2$

$\lambda\mathbf{e}_1$

$\mathcal{N}(\mathbf{X}^0)$ **X**

$\mathbf{X}^0$

$\lambda\mathbf{u}_2$

$\boldsymbol{\xi}$

$\lambda\mathbf{u}_3$ $\lambda\mathbf{u}_1$

local reference

FIGURE 7.2

material. We shall therefore impose an additional continuity requirement.
If two neighbourhoods $\mathcal{N}(\mathbf{X}^0)$ and $\mathcal{N}(\mathbf{X}^1)$ overlap, then we shall require
that their two local reference configurations can be fitted together *without
any rotation* in such a way that the particles of the overlap region occupy the
same position in the two reference neighbourhoods. Starting with an
arbitrary orientation for the local reference at $\mathbf{X}^0$, we can now construct the
reference configurations at points along any simple arc through $\mathbf{X}^0$, by first
selecting a sequence of overlapping neighbourhoods along the arc, and
choosing configurations in a continuous way for adjacent neighbourhoods.
There still remains the question of whether the continuations along two
paths from $\mathbf{X}^0$ to some other point $\mathbf{X}$ result in the same orientation of the
local reference configuration at $\mathbf{X}$. For the moment we shall *assume* that
this is so, returning in §7.4 to discuss the point further.

We suppose now that the two Cartesian coordinate systems (x_i) and
(ξ_k) have the same orthonormal triad of basis vectors, which we denote by
$\{\mathbf{u}_k: k = 1, 2, 3\}$. Consider a small material element occupying a position
between $\boldsymbol{\xi}$ and $\boldsymbol{\xi} + \lambda\mathbf{u}_1$ in an appropriate local reference configuration, with
λ being infinitesimal. In the current state it occupies a position between
$\mathbf{x}$ and, say, $\mathbf{x} + \lambda\mathbf{e}_1$ (see Fig. 7.1), where

$$\lambda\mathbf{e}_1 = \lambda\frac{\partial\mathbf{x}}{\partial\xi_1} = \lambda\frac{\partial x_i}{\partial\xi_1}\mathbf{u}_i.$$

Thus we can construct a triad of vectors $(\mathbf{e}_1, \mathbf{e}_2, \mathbf{e}_3)$ at each point $\mathbf{x}$ in the
current state, such that

$$\mathbf{e}_k = \frac{\partial x_i}{\partial \xi_k} \mathbf{u}_i. \tag{7.33}$$

If the material is truly crystalline, then we can think of the triad $\{\mathbf{u}_k\}$ (cf. §7.1c) as a set of lattice vectors in the unstrained state of the crystal and of $\{\mathbf{e}_k\}$ as being the corresponding deformed triad in the current state. The mapping $\mathbf{x} \to \boldsymbol{\xi}$ from current state to local reference configuration corresponds to the mapping $\mathbf{e}_k \to \mathbf{u}_k$ between lattice vectors in the sense that the material element $(\mathbf{x}, \mathbf{x} + \lambda \mathbf{e}_k)$ maps onto $(\boldsymbol{\xi}, \boldsymbol{\xi} + \lambda \mathbf{u}_k)$. It is helpful in connecting the present discussion with earlier chapters to preserve some of the language appropriate to crystals, and we shall continue to refer to the triads as lattice vectors, and to the above mapping as one between a deformed crystal and a perfect reference crystal. It should be emphasized, however, that the construction of a pseudo-crystalline structure is applicable to any elastic continuum whether or not there is any real underlying crystallinity.

An additional advantage in this choice of language is that it coincides with that in which most of the work on this subject has been written. In the fullest exposition, by Bilby,[5] the starting point of the theory is the supposition of a field of lattice triads $\{\mathbf{e}_k(x)\}$ defined over the body. We shall adopt Bilby's notation and set

$$\partial x_i / \partial \xi_k = (e/u)_{ki} \tag{7.34a}$$

so that

$$\mathbf{e}_k = (e/u)_{ki} \mathbf{u}_i. \tag{7.35a}$$

The field of matrices $(e/u)_{ki}$, which appears as fundamental in Bilby's work, is from the present viewpoint defined in terms of a certain deformation gradient. It is also convenient to define the inverse matrix of (e/u):

$$(u/e)_{ik} = \partial \xi_k / \partial x_i, \tag{7.34b}$$

where

$$\mathbf{u}_i = (u/e)_{ik} \mathbf{e}_k \tag{7.35b}$$

and

$$(u/e)_{ik}(e/u)_{kl} = (e/u)_{ik}(u/e)_{kl} = \delta_{il}. \tag{7.36}$$

The spatial and material strain tensors as measured from the local reference configurations can be defined as in equations (2.49) and (2.50):

$$e_{ik} = \frac{1}{2}\left(\delta_{ik} - \frac{\partial \xi_l}{\partial x_i}\frac{\partial \xi_l}{\partial x_k}\right) = \tfrac{1}{2}[\delta_{ik} - (u/e)_{il}(u/e)_{kl}],$$

$$E_{ik} = \frac{1}{2}\left(\frac{\partial x_l}{\partial \xi_i}\frac{\partial x_l}{\partial \xi_k} - \delta_{ik}\right) = \tfrac{1}{2}[(e/u)_{il}(e/u)_{kl} - \delta_{ik}]. \tag{7.37}$$

For the same reasons as before (cf. §2.7D), W must be a function of the latter strain components alone, rather than of the full deformation gradients $\{\partial x_i/\partial \xi_k\}$, in order to preserve the invariance under rotation of the spatial coordinate system. In particular, for small strains in an isotropic material, we have equation (2.8) for the stresses, with the difference that now e_{ik} is given by equation (7.37) rather than by equation (2.49). The distinction lies in the fact that (ξ_k) is not a global configuration, whereas (X_α) is.

We can also obtain the dislocation density in terms of (e/u). Let $\mathscr{C}$ be a closed circuit in the current configuration of the body with a general element of arc $\mathbf{dx} = dx_k\mathbf{u}_k$. This element corresponds in its local reference state to an element

$$d\boldsymbol{\xi} = d\xi_k\mathbf{u}_k = \frac{\partial \xi_k}{\partial x_l}dx_l\mathbf{u}_k = (u/e)_{lk}dx_l\mathbf{u}_k.$$

If we similarly map each element of $\mathscr{C}$ onto its reference state and join the mapped elements together, we get an arc which in general does not close. In fact, the closure failure is just what we defined in Chapter 1 to be the (true) Burgers vector of the dislocations threading $\mathscr{C}$. This is given by

$$-\mathbf{B}(\mathscr{C}) = \oint_{\mathscr{C}} (u/e)_{lk}dx_l\mathbf{u}_k = \int_S \varepsilon_{jpk}\partial_p(u/e)_{kl}n_jdS\,\mathbf{u}_l,$$

using Stokes' theorem to transform to an integral over any surface bounded by $\mathscr{C}$. If $\mathscr{C}$ is infinitesimal and planar we get approximately

$$-\mathbf{B}(\mathscr{C}) \approx \varepsilon_{jpk}\partial_p(u/e)_{kl}n_jdS\,\mathbf{u}_l.$$

In order to relate this to the dislocation density as defined by equation (7.3) we need the local Burgers vector, $\mathbf{b}(\mathscr{C})$, which is obtained by mapping $\mathbf{B}(\mathscr{C})$ back onto the current configuration, treating $\mathbf{B}$ as a material element. The mapping is given by $\mathbf{u}_l \rightarrow \mathbf{e}_l$, so that

$$-\mathbf{b}(\mathscr{C}) = \varepsilon_{jpk}\partial_p(u/e)_{kl}n_jdS\,\mathbf{e}_l = -a_{kj}n_jdS\mathbf{u}_k$$

by comparison with equation (7.3). Hence

$$a_{mj} = -\varepsilon_{jpk}\partial_p(u/e)_{kl}(e/u)_{lm} = \varepsilon_{jpk}(u/e)_{kl}\partial_p(e/u)_{lm} \qquad (7.38)$$

after using equation (7.36).

The connection between this non-linear result and our earlier equation (7.13) is indicated in Problem 8.

B *Geometrical structures on a dislocated crystal*

One of the earliest observations in the theory of continuous distributions of dislocations[1,2] was that it is possible to define in a very natural way a parallelism for a dislocated crystal. The connection is not symmetric, so that the resulting geometry is non-Riemannian, and the torsion is directly related to the dislocation density. The connection is, however, flat, at least for the case so far considered, although we shall introduce further generalizations in §7.4 which involve a non-zero curvature. These results, which we shall now outline, provide a unifying background to the theory, and lead readily to a formulation of the internal stress problem.

Let $v_l u_l$ and $v_l'u_l$ be two vectors at $\mathbf{x}$ and $\mathbf{x}'$ respectively. We say that they are parallel if they correspond to the same number of local lattice steps – or equivalently if when mapped onto the two local reference crystals they give vectors which are parallel in the Euclidean sense. This means that

$$v_l(u|\mathbf{x}|e)_{lk} = v_l'(u|\mathbf{x}'|e)_{lk}, \tag{7.39}$$

i.e.

$$v_l' = v_m(u|\mathbf{x}|e)_{mk}(e|\mathbf{x}'|u)_{kl},$$

where $(u|\mathbf{x}|e)$ means (u/e) evaluated at $\mathbf{x}$. Taking $\mathbf{x}' = \mathbf{x} + dx_k\mathbf{u}_k$, and assuming the matrices to be differentiable, we get

$$dv_l = v_l' - v_l = v_m(u|\mathbf{x}|e)_{mk}[(e|\mathbf{x}|u)_{kl} + dx_p\partial_p(e|\mathbf{x}|u)_{kl}] - v_l$$

$$= v_m(u|\mathbf{x}|e)_{mk}\partial_p(e|\mathbf{x}|u)_{kl}dx_p.$$

Thus the parallelism corresponds to a linear connection,

$$dv_l = -L_{ljm}v_jdx_m, \tag{7.40}$$

where

$$L_{ljm} = -(u/e)_{jk}\partial_m(e/u)_{kl}. \tag{7.41}$$

The torsion tensor associated with this connection is, by definition,

$$T_{ljm} = \tfrac{1}{2}(L_{ljm} - L_{lmj}), \tag{7.42}$$

that is, the antisymmetric part of the connection. Then using equation (7.38) we see that

$$T_{jmn}\varepsilon_{mnk} = L_{jmn}\varepsilon_{mnk} = a_{jk}. \tag{7.43}$$

Hence we have also the inverse relation,

$$T_{jmn} = \tfrac{1}{2}a_{jk}\varepsilon_{kmn}. \tag{7.44}$$

Consequently the torsion and dislocation density are equivalent quantities in that each determines the other.

Associated with any connection such as $\mathbf{L}$ there is a curvature tensor defined as follows. Let a vector $v_l u_l$ at $\mathbf{x}$ be translated parallel to itself first to $\mathbf{x} + d\mathbf{x}^{(1)}$ and then to $\mathbf{x} + d\mathbf{x}^{(1)} + d\mathbf{x}^{(2)}$. The successive changes in the components of the vector are to

$v_l - L_{ljm}(\mathbf{x})v_j dx_m{}^{(1)}$

and

$v_l - L_{ljm}(\mathbf{x})v_j dx_m{}^{(1)} - L_{ljm}(\mathbf{x}+\mathbf{dx}^{(1)})\,[v_j - L_{jki}(\mathbf{x})v_k dx_i{}^{(1)}]dx_m{}^{(2)}$

$\qquad = v_l - L_{ljm}(\mathbf{x})v_j(dx_m{}^{(1)} + dx_m{}^{(2)}) - [\partial_i L_{lkm} - L_{ljm}(\mathbf{x})L_{jki}(\mathbf{x})]v_k dx_i{}^{(1)}dx_m{}^{(2)}.$

If the vector had been translated first through $\mathbf{dx}^{(2)}$ and then through $\mathbf{dx}^{(1)}$ we would obtain a similar expression with (1) and (2) interchanged. The difference between the two results can be written as $-L_{lnim}v_n dx_i{}^{(1)}dx_m{}^{(2)}$, where

$$L_{lnim} = \partial_i L_{lnm} - \partial_m L_{lni} + L_{lji}L_{jnm} - L_{ljm}L_{jni}. \qquad (7.45)$$

L_{lnim} is called the Riemann-Christoffel curvature tensor.

In the present case the parallelism was defined in terms of the crystal lattice, which was earlier assumed to be uniquely defined throughout the body (i.e. we assumed that the mapping $\mathbf{x} \to \boldsymbol{\xi}$ onto local reference con-figurations can be specified uniquely and continuously throughout the body). Thus the vector at $\mathbf{x}+\mathbf{dx}^{(1)}+\mathbf{dx}^{(2)}$ has to be independent of the path by which this point is reached from $\mathbf{x}$ – it is the unique vector

$v_m(u|\mathbf{x}|e)_{mk}(e|\mathbf{x}+\mathbf{dx}^{(1)}+\mathbf{dx}^{(2)}|u)_{kl}\mathbf{u}_l$

(cf. eq. (7.39)). Hence the particular connection for a dislocated lattice has to have zero curvature. In fact it is easy to verify directly from equations (7.41) and (7.45) that $L_{lnim} \equiv 0$. We shall later describe an extension of the theory which removes this curvature restraint.

We want to phrase the theory entirely in terms of strain and dislocation density – in particular to write the curvature condition in these terms. The first step is to obtain the connection in terms of e_{kl} and a_{kl}.[13] From equations (7.37) and (7.41),

$$\partial_m e_{kl} = -\tfrac{1}{2}[\partial_m(u/e)_{kn}(u/e)_{ln} + (u/e)_{kn}\partial_m(u/e)_{ln}] = -\tfrac{1}{2}(L'_{lkm}+L'_{klm}), \qquad (7.46)$$

where

$$L'_{lkm} = L_{ikm}(u/e)_{in}(u/e)_{ln} = L_{ikm}(\delta_{il}-2e_{il}). \qquad (7.47)$$

If we now define the Christoffel symbol from e_{kl} in the usual way:

$$e_{ijk} = \tfrac{1}{2}(\partial_i e_{jk} + \partial_j e_{ik} - \partial_k e_{ij}) \qquad (7.48)$$

and a tensor $\mathbf{T}'$ in a similar way to $\mathbf{L}'$:

$$T'_{lkm} = T_{ikm}(\delta_{il}-2e_{il}),$$

it follows that

$$L'_{kij} = 2e_{ijk} + (T'_{kij}+T'_{ijk}-T'_{jki}). \qquad (7.49)$$

This is the required condition, giving $\mathbf{L}$ in terms of the strain and dislocation density. Substituting it into equation (7.45) gives the curvature condition as a constraint on these last two tensors.

It is clear that this general non-linear condition is going to be fairly complicated, and applications that have so far been made have been restricted to the small-strain approximation. We shall from now on assume that e_{kl} and L_{klm} are of first order of smallness, and that all second-order terms can be neglected (note that this does not require the matrix elements $(u/e)_{kl} - \delta_{kl}$ to be small, since there may still be large rotations present). Then, replacing **T** by **a**, we get after some manipulation, from equation (7.49),

$$L_{kij} = \varepsilon_{kiq}(-a_{jq} + \tfrac{1}{2}a\delta_{jq} + \varepsilon_{nrq}\partial_n e_{jr}) - \partial_j e_{ik}, \tag{7.50}$$

where $a = a_{pp}$.

In this approximation,

$$L_{lnim} \approx \partial_i L_{lnm} - \partial_m L_{lni} = 0.$$

The last term in equation (7.50) does not contribute, and L_{lnim} is antisymmetric in both pairs (ln and im) of indices. It is then sufficient to consider the contraction

$$L_{pq} = -\tfrac{1}{4}\varepsilon_{pln}\varepsilon_{qim}L_{lnim} = \varepsilon_{qim}\partial_i(a_{mp} - \tfrac{1}{2}a\delta_{mp} - \varepsilon_{nrp}\partial_n e_{mr}). \tag{7.51}$$

The part of L_{pq} which is antisymmetric in p and q, when set equal to zeros yields the condition (7.8). The symmetric part becomes equation (7.9). Thus the compatibility conditions and conservation equation for Burgers vector, are equivalent to the vanishing of the curvature of the lattice connection.

We proved in §2.5 that any non-singular matrix has a polar decomposition – that is, it can be written as the product of an orthogonal matrix and a symmetric matrix. In particular there exists a symmetric matrix **f** and an orthogonal matrix $\mathbf{\Omega}$ such that

$$(\mathbf{u}/\mathbf{e}) = (\mathbf{1} - \mathbf{f})\mathbf{\Omega}^T. \tag{7.52}$$

Denoting by **e** the spatial strain tensor and using equation (7.37) gives

$$(\mathbf{u}/\mathbf{e})\,(\mathbf{u}/\mathbf{e})^T = \mathbf{1} - 2\mathbf{e} = (\mathbf{1} - \mathbf{f})^2.$$

If the strain **e** is small, then $\mathbf{f} \approx \mathbf{e}$. Hence

$$(\mathbf{e}/\mathbf{u}) \approx \mathbf{\Omega}(\mathbf{1} + \mathbf{e}). \tag{7.53}$$

The decompositions (7.52) and (7.53) give useful expressions for the dislocation density.[14] Assuming that the strain and gradients of both strain and rotation ($\mathbf{\Omega}$) are small, we obtain from equation (7.38), to first order,

$$a_{mj} = \varepsilon_{jpk}(\Omega_{lk}\partial_p\Omega_{lm} + \partial_p e_{km}). \tag{7.54}$$

If we eliminate $\mathbf{\Omega}$ from these relations we retrieve equations (7.8) and (7.9) after neglecting second-order terms.

Equation (7.54) closely resembles the result (7.7). However, in deriving the earlier equation we made the additional assumption that the rotation itself is small. Equation (7.54) holds for large rotations. It reduces to equation (7.7) when $\Omega \approx 1$.

c Displacements and coordinates

We have seen before that a material element $dx_k \mathbf{u}_k$ in the real crystal corresponds to an element $d\xi_k \mathbf{u}_k$ in the reference crystal, where

$$d\xi_k = dx_l(u/e)_{lk}. \tag{7.55}$$

We now wish to examine the possible existence of a global deformation from the reference to the real crystal – that is a differentiable mapping $\xi_k = \xi_k(x_l)$ which sets the particles of the reference crystal in 1–1 correspondence with those of the real one. If such a mapping exists, then

$$(u/e)_{lk} = \partial \xi_k / \partial x_l, \tag{7.56}$$

and the inverse matrix is

$$(e/u)_{km} = \partial x_m / \partial \xi_k.$$

Then from equation (7.41)

$$L_{ljm} = -\frac{\partial^2 x_l}{\partial \xi_k \partial \xi_i} \frac{\partial \xi_i}{\partial x_m} \frac{\partial \xi_k}{\partial x_j}. \tag{7.57}$$

This connection is symmetric in (j, m), hence has zero torsion, and corresponds to zero dislocation density.

Conversely, if the dislocation density vanishes, then there exists a global deformation. For, from equations (7.38) and (7.44),

$$T_{ljm} = \tfrac{1}{2}[\partial_m(u/e)_{jk} - \partial_j(u/e)_{mk}] (e/u)_{kl}$$

and, if this vanishes, then

$$\partial_m(u/e)_{jk} = \partial_j(u/e)_{mk}.$$

This implies the existence of a function $\xi_k(x_l)$ such that equation (7.56) holds, which in turn implies that equation (7.55) is integrable to the global mapping

$$\xi_k = \xi_k(x_l).$$

In spite of these results, well-defined *lattice curves* do exist, even in a continuously dislocated crystal. Suppose, for example, that we take the lattice vector $\mathbf{e}_1$ and then define the corresponding lattice curve to be such that it is tangent to $\mathbf{e}_1$ everywhere along its length. Its equation is

$$d\mathbf{x}/dt = \mathbf{e}_1 \quad \text{or} \quad dx_k/dt = (e/u)_{1k}, \tag{7.58}$$

where t is an appropriate parametrization. By the existence and uniqueness theorem for systems of ordinary differential equations, there is one and only one solution curve through each point $((e/u)_{kl}$ have already been assumed to be continuously differentiable functions of position).

Differentiating equation (7.58), and using equation (7.41) gives

$$\frac{d^2 x_k}{dt^2} = \partial_l (e/u)_{1k} \frac{dx_l}{dt} = -(e/u)_{1j} L_{kjl} \frac{dx_l}{dt} = -L_{kjl} \frac{dx_j}{dt} \frac{dx_l}{dt}. \tag{7.59}$$

This second-order equation is satisfied by all the lattice curves, not only those which follow e_1. It shows that the lattice curves are geodesic with respect to the connection L – that is, the tangent vectors at all points of a lattice curve are parallel to one another.

D *Plane strain: large rotations*

We wish to reconsider the deformations studied in §7.2c but now allowing finite lattice rotations to occur. We shall still assume that the strain and the gradients of both strain and rotation are small. In such a case the (e/u) matrices will not be approximately equal to 1, and it is not useful to express the deformation as in equation (7.23). Instead, the polar decomposition (7.53) must be used. The non-zero strains will then be e_{11}, e_{12}, and e_{22}, and the rotation will have the form

$$\Omega = \begin{pmatrix} \cos\theta & \sin\theta & 0 \\ -\sin\theta & \cos\theta & 0 \\ 0 & 0 & 1 \end{pmatrix}. \tag{7.60}$$

The angle θ depends only on x_1 and x_2 and although θ itself may be large, we shall assume that the derivatives $\partial_k \theta$ are small.

The components of dislocation density can be obtained from equation (7.54). The only non-zero components are a_{13} and a_{23}, and these turn out to be given by equations (7.26), just as in the small-rotation case.

Again the simplest way in which such deformations can occur is through the operation of two systems of edge dislocations. Now, however, we cannot take the two Burgers vectors to lie along the coordinate directions because of the large lattice rotations. We must take them along the lattice directions e_1 and e_2, which, to zero order, are obtained by rotating u_1 and u_2 through θ (as in Fig. 7.3). If the two types of dislocations shown have densities $A^{(1)}$ and $A^{(2)}$ (these are the numerical densities multiplied by the magnitudes of the Burgers vectors), then

$$a_{13} = A^{(1)} \cos\theta + A^{(2)} \sin\theta, \qquad a_{23} = A^{(1)} \sin\theta - A^{(2)} \cos\theta. \tag{7.61}$$

As earlier, we shall examine the strain-free case more closely. Combining equations (7.26) and (7.61) we then obtain

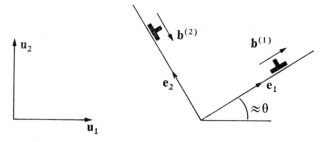

FIGURE 7.3

$$\partial_1\theta = -A^{(1)}\cos\theta - A^{(2)}\sin\theta, \qquad \partial_2\theta = A^{(1)}\sin\theta + A^{(2)}\cos\theta. \quad (7.62)$$

For given densities $A^{(1)}$ and $A^{(2)}$, suitably conditioned to produce no strains, these equations may be integrated to give the lattice rotation. Having solved for $\theta(x_1, x_2)$, it is then possible to find the lattice curves. The e_1 curves, for example, are given by equation (7.58), which can be written as

$$dx_2/dx_1 = \tan\theta(x_1, x_2). \qquad (7.63)$$

For plane strain, this curve gives also the trace of the glide-planes of the $A^{(1)}$ dislocations on the x_1, x_2-coordinate plane. The e_2 curves may be found similarly.

We shall give several examples which we shall approach via an inverse procedure. We shall suppose certain forms for the two sets of lattice curves, and then calculate the dislocation density needed to produce such a state of strain-free lattice rotation. First of all, suppose that the lattice curves make fixed angles, say β and $\tfrac{1}{2}\pi - \beta$, with the radial line from the origin (Fig. 7.4). Then

$$\theta = \tan^{-1}(x_2/x_1) - \beta \qquad (7.64)$$

for points (x_1, x_2) on the e_1 curve.

From equation (7.62),

$$A^{(1)} = -\cos\theta\cdot\partial_1\theta - \sin\theta\cdot\partial_2\theta = \sin\beta/r,$$
$$A^{(2)} = -\sin\theta\cdot\partial_1\theta + \cos\theta\cdot\partial_2\theta = \cos\beta/r. \qquad (7.65)$$

The e_1, e_3 glide-planes, from equation (7.63), are in this case spirals of the type

$$r = r_0 \exp(-\alpha \cot\beta)$$

in terms of polar coordinates (r, α).

A special case of this example arises when $\beta = \pi/2$. Then $A^{(2)} = 0$ and $A^{(1)} = 1/r$. The e_2 lattice lines are radial and the e_1 lines become circles,

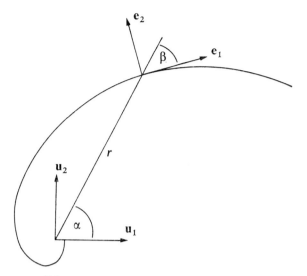

FIGURE 7.4

centred at the origin. This simplest example of lattice bending is produced by the operation of a single system of edge dislocations.

A third example, which was used by Nye[3] to explain the plastic bending of hexagonal crystals, occurs when the e_1 lattice curves are the system of involutes of a circle. Taking the circle to have radius a and to be centred at the origin, we have the geometry of Figure 7.5. The involutes are the orthogonal trajectories to the tangents to the circle. They can be regarded as drawn by the end of a string which is unwound from around the circle. If (r, α) are polar coordinates of a point P, we have, since PT is normal to the involute at P,

$$\theta = \alpha - \chi - \tfrac{1}{2}\pi, \tag{7.66}$$

and since PT is tangent to the circle at T,

$$a = r \sin \chi. \tag{7.67}$$

Hence

$$\theta - \tan^{-1}\frac{x_2}{x_1} - \sin^{-1}\frac{a}{(x_1{}^2 + x_2{}^2)^{1/2}} - \frac{\pi}{2}. \tag{7.68}$$

Having found the rotation as a function of x_1 and x_2, the dislocation densities on the two slip systems can be found from equation (7.62). Again it turns out that $A^{(2)}$ is zero, so that only the one slip system is needed. The density on that system is

$$A^{(1)} = (r^2 - a^2)^{-1/2}. \tag{7.69}$$

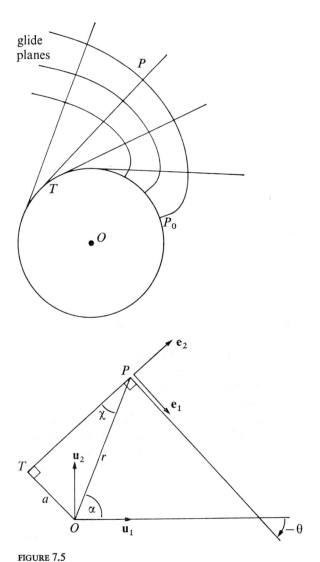

FIGURE 7.5

7.4
DISCLINATIONS

A *Physical considerations*

In §7.3 we made the assumption that a unique local reference configuration could be found for each neighbourhood of the body by constructing the configurations in a continuous way along arbitrary curves lying in the body.

We shall now remove this restriction, in the process extending the theory to include new physical phenomena.

We continue to suppose that local reference configurations exist for each neighbourhood and that they can be chosen in a continuous way along any piecewise smooth curve in the body which does not intersect itself. However, if we continue the configurations around any closed curve, then we no longer suppose that the reference configuration that we end up with is the same as the one we started out with. The local reference configurations are by definition in the natural state, so that the change in going around a circuit can only consist of a rigid-body rotation. Furthermore, because of the continuity of reference configurations around the circuit, the rotation must belong to the symmetry group of the material concerned. (For example, with a cubic material, we might select local reference configurations in which the coordinate basis $(\mathbf{u}_1, \mathbf{u}_2, \mathbf{u}_3)$ corresponds in every case to the cube axes. Then any two reference configurations for a particular neighbourhood can differ only by an element of the cubic symmetry group.) In particular, for an isotropic material, any rotation might occur.

Suppose that for a circuit $\mathscr{C}$, starting and finishing at a point $\mathbf{x}$ in the current configuration of the body, the reference configurations are rotated by an orthogonal transformation $\mathbf{R}(\mathscr{C})$. In other words, if $\mathbf{d\xi} = d\xi_k \mathbf{u}_k$ is the vector corresponding to a certain material element at $\mathbf{x}$ in the initial reference configuration at $\mathbf{x}$ and $\mathbf{d\xi}' = d\xi_k' \mathbf{u}_k$ is the vector corresponding to the same material element in the final reference configuration at $\mathbf{x}$, then

$$\mathbf{d\xi}' = \mathbf{R}(\mathscr{C})\mathbf{d\xi},$$

or, in component notation,

$$d\xi_k' = R_{kl}(\mathscr{C})d\xi_l. \tag{7.70}$$

From equations (7.34) and (7.35), the changes in the lattice triad $\{\mathbf{e}_k\}$ and the two transformation matrices that result from following the circuit can be calculated, the result being

$$\mathbf{e}_k' = R_{kl}(\mathscr{C})\mathbf{e}_l, \tag{7.71}$$

$$(e/u)_{ki}' = R_{kl}(\mathscr{C})\,(e/u)_{li}, \tag{7.72}$$

$$(u/e)_{il}' = (u/e)_{ik}R_{lk}(\mathscr{C}). \tag{7.73}$$

It follows then from equation (7.37) that the *spatial* strain tensor, e_{ik}, is unchanged. Therefore there still exists a unique strain tensor at each point, independent of the path along which we have continued the local configurations in order to reach the point in question.

Before proceeding any further, we should ascertain whether such a situation as this corresponds to any physically realistic occurrence. The

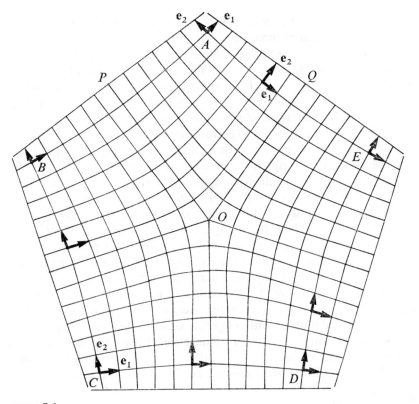

FIGURE 7.6

answer is in the affirmative: by this generalization we allow the theory to include dislocations of rotation (or disclinations) as well as dislocations of translation. Consider for example the configuration of a cubic crystal illustrated in Figure 7.6, which is one of the three basic rotation dislocations. If the quadrant POQ of material were to be removed, and the faces OP and OQ joined together, we would have a perfect crystal. In the dislocated state, these two faces are rotated with respect to one another, rather than translated as for the dislocations dealt with previously. Now, let the circuit $\mathscr{C}$ be $ABCDEA$ and let the lattice vectors $\mathbf{e}_1$ and $\mathbf{e}_2$ be initially as shown at A. Moving these in a continuous way around $\mathscr{C}$, we find that they do not return to their original values after completing the circuit – the difference being given by equation (7.71) in fact.

For a given direction of the dislocation line, there are three basic rotation dislocations in a simple cubic material, corresponding to 90° rotations about the three cube axes. A similar result holds in isotropic materials except that the angle of rotation is unrestricted. In the terms of Figure 7.6 this is matched by an arbitrariness in the angle POQ of the wedge of inserted material.

B *Connection, dislocation density, and curvature*[7,15,10]

Starting with a triad $\{e_k(x)\}$ of lattice vectors at the point x, defined by equation (7.33) using one of the possible local reference configurations at x, we can construct corresponding triads at points on simple curves through x. In particular we can find the triad $\{e_k(x+dx)\}$ at a neighbouring point $x+dx$ by using the straight-line segment between the two points. Defining $de_k = e_k(x+dx) - e_k(x)$ and assuming that the change de_k is of the same order as dx, then there must exist a set of coefficients J_{klm} such that

$$de_k = J_{klm}e_l(x)dx_m, \qquad (7.74)$$

since $\{e_l(x)\}$ form a basis. The J_{klm} will depend in general on the point x, on the particular triad we choose to start off with, and also on the direction cosines, $dx_m/|dx|$, of the small line segment. However, we shall make the assumption that J_{klm} do not depend on these direction cosines. This will mean that if we move from x to $x+dx$ along any path whose length is of the same order of smallness as $|dx|$, then equation (7.74) still gives the change in lattice vectors to first order in $|dx|$.

Introducing the transformation matrix, $(e/u)_{kp} = e_k \cdot u_p$, as before, equation (7.74) can be written in the equivalent form

$$d(e/u)_{kp} = J_{klm}(e/u)_{lp}dx_m. \qquad (7.75)$$

This equation gives the change in $(e/u)_{kp}$ along any smooth curve through x; regarded as a system of ordinary differential equations, it is integrable along any such curve. It is integrable throughout space if and only if the right-hand side equals $\partial_m(e/u)_{kp}dx_m$, that is

$$J_{klm} = \partial_m(e/u)_{kp}(e/u)_{pl}. \qquad (7.76)$$

If this equation holds, then we have again the situation studied in the previous section.

Along any arc through x we can construct a unique set of lattice vectors, given the starting values $\{e_k(x)\}$, and we can use these vectors to define a parallelism along the arc. We say that two vectors $v = v_l u_l$ at x and $v' = v_l' u_l$ at x' on the arc are parallel if they have the same components with respect to the two lattice triads at x and x' respectively. This implies equation (7.39) exactly as before. But now if we take $x' = x+dx$ and let $v_l' - v_l = dv_l$, then in view of equation (7.75) we have

$$dv_l = v_j(u|x|e)_{jk}J_{kpm}(e|x|u)_{pl}dx_m.$$

Thus we again have a linear connection, as in equation (7.40) but with

$$L_{ljm} = -(u/e)_{jk}J_{kpm}(e/u)_{pl}. \qquad (7.77)$$

This reduces to equation (7.41) when equation (7.76) holds.

As remarked earlier, the starting triad $\{e_k(\mathbf{x})\}$ is to some extent arbitrary, and it is important to verify that this indefiniteness produces no resulting arbitrariness in the connection. Using equations (7.71)–(7.73), we see that any two sets of starting triads must be related by an equation $e_k{}' = R_{kl}e_l$ with (R_{kl}) some orthogonal matrix. The corresponding transformation matrices are related by $(e/u)'_{ki} = R_{kl}(e/u)_{li}$ and so from equation (7.75) $J'_{klm} = R_{kp}R_{ln}J_{pnm}$. Combining these results with equation (7.77) now shows that the elements L_{ljm} are invariant with respect to a change in starting triad.

Equations (7.46)–(7.49) still hold within the present more general framework. Since the spatial strain tensor e_{ik} is still well-defined, it follows from equation (7.37) that

$$d(e_{ik}) = \partial_m e_{ik}dx_m = -\tfrac{1}{2}[d(u/e)_{il}(u/e)_{kl}+(u/e)_{il}d(u/e)_{kl}]$$

$$= \tfrac{1}{2}[(u/e)_{kl}(u/e)_{in}J_{nlm}+(u/e)_{kn}(u/e)_{il}J_{nlm}]\,dx_m$$

$$= -\tfrac{1}{2}(L'_{ikm}+L'_{kim})\,dx_m \tag{7.78}$$

after using definition (7.47) and equation (7.77). Comparing coefficients of each dx_m now establishes equation (7.46). The torsion T_{ljm} is defined as the antisymmetric part of the connection, as in equation (7.42), and T'_{ljm} is derived from it as before. Then equation (7.49) is obtained using the same manipulations as in §7.3.

We are now in a position to define the dislocation density tensor, essentially as before in terms of the closure failure of Burgers circuits. For simplicity we consider a parallelogram-shaped circuit $ABCDA$ which consists of steps $\mathbf{x} \to \mathbf{x}+\mathbf{dx}^{(1)} \to \mathbf{x}+\mathbf{dx}^{(1)}+\mathbf{dx}^{(2)} \to \mathbf{x}+\mathbf{dx}^{(2)} \to \mathbf{x}$. The steps are mapped onto local reference crystals as follows:

$$\overrightarrow{AB} = dx_k{}^{(1)}\mathbf{u}_k \to dx_k{}^{(1)}(u|\mathbf{x}|e)_{kl}\mathbf{u}_l,$$

$$\overrightarrow{BC} = dx_k{}^{(2)}\mathbf{u}_k \to dx_k{}^{(2)}(u|\mathbf{x}+\mathbf{dx}^{(1)}|e)_{kl}\mathbf{u}_l$$

$$= dx_k{}^{(2)}[(u|\mathbf{x}|e)_{kl}-(u|\mathbf{x}|e)_{kn}J_{nlm}dx_m{}^{(1)}]\mathbf{u}_l$$

(using equations (7.75) and (7.36)); similar expressions hold for $\overrightarrow{AD}$ and $\overrightarrow{DC}$, with $\mathbf{dx}^{(1)}$ and $\mathbf{dx}^{(2)}$ interchanged. The difference between the maps of $\overrightarrow{ABC}$ and $\overrightarrow{ADC}$ gives the true Burgers vector

$$-\mathbf{B} = -(u/e)_{kn}J_{nlm}(dx_m{}^{(1)}dx_k{}^{(2)}-dx_k{}^{(1)}dx_m{}^{(2)})\mathbf{u}_l.$$

Mapping back gives the local Burgers vector in equation (7.3)

$$-\mathbf{b} \equiv -a_{pj}n_jdS\,\mathbf{u}_p$$

$$= -(u/e)_{kn}J_{nlm}(e/u)_{lp}(dx_m{}^{(1)}dx_k{}^{(2)}-dx_k{}^{(1)}dx_m{}^{(2)})\mathbf{u}_p.$$

Hence

$$a_{pj} = L_{pkm}\varepsilon_{kmj} = T_{pkm}\varepsilon_{kmj}, \tag{7.79}$$

where the torsion $\mathbf{T}$ is defined as in equation (7.42). Thus we have the same relationship between the dislocation density and the connection as in the earlier theory.

The most significant change in the equations describing the present situation from those of §7.3 is that the curvature tensor, defined by equation (7.45), is no longer zero. The curvature tensor measures the change induced in a vector by parallel transport around a small closed circuit, and since the lattice vectors are changed by such a process, so will be any other vector. A vector $v_l\mathbf{u}_l$, if moved by parallel displacement around the parallelogram circuit $\mathbf{x} \rightarrow \mathbf{x}+\mathbf{dx}^{(1)} \rightarrow \mathbf{x}+\mathbf{dx}^{(1)}+\mathbf{dx}^{(2)} \rightarrow \mathbf{x}+\mathbf{dx}^{(2)} \rightarrow \mathbf{x}$, changes to $v_l'\mathbf{u}_l$, where

$$v_l' = v_l - L_{lnim}v_n dx_i^{(1)} dx_m^{(2)}.$$

Taking the vector $v_l\mathbf{u}_l$ to be the lattice vector $\mathbf{e}_p$, we have

$$(e/u)'_{pl} = (e/u)_{pl} - L_{lnim}(e/u)_{pn} dx_i^{(1)} dx_m^{(2)}. \tag{7.80}$$

Comparing this equation with equation (7.72) we obtain the relationship between the rotation of local reference configurations and the curvature tensor:

$$(u/e)_{np}R_{pm}(\mathscr{C})\,(e/u)_{ml} = \delta_{ln} - L_{lnim}dx_i^{(1)} dx_m^{(2)}. \tag{7.81}$$

It follows from its definition (7.45) that L_{lnim} is antisymmetric in the pair of indices i and m. Therefore we can replace $dx_i^{(1)}dx_m^{(2)}$ in equation (7.81) in favour of the vector element of area of the parallelogram, $n_q dS = \varepsilon_{qim}dx_i^{(1)}dx_m^{(2)}$. Secondly we can express the small rotation $R_{pm}(\mathscr{C})$ in terms of its corresponding infinitesimal rotation vector $\omega_j(\mathscr{C})$ as follows:

$$R_{pm}(\mathscr{C}) = \delta_{pm} + \varepsilon_{pmj}\omega_j(\mathscr{C}).$$

With these changes, equation (7.81) may be written in the form

$$(u/e)_{np}\varepsilon_{pmj}(e/u)_{ml}\omega_j(\mathscr{C}) = -\tfrac{1}{2}L_{lnim}\varepsilon_{imq}n_q dS. \tag{7.82}$$

It is apparent from this equation that the curvature tensor L_{lnim} is connected directly with the density of disclinations, since it is the disclinations which determine the rotation vector $\omega_j(\mathscr{C})$ associated with each infinitesimal Burgers circuit. In fact, we may, if we wish, define a third-order tensor d_{lnq}, called the *density of disclination tensor*, by means of the equation

$$(u/e)_{np}\varepsilon_{pmj}(e/u)_{ml}\omega_j(\mathscr{C}) = d_{lnq}n_q dS,$$

in which case equation (7.82) becomes

$$d_{lnq} = -\tfrac{1}{2}L_{lnim}\varepsilon_{imq}, \qquad L_{lnim} = -d_{lnq}\varepsilon_{qim}. \tag{7.83}$$

In writing the second form, we have again used the fact that $L_{lnim} = -L_{lnmi}$. Equation (7.83) is the new curvature condition, replacing the earlier condition of vanishing curvature which was found to hold when disclinations are absent.

Again we are interested in phrasing the curvature condition (7.83) in terms of the dislocation density and strain tensors. This is achieved by substituting from equation (7.49) into equation (7.45) and then using the resulting expression for the curvature in equation (7.83). The resulting equation is algebraically complicated and, as before, we shall restrict attention to small strains and dislocation densities. In this case equations (7.50) and (7.51) are again obtained, the only difference being that now the left-hand side of the second equation is non-zero. Thus in place of equations (7.8) and (7.9) we now have that

$$\partial_p a_{jp} = -\varepsilon_{jpq} L_{pq}, \tag{7.84}$$

$$\varepsilon_{qim}\varepsilon_{pnr}\partial_i\partial_n e_{mr} = (\varepsilon_{qim}\partial_i a_{mp} - L_{pq})^{(S)}. \tag{7.85}$$

From equation (7.83) it follows that the contracted tensor L_{pq} is given by

$$d_{lnq} = \varepsilon_{lnp} L_{pq}, \qquad L_{pq} = \tfrac{1}{2}\varepsilon_{pln} d_{lnq}. \tag{7.86}$$

When disclinations are present in the medium, we can formulate the internal stress problem by supposing that a_{jp} and d_{lnq} (or equivalently L_{pq}) are prescribed functions giving respectively the densities of dislocations and disclinations. Equation (7.84) is a condition of consistency of these two densities, and equation (7.85) determines the 'incompatibility' of the elastic strain. The general solution of this problem will be obtained in Chapter 8 for an isotropic medium.

PROBLEMS

1   From equation (7.7), show that $a_{j1} = 0$ if and only if there exists a set of single-valued quantities $\{u_i\}$ such that $e_{pj} = \tfrac{1}{2}(\partial_p u_j + \partial_j u_p)$ and $\omega_{pj} = \tfrac{1}{2}(\partial_p u_j - \partial_j u_p)$.

2   Consider a distribution of dislocations with Burgers vector $\tfrac{1}{2}[1\bar{1}0]$ gliding on planes parallel to (111) in a face-centred cubic crystal. Suppose that there are N lines per unit area. Taking axes along the cube directions find the dislocation density tensor when the dislocations are (a) all pure screws and (b) all pure edges.

3   By integrating equation (7.8) over a volume and using the divergence theorem, interpret this equation as a conservation law of Burgers vector.

4   Show how the parametrization t of the lattice curves is related to the arc-length along the curve in order that equation (7.14) should hold. Obtain the generalization of equation (7.14) when an arbitrary parametrization is used.

5   Eliminate θ from equations (7.26) and relate the resulting equation to equation (7.9). Show that the equation so obtained is sufficient to guarantee

the existence of a function $\theta(x_1, x_2)$ such that equations (7.26) hold. What becomes of equations (7.8) and (7.10) for plane strain?

6 Show that the condition $\partial_2 a_{13} - \partial_1 a_{23} = 0$ implies that any plane deformation is zero in the absence of tractions on the external boundary of the body. [Show first the existence of a single-valued displacement field and then use the uniqueness theorem of classical elasticity.]

7 Show how equations (7.27) arise as a special case of equation (7.28).

8 Express the matrices $(e/u)_{kl}$ and $(u/e)_{kl}$ in terms of α_{kl} in the case of small deformations. Show that equation (7.38) reduces to equation (7.13).

9 Verify that, for the connection given in equation (7.41), the curvature tensor is identically zero.

10 Obtain the curvature condition in terms of the strain and dislocation density tensors without making the small-strain approximation.

11 Show that the polar equation of the e_1-lines such as P_0P for the example in Figure 7.5 is

$$\left(\frac{r^2}{a^2} - 1\right)^{1/2} + \sin^{-1}\frac{a}{r} - \alpha = \text{const.}$$

Find the equation of the e_2-lines.

12 The e_1-lines in a strain-free state of plane strain are the family of hyperbolas $kx_1{}^2 - x_2{}^2 = \lambda$, where $k > 0$ is a constant and λ is the parameter which varies from one member of the family to another. Find the equations of the e_2-lines and the dislocation densities $A^{(1)}$ and $A^{(2)}$.

13 The elastic strain e_{ij} in a linear, isotropic, thermoelastic medium is given by[16]

$$e_{ij} = \tfrac{1}{2}(\partial_i u_j + \partial_j u_i) - \alpha T \delta_{ij}.$$

Here u_i is the displacement from the particle positions in some reference configuration and $T(\mathbf{x})$ is the difference in temperature between the deformed and reference configurations. The expansion coefficient is denoted by α. Show that this strain corresponds to an equivalent dislocation density $a_{ij} = -\alpha \varepsilon_{ijk} \partial_k T$.

REFERENCES

1 Kondo, K., Proc. 2nd Japan Congr. Appl. Mechs. (Science Council of Japan, 1953), p. 41
2 Bilby, B.A., Bullough, R., and Smith, E., Proc. Roy. Soc. A231 (1955), 263
3 Nye, J.F., Acta Metall. 1 (1953), 153
4 Kondo, K. (ed.), *Memoirs of unifying study of basic problems in engineering by means of geometry*, vols. I, II, III (Tokyo: Gakujutsu Bunken Fukyu-kai, 1955, 1958, 1962)
5 Bilby, B.A., Progr. Solid Mechs. 1 (1960), 329
6 Kröner, E., Arch. Rat. Mech. Anal. 4 (1960), 273
7 Kröner, E., *Kontinuumstheorie der Versetzungen und Eigenspannungen* (Berlin: Springer, 1958)
8 Seeger, A., Phys. Stat. Solidi 1 (1961), 667
9 Truesdell, C., and Noll, W., *Handbuch der Physik* III/3 (Berlin: Springer, 1965), sections 34, 44.
10 Noll, W., Arch. Rat. Mech. Anal. 27 (1967), 1
11 Bilby, B.A., Bullough, R., Gardner, L.R.T., and Smith, E., Proc. Roy. Soc. A244 (1958), 538

12 Fox, N., Quart. J. Mech. and Appl. Math. **21** (1968), 67
13 Kröner, E., and Seeger, A., Arch. Rat. Mech. Anal. **3** (1959), 97
14 Lardner, R.W., Int. J. Engng. Sci. **7** (1969), 417
15 Bilby, B.A., and Gardner, L.R.T., Ph.D. Thesis (L.R.T. Gardner),
   Sheffield (1958)
16 Sokolnikoff, I.S., *The mathematical theory of elasticity* (New York:
   McGraw-Hill, 1956), p. 359

8
INTERNAL STRESSES

8.1
TENSOR POTENTIALS

One class of problems that arise in dislocation theory involves finding the distribution of 'internal' stress due to a given arrangement of dislocations. With continuous distributions this means that the dislocation density tensor is a specified function of position. Assuming that the material is linear, the equations describing the situation are the equilibrium equations, stress-strain relations, and the strain incompatibility equation:

$$\partial_j \sigma_{ij} = 0, \tag{8.1}$$

$$\sigma_{ij} = c_{ijkl} e_{kl}, \tag{8.2}$$

$$\varepsilon_{ijk} \varepsilon_{lmn} \partial_j \partial_m e_{kn} = -\eta_{il}. \tag{8.3}$$

In the last equation η is given in terms of the dislocation density by equation (7.9), or by equation (7.85) in the event that rotation dislocations are present. For isotropy the elastic modulus tensor is given by equation (2.5). In order to solve this system of equations, Kröner has used an elegant extension of vector potential theory which we now describe.[1-3]

We shall use the vector operator $\mathbf{V}$ in its familiar senses of grad, div, and curl when operating on scalar and vector fields. That is, if ϕ is a (differentiable) scalar field, $\mathbf{V}\phi$ denotes the vector field whose ith component is $\partial_i \phi$ with respect to Cartesian coordinates; if $\mathbf{v}$ is a vector field, then $\mathbf{V} \cdot \mathbf{v}$ denotes the scalar field $\partial_j v_j$, and $\mathbf{V} \wedge \mathbf{v}$ denotes the vector field whose ith component is $\varepsilon_{ijk} \partial_j v_k$, with ε_{ijk} being the usual permutation tensor. But now we extend the notation to include tensor derivatives and derivatives of tensors. We denote by $\mathbf{VV}\phi$ the tensor field whose Cartesian components are $\partial_i \partial_j \phi$ and by $\mathbf{Vv}$ the tensor field with components $\partial_i v_j$ (referred to as grad grad ϕ and grad $\mathbf{v}$). The transpose of $\mathbf{Vv}$ is denoted by $\mathbf{vV}$ and we define the deformation of $\mathbf{v}$, written def $\mathbf{v}$, to be the symmetric part of $\mathbf{Vv}$. Thus

$$(\mathbf{Vv})_{ij} = \partial_i v_j, \qquad (\mathbf{vV})_{ij} = \partial_j v_i, \qquad (\text{def } \mathbf{v})_{ij} = \tfrac{1}{2}(\partial_i v_j + \partial_j v_i). \tag{8.4}$$

Given a differentiable tensor field $\mathbf{T}$ we can define two vector fields which are divergences of $\mathbf{T}$ analogous to the vector divergence, as follows:

$$(\mathbf{V} \cdot \mathbf{T})_j = \partial_i T_{ij}, \qquad (\mathbf{T} \cdot \mathbf{V})_i = \partial_j T_{ij}. \qquad (8.5)$$

If $\mathbf{T}$ is symmetric, then of course there is no difference between these two. We shall use the notation $\mathbf{V} \cdot \mathbf{T} \cdot \mathbf{V}$ to denote the scalar field $\mathbf{V} \cdot (\mathbf{T} \cdot \mathbf{V}) = \partial_i \partial_j T_{ij}$, Finally, we can define the curl of a tensor in two ways also, depending on which index of the tensor is operated on. Thus

$$(\mathbf{V} \wedge \mathbf{T})_{il} = \varepsilon_{ijk} \partial_j T_{kl}, \qquad (\mathbf{T} \wedge \mathbf{V})_{il} = \varepsilon_{ljk} \partial_j T_{ik}. \qquad (8.6)$$

(Note that some authors define the second quantity with the opposite sign.) If we perform a double curl operation, once on each index of $\mathbf{T}$, we obtain an important tensor called the incompatibility of $\mathbf{T}$ and written inc $\mathbf{T}$. In terms of components,

$$(\mathrm{inc}\ \mathbf{T})_{il} = (\mathbf{V} \wedge \mathbf{T} \wedge \mathbf{V})_{il} = \varepsilon_{ijk} \varepsilon_{lmn} \partial_j \partial_m T_{kn}. \qquad (8.6')$$

The relevance of these definitions to continuum mechanics, and dislocation theory in particular, is immediately obvious. If $\mathbf{u}$ is the displacement field of classical elasticity, then the strain tensor $\mathbf{e}$ is equal to def $\mathbf{u}$. The equilibrium equation (8.1) is $\mathbf{V} \cdot \boldsymbol{\sigma} = 0$. Equation (8.3) takes the form inc $\mathbf{e} = -\boldsymbol{\eta}$. Therefore it is of some interest to extend the well-known theorems for vector fields to tensors, and we begin by quoting, without proof, the standard results that we shall need (see for example reference 4).

Lemma 1

(a) $\mathbf{V} \wedge (\mathbf{V}\phi) \equiv \mathbf{0}, \qquad \mathbf{V} \cdot (\mathbf{V} \wedge \mathbf{v}) \equiv 0.$

(b)   For any vector field $\mathbf{v}$, defined and continuously differentiable through all space and tending to zero at infinity at least as fast as $r^{-(1+\alpha)}$ for some $\alpha > 0$, there exist scalar and vector fields ϕ and $\mathbf{A}$ (called potentials) such that

$\mathbf{v} = \mathbf{V}\phi + \mathbf{V} \wedge \mathbf{A}.$

$\mathbf{A}$ is not uniquely determined, and can be selected to satisfy some auxiliary constraint which is usually chosen to be the condition $\mathbf{V} \cdot \mathbf{A} = 0$.

(c)   If $\mathbf{V} \cdot \mathbf{v} = 0$, then in the above representation $\phi = 0$ and $\mathbf{v} = \mathbf{V} \wedge \mathbf{A}$.

(d)   If $\mathbf{V} \wedge \mathbf{v} = \mathbf{0}$, then $\mathbf{A} = \mathbf{0}$ and $\mathbf{v}$ has the representation $\mathbf{v} = \mathbf{V}\phi$.

The following two results, the first of which is obvious, are the generalizations of parts (a) and (b) of Lemma 1.

Lemma 2

$$\mathbf{V} \cdot (\mathrm{inc}\ \mathbf{T}) = (\mathrm{inc}\ \mathbf{T}) \cdot \mathbf{V} \equiv \mathbf{0}, \qquad \mathrm{inc}(\mathrm{def}\ \mathbf{v}) \equiv \mathbf{0}. \qquad (8.7)$$

Lemma 3

If **T** is a continuously differentiable tensor field which tends to zero at infinity at least as fast as $r^{-2-\alpha}$ for some $\alpha > 0$, then there exist a scalar ϕ, vectors **b** and **c**, and a tensor **B** such that

$$\mathbf{T} = \nabla\nabla\phi + \nabla\wedge\mathbf{b}\nabla + \nabla\mathbf{c}\wedge\nabla + \text{inc } \mathbf{B}. \tag{8.8}$$

Furthermore the subsidiary conditions

$$\nabla\cdot\mathbf{b} = \nabla\cdot\mathbf{c} = 0, \qquad \nabla\cdot\mathbf{B} = \mathbf{B}\cdot\nabla = 0 \tag{8.9}$$

can be imposed. (Note that the two middle terms in equation (8.8) are unambiguous since for instance $\nabla\wedge(\mathbf{b}\nabla) = (\nabla\wedge\mathbf{b})\nabla$, each having an *il*-component equal to $\varepsilon_{ijk}\partial_j\partial_l b_k$.)

Proof For fixed j, the three numbers (T_{ij}) can be regarded as defining a vector field. Using Lemma 1(b), there exist scalar and vector potentials for this field:

$$T_{ij} = \partial_i a_j + \varepsilon_{ilk}\partial_l A_{kj},$$

where j is a parameter throughout. This can be written

$$\mathbf{T} = \nabla a + \nabla\wedge\mathbf{A}. \tag{8.10}$$

Furthermore we can choose **A** in such a way that $\nabla\cdot\mathbf{A} = \mathbf{0}$.

Similarly for fixed k, the three numbers (A_{kj}) form a vector field, for which two potentials exist. Thus **A** can be written

$$\mathbf{A} = \mathbf{b}\nabla + \mathbf{B}\wedge\nabla,$$

where we can require also that $\mathbf{B}\cdot\nabla = \mathbf{0}$. Finally, for the vector field **a**, there exist potentials ϕ and **c** such that

$$\mathbf{a} = \nabla\phi + \nabla\wedge\mathbf{c}$$

with $\nabla\cdot\mathbf{c} = 0$. If we now substitute these representations for **A** and **a** into equation (8.10), we obtain equation (8.8) directly.

Of the subsidiary conditions (8.9) it remains to show that $\nabla\cdot\mathbf{b}$ and $\nabla\cdot\mathbf{B}$ are zero. Since $\nabla\cdot\mathbf{A} = \mathbf{0}$, we have

$$(\nabla\cdot\mathbf{b})\nabla + (\nabla\cdot\mathbf{B})\wedge\nabla = \mathbf{0},$$

or in other words the sum of the gradient of $\nabla\cdot\mathbf{b}$ and the curl of $\nabla\cdot\mathbf{B}$ is identically zero. Taking the divergence we get $\nabla^2(\nabla\cdot\mathbf{b}) = 0$. Thus $\nabla\cdot\mathbf{b}$ is a harmonic function, tending to zero at infinity, and hence is identically zero. This leaves the vector field $\nabla\cdot\mathbf{B}$ having a zero curl. Since its divergence is zero also (as follows from the fact, proved earlier, that $\mathbf{B}\cdot\nabla = \mathbf{0}$), and it tends to zero at infinity, then it must be identically zero, as required.

The four potentials, ϕ, **b**, **c**, and **B**, can be expressed explicitly in terms of **T** by means of integral formulae analogous to those of vector field theory. By taking appropriate divergences and curls of equation (8.8) and using the identity for any vector or tensor field **X**,

$$\mathbf{V} \wedge (\mathbf{V} \wedge \mathbf{X}) = \mathbf{V}(\mathbf{V} \cdot \mathbf{X}) - \nabla^2 \mathbf{X}, \tag{8.11}$$

we obtain the results

$$\nabla^4 \phi = \mathbf{V} \cdot \mathbf{T} \cdot \mathbf{V}, \qquad \nabla^4 \mathbf{b} = -\mathbf{V} \wedge \mathbf{T} \cdot \mathbf{V},$$
$$\nabla^4 \mathbf{c} = -\mathbf{V} \cdot \mathbf{T} \wedge \mathbf{V}, \qquad \nabla^4 \mathbf{B} = \text{inc } \mathbf{T}. \tag{8.12}$$

Here ∇^4 denotes the biharmonic operator $\nabla^2 \nabla^2$. The Green's function for ∇^4 in the whole of space is $-|\mathbf{r} - \mathbf{r}'|/8\pi$, so that

$$\phi(\mathbf{r}) = -\frac{1}{8\pi} \int \mathbf{V}' \cdot \mathbf{T}(\mathbf{r}') \cdot \mathbf{V}' \, |\mathbf{r} - \mathbf{r}'| \, dV',$$

$$\mathbf{b}(\mathbf{r}) = \frac{1}{8\pi} \int \mathbf{V}' \wedge \mathbf{T}(\mathbf{r}') \cdot \mathbf{V}' \, |\mathbf{r} - \mathbf{r}'| \, dV', \tag{8.13}$$

$$\mathbf{c}(\mathbf{r}) = \frac{1}{8\pi} \int \mathbf{V}' \cdot \mathbf{T}(\mathbf{r}') \wedge \mathbf{V}' \, |\mathbf{r} - \mathbf{r}'| \, dV',$$

$$\mathbf{B}(\mathbf{r}) = -\frac{1}{8\pi} \int \text{inc } \mathbf{T}(\mathbf{r}') \, |\mathbf{r} - \mathbf{r}'| \, dV'.$$

It is readily verified that under the supposed behaviour of **T** at infinity these solutions satisfy the conditions (8.9).

In continuum mechanics, many of the tensors we have to deal with are symmetric, and in such cases the representation in terms of potentials become simpler. The transpose of equation (8.8) is

$$\mathbf{T}^T = \mathbf{V}\mathbf{V}\phi + \mathbf{V}\mathbf{b} \wedge \mathbf{V} + \mathbf{V} \wedge \mathbf{c}\mathbf{V} + \mathbf{V} \wedge \mathbf{B}^T \wedge \mathbf{V}.$$

If we add this to equation (8.8) and divide by 2, we get an alternative expression for **T** when it is symmetric:

$$\mathbf{T} = \mathbf{V}\mathbf{V}\phi + \tfrac{1}{2}\mathbf{V} \wedge (\mathbf{b}+\mathbf{c})\mathbf{V} + \tfrac{1}{2}\mathbf{V}(\mathbf{b}+\mathbf{c}) \wedge \mathbf{V} + \mathbf{V} \wedge \mathbf{B}' \wedge \mathbf{V}$$
$$= \text{def } \mathbf{v} + \text{inc } \mathbf{B}', \tag{8.14}$$

where $\mathbf{v} = \mathbf{V}\phi + \mathbf{V} \wedge (\mathbf{b}+\mathbf{c})$ and $\mathbf{B}' = \tfrac{1}{2}(\mathbf{B}+\mathbf{B}^T)$. Thus we have

Lemma 4

If **T** is a symmetric tensor field, of order $r^{-(2+\alpha)}$ at infinity, then there exists a vector field **v** and a symmetric tensor field **B'** such that equation (8.14) holds. Furthermore we can choose $\mathbf{V} \cdot \mathbf{B}' = 0$.

As in the general case, there are integral formulae for the potentials **v** and **B′**, and in fact **B′** is still given by the fourth of equations (8.13). For **v**, taking the divergence of equation (8.14), we obtain

$$\mathbf{V} \cdot \mathbf{T} = \mathbf{V} \cdot (\text{def } \mathbf{v}) = \tfrac{1}{2}[\nabla^2 \mathbf{v} + \mathbf{V}(\mathbf{V} \cdot \mathbf{v})].$$

Taking a second divergence gives

$$\mathbf{V} \cdot \mathbf{T} \cdot \mathbf{V} = \nabla^2 (\mathbf{V} \cdot \mathbf{v}).$$

Therefore

$$\nabla^4 \mathbf{v} = 2\nabla^2 (\mathbf{V} \cdot \mathbf{T}) - \mathbf{V}[\nabla^2 (\mathbf{V} \cdot \mathbf{v})] = 2\nabla^2 (\mathbf{V} \cdot \mathbf{T}) - \mathbf{V}(\mathbf{V} \cdot \mathbf{T} \cdot \mathbf{V}). \tag{8.15}$$

Again the solution can be expressed using the Green's function for ∇^4, the result in terms of components being

$$v_i(\mathbf{r}) = -\frac{1}{8\pi} \int [2\nabla'^2 \partial_j' T_{ji}(\mathbf{r}') - \partial_i' \partial_j' \partial_k' T_{jk}(\mathbf{r}')] \, |\mathbf{r} - \mathbf{r}'| \, dV'. \tag{8.16}$$

For tensor fields also, there are special cases analogous to parts (c) and (d) of Lemma 1. The first arises when **T** has zero incompatibility, in which case **B** = **0** from equation (8.13). Sticking to symmetric tensors, we then have the following result:

Lemma 5
If the incompatibility of a symmetric tensor vanishes, then it can be written as the deformation of a vector.

This theorem is no more than the statement, familiar from classical elasticity, that the strain compatibility equations are sufficient for the existence of a displacement field from which the strain can be derived. We have seen that the resulting displacement field is expressible as a volume integral in terms of the strain via equation (8.16). However, it is well known that an equivalent, and generally simpler, integral representation is available in this case, called the Cesaro integral. This states that

$$\mathbf{v}(\mathbf{r}) = \mathbf{v}_0 + \mathbf{w}_0 \wedge (\mathbf{r} - \mathbf{r}_0) + \int_{r_0}^{r} \{\mathbf{T}(\mathbf{r}') - (\mathbf{r} - \mathbf{r}') \wedge [\mathbf{V}' \wedge \mathbf{T}(\mathbf{r}')]\} \cdot d\mathbf{r}', \tag{8.17}$$

where $\mathbf{v}_0$ and $\mathbf{w}_0$ are arbitrary constant vectors, $\mathbf{r}_0$ is an arbitrary point, and the integral is evaluated along any path from $\mathbf{r}_0$ to $\mathbf{r}$. It is readily verified that when inc **T** = **0** this integral is path-independent (see Prob. 8.5), and clearly def **v** = **T**. Besides the reduction in the dimensionality of the integral, Cesaro's formula has the advantage over equation (8.16) of not requiring **T** to have a certain type of behaviour at infinity.

The second special case arises when **T** has zero divergences, $\mathbf{V} \cdot \mathbf{T} = \mathbf{T} \cdot \mathbf{V} = \mathbf{0}$. Then, from equation (8.13), ϕ, **b**, and **c** are all zero and we are left with

K

$\mathbf{T} =$ inc $\mathbf{B}$. Similarly equation (8.16) gives $\mathbf{v} = \mathbf{0}$ when T is symmetric. Thus we have

Lemma 6

If $\mathbf{T}$ is a tensor, of order $r^{-(2+\alpha)}$ at infinity, such that $\mathbf{V} \cdot \mathbf{T} = \mathbf{T} \cdot \mathbf{V} = \mathbf{0}$, then there exists a tensor $\mathbf{B}$ such that $\mathbf{T} =$ inc $\mathbf{B}$. Furthermore we can require that $\mathbf{V} \cdot \mathbf{B} = \mathbf{B} \cdot \mathbf{V} = \mathbf{0}$. If $\mathbf{T}$ is symmetric, it is sufficient to require that $\mathbf{V} \cdot \mathbf{T} = \mathbf{0}$. Then $\mathbf{B}$ is also symmetric, and satisfies $\mathbf{V} \cdot \mathbf{B} = \mathbf{0}$. In both cases, $\mathbf{B}$ is given explicitly in the fourth of equations (8.13).

So far we have always taken the subsidiary conditions as the vanishing of the divergences of the vector and tensor potentials. In fact this is only one among many possible conditions, and in the application of these results to elasticity it is more convenient to take a different one. Referring specifically to the situation in the last lemma, with $\mathbf{T}$ and $\mathbf{B}$ both symmetric, in place of $\mathbf{V} \cdot \mathbf{B} = \mathbf{0}$ we can impose the constraint $\mathbf{V} \cdot \mathbf{C} = \mathbf{0}$, where $\mathbf{C}$ is the tensor with components $C_{ij} = B_{ij} + kB_{ll}\delta_{ij}$ and k is any constant different from -1. The proof of this is as follows.

Let $\mathbf{B}$ be any potential such that $\mathbf{T} =$ inc $\mathbf{B}$, so that in general $\mathbf{V} \cdot \mathbf{C} \neq \mathbf{0}$. We can add to $\mathbf{B}$ the deformation of any vector field $\mathbf{v}$ and it will still be a potential for $\mathbf{T}$, by virtue of equation (8.7). For this new potential, the new C-tensor is $C_{ij}' = C_{ij} + \frac{1}{2}(\partial_i v_j + \partial_j v_i) + k\partial_k v_k \delta_{ij}$. Then we shall have $\mathbf{V} \cdot \mathbf{C}' = \mathbf{0}$ provided $\mathbf{v}$ satisfies the equations

$$\nabla^2 v_j + (1+2k)\partial_j \partial_i v_i = -2\partial_i C_{ij}.$$

Thus if we can show that a solution of these equations always exists, the proof will be complete. Now if a solution $\mathbf{v}$ exists, it can, by Lemma 1(b), be expressed in terms of potentials in the form $\mathbf{v} = \nabla\phi + \mathbf{V} \wedge \mathbf{A}$, with $\mathbf{V} \cdot \mathbf{A} = 0$. The differential equation for $\mathbf{v}$ then becomes, after substituting this representation for $\mathbf{v}$,

$$2(1+k)\mathbf{V}(\nabla^2\phi) + \mathbf{V} \wedge (\nabla^2\mathbf{A}) = -2\mathbf{V} \cdot \mathbf{C}.$$

If we now take the divergence and the curl of this equation and use the identity (8.11), we obtain

$$(1+k)\nabla^4\phi = -\mathbf{V} \cdot \mathbf{C} \cdot \mathbf{V}, \qquad \nabla^4\mathbf{A} = \mathbf{V} \cdot \mathbf{C} \wedge \mathbf{V}.$$

These equations may now be solved for ϕ and $\mathbf{A}$ (cf. eqs. (8.12) and (8.13)) and hence $\mathbf{v}$ may be found.

Finally, we mention the fact, which will prove useful later on, that for any tensor $\mathbf{T}$ we have the identity

$$(\text{inc } \mathbf{T})_{il} = (\nabla^2 T_{kk} - \partial_i \partial_k T_{ik})\delta_{il} + \partial_i \partial_k T_{lk} + \partial_l \partial_k T_{ki} - \nabla^2 T_{li} - \partial_i \partial_l T_{kk}. \qquad (8.18)$$

(For a proof, see Prob. 8.2). The right-hand side here is the form in which the compatibility conditions of classical elasticity are usually written.

8.2
APPLICATION TO DISLOCATIONS

A *Solution for the stress-field*

The stress tensor $\boldsymbol{\sigma}$ is symmetric and has zero divergence for equilibrium in the absence of body forces. Hence by Lemma 6 there exists a symmetric tensor $\mathbf{B}$, called the stress function, such that $\boldsymbol{\sigma} = \text{inc } \mathbf{B}$. We shall consider only isotropic materials, for which the stress-strain relation has the forms in equations (3.8) and (2.9). In this case it is convenient to use in place of $\mathbf{B}$ a tensor $\boldsymbol{\chi}$ defined as

$$\chi_{ij} = \frac{1}{2\mu}\left(-B_{ij} + \frac{\nu}{1+2\nu} B_{kk}\delta_{ij}\right),\tag{8.19}$$

where ν is Poisson's ratio. The inverse relation is

$$B_{ij} = -2\mu\left(\chi_{ij} + \frac{\nu}{1-\nu}\chi_{kk}\delta_{ij}\right),\tag{8.20}$$

which can be obtained by first setting $i = j$ and summing in equation (8.19) to give B_{kk} in terms of χ_{kk}. Thus we have

$$\boldsymbol{\sigma} = -2\mu \text{ inc}\left(\chi_{ij} + \frac{\nu}{1-\nu}\chi_{kk}\delta_{ij}\right).\tag{8.21}$$

If we use equation (8.18) we can obtain an expanded version of this formula. In general, this is rather unwieldy, but if $\nabla\cdot\boldsymbol{\chi} = 0$ it becomes much simpler:

$$\sigma_{il} = 2\mu\left(\nabla^2\chi_{il} - \frac{1}{1-\nu}(\nabla^2\chi_{kk}\delta_{il} - \partial_i\partial_l\chi_{kk})\right).\tag{8.22}$$

But, as follows from our discussion after Lemma 6 in §8.1, we can in fact always find a potential $\boldsymbol{\chi}$ with zero divergence such that the stress-field is given by equation (8.21). (The proviso there that $k \neq -1$ is reflected here in the restriction that $\nu \neq -1$.) The corresponding strain tensor is readily seen to be

$$e_{il} = -\nabla^2\chi_{kk}\delta_{il} + \nabla^2\chi_{il} + \frac{1}{1-\nu}\partial_i\partial_l\chi_{kk}.\tag{8.23}$$

Then if we substitute this result into the incompatibility relations (8.3), we obtain the remaining conditions of the problem as certain differential equations for the components χ_{ij}. Using equation (8.18) again and the fact that $\partial_i \chi_{ij} = 0$, these are seen to be simply

$$\nabla^4 \chi_{ij} = \eta_{ij}. \tag{8.24}$$

The solution of this equation is obtained, as were equations (8.13), by using the Green's function for the biharmonic operator, and is

$$\chi_{ij}(\mathbf{r}) = -\frac{1}{8\pi} \int \eta_{ij}(\mathbf{r}') |\mathbf{r} - \mathbf{r}'| \, dV'. \tag{8.25}$$

For consistency we must show finally that this solution satisfies the assumed subsidiary condition. From equations (7.73), (7.72), and (7.20), it follows that $\partial_j \eta_{ij} = 0$; in fact, if this condition were not satisfied, equation (8.3) would have no solution. Then differentation of equation (8.25) followed by an integration by parts (using the divergence theorem) yields the desired result that $\partial_i \chi_{ij} = 0$.

In the absence of rotation dislocations, $\boldsymbol{\eta}$ is given by the right-hand side of equation (7.22):

$$\eta_{ij} = -(\varepsilon_{jmn}\partial_m a_{ni})^{(s)}. \tag{8.26}$$

Substituting this into equation (8.25), using the divergence theorem to switch the derivative over to $|\mathbf{r} - \mathbf{r}'|$, and then changing to a derivative with respect to $\mathbf{r}$ rather than $\mathbf{r}'$, we obtain finally

$$\chi_{ij}(\mathbf{r}) = \frac{1}{8\pi} \left(\varepsilon_{jmn} \int a_{ni}(\mathbf{r}') \partial_m |\mathbf{r} - \mathbf{r}'| \, dV' \right)^{(s)}. \tag{8.27}$$

In the process of arriving at this result we have set a surface term at infinity equal to zero. A sufficient condition for this to be valid is that $a_{ni}(\mathbf{r}) \sim r^{-(3+\alpha)}$ at large distances, a condition which also suffices to guarantee the convergence of the integral in equation (8.27).

B *Stress-field of a single loop*

The expressions in the previous section, which were proved for arbitrary continuous distributions of dislocation, can be used to derive results for discrete dislocation lines by allowing the dislocation density to be concentrated in the neighbourhood of the lines in question. As long as we are concerned with distances from the dislocation line which are large compared with the diameter of the region in which the dislocation density is non-zero, then the actual structure of this core region is of no consequence. We might just as well give a_{ij} a δ-function type of behaviour near the dislocation line as include any spread of dislocation density throughout the core; in either case we shall reproduce the results proved earlier in §3.6. This ceases to be true, however, when we consider points close to the dislocation line. By allowing the dislocation to be distributed throughout some core region rather than concentrated on a single line we remove the stress singularity and also obtain a finite strain energy without having to introduce some inner cut-off at $r = r_0$ in the energy integral. In fact, we shall be able to determine an effective value for r_0 in terms of the assumed structure of the dislocation core.

In this section we shall consider a dislocation line which forms a single closed loop, Γ. We suppose that the total Burgers vector $\mathbf{b}$ is distributed over a thin filament around Γ. The volume element dV' in equation (8.27) can be written as $dS' \cdot dl'$ with dS' an element of area of the filament cross-section and dl' an element of length along Γ. Comparing with equation (7.1) we have

$$dl' \int a_{ni}(\mathbf{r}')dS' = b_n dl_i',$$
(8.28)

where $dl_i' = dl'\xi_i'$ is a vector element along Γ.

In equation (8.27), when the field-point $\mathbf{r}$ is far away from the dislocation line, $|\mathbf{r}-\mathbf{r}'|$ is roughly the same for all points $\mathbf{r}'$ in any cross-section of the core, and the integration over dS' can be performed. Using equation (8.28) we have

$$\chi_{ij}(r) = \frac{1}{8\pi} (\varepsilon_{jmn}b_n \oint_\Gamma \partial_m |\mathbf{r}-\mathbf{r}'| \, dl_i')^{(s)}.$$
(8.29)

The stress and strain fields can be obtained immediately by substituting into equations (8.22) and (8.23). The first of these is

$$\sigma_{il}(r) = \frac{\mu}{4\pi} \varepsilon_{kmn} b_n \oint_\Gamma \left(\partial_m \nabla^2 |\mathbf{r}-\mathbf{r}'| \frac{1}{2} (\delta_{kl} dl_i' + \delta_{ki} dl_l') \right.$$
$$\left. + \frac{1}{1-\nu} (\partial_m \partial_i \partial_l - \delta_{il} \partial_m \nabla^2) |\mathbf{r}-\mathbf{r}'| \, dl_k' \right). \quad (8.30)$$

This result is identical with the formula of Peach and Koehler[5] which we proved in §3.6B using the Green's function method.

c Interaction energies of closed loops

We are particularly interested in the total strain energy of various states of internal stress. Having obtained the stress-field for an arbitrary distribution of dislocation density, we are in a position, in principle at least, to calculate the energy. We shall consider first the interaction energy, as defined in equation (2.32), between two states of internal stress which we label A and B. They have strain tensors $e_{ij}^{(A)}$, $e_{ij}^{(B)}$, stresses $\sigma_{ij}^{(A)}$, $\sigma_{ij}^{(B)}$ and corresponding stress-functions $B_{ij}^{(A)}$, $B_{ij}^{(B)}$ or $\chi_{ij}^{(A)}$, $\chi_{ij}^{(B)}$, and incompatibilities $\eta_{ij}^{(A)}$ and $\eta_{ij}^{(B)}$. Then

$$U_{\text{int}}^{(AB)} = \int e_{ij}^{(A)}\sigma_{ij}^{(B)}dV = \int e_{ij}^{(A)}\varepsilon_{ikl}\varepsilon_{jmn}\partial_k\partial_m B_{ln}^{(B)}dV$$

$$= \int \varepsilon_{ikl}\varepsilon_{jmn}\partial_k\partial_m e_{ij}^{(A)}B_{ln}^{(B)}dV,$$

after integrating by parts and neglecting the surface terms at infinity. If we now substitute from equations (8.3) and (8.20) we have

$$U_{\text{int}}^{(AB)} = 2\mu \int \eta_{ln}^{(A)}\left(\chi_{ln}^{(B)} + \frac{\nu}{1-\nu}\chi_{kk}^{(B)}\delta_{ln}\right)dV. \tag{8.31}$$

The surface terms at infinity will vanish if for both A and B fields the stresses tend to zero at least as fast as $r^{-(3/2+\alpha)}$ for some $\alpha > 0$. For dislocation loops, equation (8.30) shows that the stresses are $O(r^{-2})$ at large distances, so that equation (8.31) is a valid expression for the interaction energy between two such loops.

Considering this case now, in which the two states correspond to dislocation loops $\Gamma^{(A)}$ and $\Gamma^{(B)}$, we substitute from equation (8.26) for $\eta^{(A)}$, integrate once by parts, and perform the integration across the filament cross-section of the loop $\Gamma^{(A)}$ using equation (8.28). This procedure, valid if the distance between the two loops is large compared to the core radius, leads to

$$U_{\text{int}}^{(AB)} = 2\mu\varepsilon_{jmn}b_n^{(A)}\oint_{\Gamma^{(A)}} \partial_m\left(\chi_{ij}^{(B)} + \frac{\nu}{1-\nu}\chi_{kk}^{(B)}\delta_{ij}\right)dl_i^{(A)}. \tag{8.32}$$

(We use $\mathbf{r}$ as the running point on $\Gamma^{(A)}$ and later $\mathbf{r}'$ as a point on $\Gamma^{(B)}$). Substituting from equation (8.29) for $\boldsymbol{\chi}^{(B)}$ we obtain the result in the form

$$U_{\text{int}}^{(AB)} = M_{nq}^{(AB)}b_n^{(A)}b_q^{(B)}, \tag{8.33}$$

where

$$M_{nq}^{(AB)} = \frac{\mu}{8\pi}\oint_{\Gamma^{(A)}}\oint_{\Gamma^{(B)}} \varepsilon_{jmn}\varepsilon_{kpq}\partial_m\partial_p|\mathbf{r}-\mathbf{r}'|\left(dl_k^{(A)}dl_j^{(B)}\right.$$

$$\left. + \delta_{jk}dl_i^{(A)}dl_i^{(B)} + \frac{2\nu}{1-\nu}dl_j^{(A)}dl_k^{(B)}\right). \tag{8.34}$$

This quantity is sometimes called the dislocation mutual inductance because of the close similarity between the interaction energy, in equation (8.33), of two dislocation loops and the corresponding result for current loops in magnetostatics.

A somewhat different formula for $U_{\text{int}}^{(AB)}$, first obtained by Blin,[6] can be obtained by transforming equation (8.34). Using the identity introduced in Problem 8.2 and equation (3.68) with f given by $|\mathbf{r}-\mathbf{r}'|$, we get

$$\oint_{\Gamma^{(A)}} \oint_{\Gamma^{(B)}} \varepsilon_{jmn}\varepsilon_{kpq}\partial_m\partial_p |\mathbf{r}-\mathbf{r}'| (dl_k^{(A)}dl_j^{(B)} + \delta_{jk}dl_i^{(A)}dl_i^{(B)} - 2dl_j^{(A)}dl_k^{(B)})$$

$$= \oint_{\Gamma^{(A)}} \oint_{\Gamma^{(B)}} \nabla^2 |\mathbf{r}-\mathbf{r}'| (2dl_q^{(A)}dl_n^{(B)} - dl_n^{(A)}dl_q^{(B)})$$

$$= \oint_{\Gamma^{(A)}} \oint_{\Gamma^{(B)}} \frac{2}{|\mathbf{r}-\mathbf{r}'|} (-2\varepsilon_{nqk}\varepsilon_{ijk}dl_i^{(A)}dl_j^{(B)} + dl_n^{(A)}dl_q^{(B)}).$$

Therefore

$$U_{\text{int}}^{(AB)} = \frac{\mu}{8\pi} \oint_{\Gamma^{(A)}} \oint_{\Gamma^{(B)}} \left(\frac{2}{|\mathbf{r}-\mathbf{r}'|} (-2b_n^{(A)}b_q^{(B)}\varepsilon_{nqk}\varepsilon_{ijk}dl_i^{(A)}dl_j^{(B)} \right.$$

$$+ b_n^{(A)}b_q^{(B)}dl_n^{(A)}dl_q^{(B)})$$

$$+ \frac{2}{1-\nu} \varepsilon_{jmn}\varepsilon_{kpq}b_n^{(A)}b_q^{(B)}\partial_m\partial_p |\mathbf{r}-\mathbf{r}'| dl_j^{(A)}dl_k^{(B)} \Big)$$

$$= + \frac{\mu}{4\pi} \oint_{\Gamma^{(A)}} \oint_{\Gamma^{(B)}} \left(-\frac{(\mathbf{b}^{(A)} \wedge \mathbf{b}^{(B)})\cdot(\mathbf{dl}^{(A)} \wedge \mathbf{dl}^{(B)})}{|\mathbf{r}-\mathbf{r}'|} \right.$$

$$+ \frac{(\mathbf{b}^{(A)}\cdot\mathbf{dl}^{(A)})\,(\mathbf{b}^{(B)}\cdot\mathbf{dl}^{(B)})}{|\mathbf{r}-\mathbf{r}'|}$$

$$+ \frac{1}{1-\nu} (\mathbf{b}^{(A)} \wedge \mathbf{dl}^{(A)})\cdot\nabla\nabla |\mathbf{r}-\mathbf{r}'|\cdot(\mathbf{b}^{(B)} \wedge \mathbf{dl}^{(B)}) \Big). \quad (8.35)$$

This is usually known as Blin's formula.

D *Self-energy of a closed loop*[1, 3]

For the self-energy of a certain state (A) of internal stress, equation (8.31) is still valid provided we replace B by A and multiply by $1/2$. However, the transition to the form (8.32) breaks down because the integration over the filament cross-section cannot be performed using equation (8.28) in this case. Thus, for example, Blin's formula cannot simply be used with $\Gamma^{(B)} = \Gamma^{(A)}$ to give the self-energy of a single loop – the resulting integral is in fact divergent. To deal with this case we have to include more detail of the dislocation structure.

Let $\mathbf{l}$ represent a general point on the dislocation axis, and at each point $\mathbf{l}$ take $\mathbf{q} = (q_1, q_2)$ as vector and coordinates in the plane perpendicular to the axis at $\mathbf{l}$ (see Fig. 8.1). Then $\mathbf{r} = \mathbf{l}+\mathbf{q}$ is a general point in the filament. Assume the structure to be such that

$$a_{ni}(\mathbf{r})\, dV = b_n\gamma(\mathbf{q})d^2\mathbf{q}\, dl_i, \quad (8.36)$$

where $\gamma(\mathbf{q})$ is a positive function such that

$$\int \gamma(\mathbf{q})d^2\mathbf{q} = 1.$$

Thus the dislocation is supposed to consist of a distribution of parallel Burgers vector, of total amount $\mathbf{b}$, whose spread is uniform along its length and is determined by the function $\gamma(\mathbf{q})$. We shall refer to $\gamma(\mathbf{q})$ as the dislocation core structure function.

When we use equation (8.36) in equation (8.27) we obtain, in place of equation (8.29),

$$\chi_{ij}(\mathbf{r}) = \frac{1}{8\pi} (\varepsilon_{jmn}b_n \int \gamma(\mathbf{q}')d^2\mathbf{q}' \oint_\Gamma \partial_m |\mathbf{r}-\mathbf{r}'| \, dl_i')^{(S)} \tag{8.37}$$

where $\mathbf{r}' = \mathbf{l}'+\mathbf{q}'$. Similarly in the expression analogous to equation (8.31) for the self-energy, we obtain, in place of equation (8.32),

$$U^{(AA)} = \mu\varepsilon_{jmn}b_n^{(A)} \int \gamma(\mathbf{q})d^2\mathbf{q} \oint_{\Gamma^{(A)}} \partial_m \left(\chi_{ij}^{(A)} + \frac{\nu}{1-\nu} \chi_{kk}^{(A)}\delta_{ij} \right) dl_i^{(A)} \tag{8.38}$$

If we now combine these two and write (dropping most of the (A) superfixes)

$$U^{(AA)} = \tfrac{1}{2}M_{nq}^{(AA)}b_n b_q, \tag{8.39}$$

we obtain, in place of equation (8.34),

$$M_{nq}^{(AA)} = \frac{\mu}{8\pi} \int \gamma(\mathbf{q})d^2\mathbf{q} \int \gamma(\mathbf{q}')d^2\mathbf{q}' \oint_\Gamma \oint_\Gamma \varepsilon_{jmn}\varepsilon_{kpq}\partial_m\partial_p |\mathbf{r}-\mathbf{r}'|$$
$$\cdot \left(dl_k dl_j' + \delta_{jk}dl_i dl_i' + \frac{2\nu}{1-\nu} dl_j dl_k' \right). \tag{8.40}$$

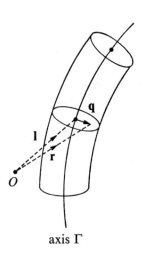

axis Γ

FIGURE 8.1

In conformity with the earlier terminology, this quantity is often called the self-inductance of the loop. Since $|\mathbf{r}-\mathbf{r}'| = |\mathbf{l}-\mathbf{l}'+\mathbf{q}-\mathbf{q}'|$, the double integral around Γ is now convergent.

It is possible in certain cases to draw conclusions from equation (8.40) without making detailed assumptions about the form of the core structure function (see for example the next section). However, there are also cases in which it is necessary, or perhaps just convenient, to assume some particular structure. In these situations several slightly different approaches have been made by various authors. First of all it is usual to assume that the dislocation will spread in the glide direction much more than in the direction normal to the glide-plane. Then taking q_1 to be the component of $\mathbf{q}$ in the glide-plane we can set $\gamma(\mathbf{q}) = \beta(q_1)\delta(q_2)$ to a reasonable approximation. The function $\beta(q_1)$ determines the distribution of Burgers vector within the glide-plane.

Perhaps the most satisfactory way of proceeding beyond this is to take $\beta(q_1)$ as given by one of the atomistic dislocation models. For example the Peierls-Nabarro model would give (cf. eqs. (4.42) and (5.99))

$$\beta(q_1) = \frac{1}{\pi}\frac{\zeta}{q_1{}^2+\zeta^2}, \tag{8.41}$$

where ζ is the dislocation width. A simpler procedure is to take $\beta(q_1)$ to be given by a step-function:

$$\beta(q_1) = \begin{cases} 1/2\zeta & \text{for } |q_1| < \zeta, \\ 0 & \text{otherwise.} \end{cases} \tag{8.42}$$

Although it is less acceptable on physical grounds, equation (8.42) has the advantage of being more amenable to calculation, without being fundamentally different from more correct expressions.

8.3
EXAMPLES

A Straight screw dislocations

Consider two parallel straight screws lying in the x_3-direction, one along the x_3-axis, the other a distance ρ away (Fig. 8.2). The interaction energy from equations (8.33) and (8.34) is

$$U^{(AB)}_{\text{int}} = M_{33}{}^{(AB)}b^{(A)}b^{(B)} \tag{8.43}$$

with

$$M_{33}{}^{(AB)} = \frac{\mu}{8\pi} \int_{\Gamma^{(A)}} \int_{\Gamma^{(B)}} \varepsilon_{jm3}\varepsilon_{kp3}\partial_m\partial_p |\mathbf{r}-\mathbf{r}'| (\delta_{jk}dl_3{}^{(A)}dl_3{}^{(B)}).$$

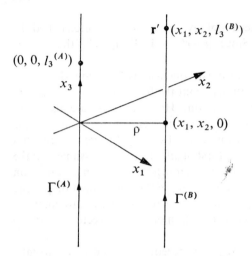

FIGURE 8.2

The integrand can be written

$$(\partial_{11}+\partial_{22})\,|\mathbf{r}-\mathbf{r}'| = \frac{2}{|\mathbf{r}-\mathbf{r}'|} + \partial_3\partial_3'\,|\mathbf{r}-\mathbf{r}'|.$$

Supposing both dislocations to occupy the length $-\tfrac{1}{2}L < l_3 < \tfrac{1}{2}L$, the second term can be integrated immediately to give

$$\left[\left(|\mathbf{r}-\mathbf{r}'|\right)_{l_3(A)=-\frac{1}{2}L}^{l_3(A)=+\frac{1}{2}L}\right]_{l_3(B)=-\frac{1}{2}L}^{l_3(B)=+\frac{1}{2}L} = -2[(L^2+\rho^2)^{1/2}-\rho].$$

The first term gives

$$2\int_{-\frac{1}{2}L}^{\cdot\frac{1}{2}L} dl_3{}^{(A)}\left[\ln\{y+(y^2+\rho^2)^{1/2}\}\right]_{-\frac{1}{2}L-l_3(A)}^{\frac{1}{2}L-l_3(A)}$$

$$= 4\left\{L\ln\left[\frac{L}{\rho}+\left(\frac{L^2}{\rho^2}+1\right)^{1/2}\right]-[(L^2+\rho^2)^{1/2}-\rho]\right\}$$

so that

$$M_{33}{}^{(AB)} = \frac{\mu}{4\pi}L\left\{2\ln\left[\frac{L}{\rho}+\left(\frac{L^2}{\rho^2}+1\right)^{1/2}\right] - 3\left(\frac{(L^2+\rho^2)^{1/2}-\rho}{L}\right)\right\}. \quad (8.44)$$

For the self-energy of a single screw, the inapplicability of formula (8.34) is demonstrated by the divergence of this expression if we allow $\rho \to 0$. In fact we must use equation (8.40). In the same way as above the energy reduces to $\tfrac{1}{2}M_{33}{}^{(AA)}b^2$, where

$$M_{33}{}^{(AA)} = \frac{\mu}{4\pi}L\int\gamma(\mathbf{q})d^2\mathbf{q}\int\gamma(q')\mathbf{d}^2\mathbf{q}'\left\{2\ln\left[\frac{L}{\rho}+\left(\frac{L^2}{\rho^2}+1\right)^{1/2}\right]\right.$$

$$\left. -3\left[\left(1+\frac{\rho^2}{L^2}\right)^{1/2}-\frac{\rho}{L}\right]\right\}$$

and $\rho = |\mathbf{q}-\mathbf{q}'|$. Supposing now that the length is much greater than the core radius ($L \gg \rho$) we obtain the approximate result

$$M_{33}{}^{(AA)} = \frac{\mu}{4\pi} L \left[2 \ln 2L - 3 - 2 \int \gamma(\mathbf{q})d^2\mathbf{q} \int \gamma(\mathbf{q}')d\mathbf{q}' \ln |\mathbf{q}-\mathbf{q}'| \right]$$

$$= \frac{\mu}{2\pi} L \left(\ln \frac{2L}{r_0} - \frac{3}{2} \right). \tag{8.45}$$

Here we have defined the effective core radius r_0 by

$$\ln r_0 = \int \gamma(\mathbf{q})d^2\mathbf{q} \int \gamma(\mathbf{q}')d^2\mathbf{q}' \ln |\mathbf{q}-\mathbf{q}'|. \tag{8.46}$$

The value of r_0 depends of course on the detailed core structure, as specified by $\gamma(\mathbf{q})$, although for any reasonable structure it will be of the same order as the width ζ. As an example, the structure of equation (8.42) gives

$$\ln r_0 = \frac{1}{4\zeta^2} \int_{-\zeta}^{\zeta} dq_1 \int_{-\zeta}^{\zeta} dq_1' \ln |q_1-q_1'| = \ln(2\zeta/e^{3/2}). \tag{8.47}$$

Some care is necessary in interpreting these results. Formulae such as equation (8.34) which we used in their derivation are only applicable to closed-loop dislocations, and their meaning is not clear if used for open lines as we have done here. In fact equations (8.44) and (8.45) are only meaningful as partial contributions to the energy from two segments of dislocation which form parts of closed loops, and the total energy would contain similar contributions from all such segments and pairs of segments. Thus it would be meaningless for example to compare equation (8.44) with equation (3.39). However, when the two screws A and B have equal and opposite Burgers vectors, they can be considered as part of the same loop, formed by joining across the top and bottom with edge dislocations. Then if $L \gg \rho$ the edges make a negligible contribution to the total energy, which then must equal

$$U_{\text{tot}} = \frac{1}{2}M_{33}{}^{(AA)}b^2 + \frac{1}{2}M_{33}{}^{(BB)}b^2 - M_{33}{}^{(AB)}b^2 \approx \frac{\mu b^2}{2\pi} L \ln \frac{\rho}{r_0} \tag{8.48}$$

for $L \gg \rho$. In this case we do obtain a meaningful result which can be compared with equation (3.41). Such a comparison shows that our present definition of r_0 in terms of the structure function agrees with the earlier core radius, at least as far as energy is concerned.

B Straight edge dislocations

For the corresponding edge problem, taking Figure 8.2 again, we can choose axes so that the dislocation along the x_3-axis has Burgers vector in the

x_1-direction, in which case we need $M_{11}{}^{(AB)}$ and $M_{12}{}^{(AB)}$. From equation (8.34)

$$M_{11}{}^{(AB)} = \frac{\mu}{8\pi} \int_{\Gamma^{(A)}} \int_{\Gamma} \left(\frac{2}{1-\nu} \partial_2 \partial_2 |\mathbf{r}-\mathbf{r}'| - \partial_3 \partial_3' |\mathbf{r}-\mathbf{r}'| \right) dl_3{}^{(A)} dl_3{}^{(B)}$$

and

$$M_{12}{}^{(AB)} = \frac{-\mu}{4\pi(1-\nu)} \int_{\Gamma^{(A)}} \int_{\Gamma^{(B)}} \partial_1 \partial_2 |\mathbf{r}-\mathbf{r}'| \, dl_3{}^{(A)} dl_3{}^{(B)}.$$

The first of these is the appropriate mutual inductance for parallel Burgers vectors, and the second for perpendicular ones. Since

$$\partial_m \partial_p |\mathbf{r}-\mathbf{r}'| = [\delta_{mp}|\mathbf{r}-\mathbf{r}'|^2 - (x_m - x_m')(x_p - x_p')]/|\mathbf{r}-\mathbf{r}'|^3,$$

we obtain, after integration,

$$M_{11}{}^{(AB)} = \frac{\mu}{2\pi(1-\nu)} L \left\{ \ln\left[\frac{L}{\rho} + \left(\frac{L^2}{\rho^2} + 1 \right)^{1/2} \right] \right.$$
$$\left. - \left[\left(1 + \frac{\rho^2}{L^2} \right)^{1/2} - \frac{\rho}{L} \right] \left(\frac{1}{2}(1+\nu) + \frac{x_2{}^2}{\rho^2} \right) \right\}$$

$$\approx \frac{\mu}{2\pi(1-\nu)} L \left(\ln\frac{2L}{\rho} - \frac{1}{2}(1+\nu) - \frac{x_2{}^2}{\rho^2} \right) \quad \text{for } L \gg \rho, \qquad (8.49)$$

$$M_{12}{}^{(AB)} = \frac{\mu}{2\pi(1-\nu)} \frac{x_1 x_2}{\rho^2} [(L^2 + \rho^2)^{1/2} - \rho] \approx \frac{\mu}{2\pi(1-\nu)} L \frac{x_1 x_2}{\rho^2} \quad \text{for } L \gg \rho. \tag{8.50}$$

For a single dislocation we must again use equation (8.40), which is equivalent, in the same way as for the screw, to averaging equation (8.49) with respect to the core structure function. Thus

$$M_{11}{}^{(AA)} = \frac{\mu}{2\pi(1-\nu)} L \int \gamma(\mathbf{q}) d^2\mathbf{q}$$
$$\cdot \int \gamma(\mathbf{q}') d^2\mathbf{q}' \left(\ln\frac{2L}{|\mathbf{q}-\mathbf{q}'|} - \frac{1}{2}(1+\nu) - \frac{(q_2 - q_2')^2}{|\mathbf{q}-\mathbf{q}'|^2} \right)$$
$$= \frac{\mu}{2\pi(1-\nu)} L \left(\ln\frac{2L}{r_0} - \frac{1}{2}(1+\nu) - \kappa \right). \tag{8.51}$$

Here r_0 is again defined by equation (8.46) and

$$\kappa = \int \gamma(\mathbf{q}) d^2\mathbf{q} \int \gamma(\mathbf{q}') d^2\mathbf{q}' \frac{(q_2 - q_2')^2}{|\mathbf{q}-\mathbf{q}'|^2}.$$

κ gives a measure of the shape of the tube of dislocation: for a dislocation spread entirely in the glide direction, $\gamma(\mathbf{q}) = \beta(q_1)\delta(q_2)$ and κ is zero; for the opposite extreme of a dislocation spread only in the climb direction, $\gamma(\mathbf{q}) = \delta(q_1)\beta(q_2)$ and $\kappa = 1$; if the structure function is cylindrically symmetric about the dislocation axis, then $\kappa = 1/2$. As remarked at the end of §8.2D, we would expect the first of these alternatives to be closest to the truth in general.

Again equations (8.49)–(8.51) have a direct interpretation only when the dislocations involved form parts of closed loops. For a pair of equal and opposite edges whose lengths are much greater than their distance of separation, we have the meaningful expression

$$U_{\text{tot}} = \tfrac{1}{2}M_{11}{}^{(AA)}b^2 + \tfrac{1}{2}M_{11}{}^{(BB)}b^2 - M_{11}{}^{(AB)}b^2$$

$$= \frac{\mu b^2}{2\pi(1-\nu)}L\left(\ln\frac{\rho}{r_0} + \frac{x_2{}^2}{\rho^2} - \kappa\right). \tag{8.52}$$

If we compare this with equation (3.46), derived earlier for the hollow-core model, we see that for the edge dislocation the two definitions of the core radius do not precisely correspond. In terms of the present r_0, the effective hollow-core radius is equal to $r_0 \exp(\kappa - \tfrac{1}{2})$. This equals r_0 when the core is cylindrically symmetric, but in general is different.

c Mixed dislocation

We consider next a straight dislocation whose Burgers vector makes an angle ψ with the line direction. Taking this latter to lie again along the x_3-axis, with the Burgers vector in the x_1x_3-plane, we have an energy

$$U = \tfrac{1}{2}M_{nq}{}^{(AA)}b_n b_q$$

$$= \tfrac{1}{2}b^2(M_{11}{}^{(AA)}\sin^2\psi + 2M_{13}{}^{(AA)}\sin\psi\cos\psi + M_{33}{}^{(AA)}\cos^2\psi).$$

From equation (8.40) it follows that M_{13} vanishes for our present geometry so that U consists simply of the sum of the energies of the separate screw and edge components of the mixed dislocation. Substituting from equations (8.45) and (8.51) we thus have

$$\frac{U}{L} = \frac{\mu b^2}{4\pi}\left[\left(\frac{\sin^2\psi}{1-\nu} + \cos^2\psi\right)\left(\ln\frac{2L}{r_0} - 1\right) - \tfrac{1}{2}\cos 2\psi - \frac{\kappa}{1-\nu}\sin^2\psi\right]. \tag{8.53}$$

It can be seen that there is considerable resemblance between this formula and equation (3.21), although once again we do not expect an exact correspondence since they represent somewhat different situations. Equation (8.53) gives the partial contribution to the energy from a finite dislocation segment in an infinite medium whereas equation (3.21) gives the total energy in a finite cylinder due to an infinite dislocation line.

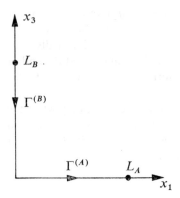

FIGURE 8.3

D *Rectangular loops*

Consider first the configuration of Figure 8.3 with a right-angled dislocation line. We shall consider the two cases of Burgers vectors in the x_3-direction (a screw-edge corner) and in the x_2-direction (an edge-edge corner). In the first case we want $M_{33}{}^{(AB)}$, which from equation (8.34) is zero, so that there is no interaction energy. In the second case we get

$$M_{22}{}^{(AB)} = -\frac{\mu}{8\pi}\frac{1+\nu}{1-\nu}\int_{L_B}^{0}\int_{0}^{L_A}\partial_1\partial_3\,|\mathbf{r}-\mathbf{r}'|\,dl_1{}^{(A)}dl_3{}^{(B)}$$

$$= \frac{\mu}{8\pi}\frac{1+\nu}{1-\nu}[(L_A+L_B)-(L_A{}^2+L_B{}^2)^{1/2}].$$

A rectangular glide-loop is defined as one lying entirely in its glide-plane (Fig. 8.4). It has a total energy consisting of the self-energies of its four sides and interaction energies of the two opposite pairs.

Combining equations (8.44), (8.45), (8.49), and (8.51) gives

$$U^{(\text{glide loop})} = \frac{\mu b^2}{2\pi}\left(L_B\ln\frac{2L_AL_B}{r_0{}^s[L_B+(L_A{}^2+L_B{}^2)^{1/2}]}\right.$$

$$\left. -\frac{2-\nu}{1-\nu}[L_A+L_B-(L_A{}^2+L_B{}^2)^{1/2}]\right)$$

$$+\frac{\mu b^2}{2\pi(1-\nu)}L_A\left(\ln\frac{2L_AL_B}{r_0{}^e[L_A+(L_A{}^2+L_B{}^2)^{1/2}]}-\kappa\right),\quad (8.54)$$

where $r_0{}^e$ and $r_0{}^s$ are the core radii for the edge and screw dislocations.

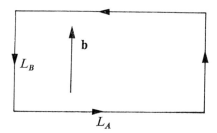

FIGURE 8.4

A rectangular prismatic loop is by definition one whose Burgers vector is perpendicular to the plane of the loop. In this case adjacent sides do contribute an interaction term. Combining this with equations (8.49) and (8.51) gives

$$U^{\text{(prismatic loop)}} = \frac{\mu b^2}{2\pi(1-\nu)}\left(L_A \ln \frac{2L_A L_B}{r_0{}^e[L_A+(L_A{}^2+L_B{}^2)^{1/2}]}\right.$$

$$+ L_B \ln \frac{2L_A L_B}{r_0{}^e[L_B+(L_A{}^2+L_B{}^2)^{1/2}]}$$

$$\left. + 2(L_A{}^2+L_B{}^2)^{1/2} - (L_A+L_B)\,(1+\kappa)\right). \quad (8.55)$$

It is interesting to examine these formulae in the case of a very elongated loop, in which limit we expect to retrieve the results obtained earlier for pairs of opposite dislocations. In this way we can obtain an estimate of the errors involved in the earlier expressions. For example, taking $L_A \gg L_B$ in equation (8.55) gives

$$U \approx \frac{\mu b^2}{2\pi(1-\nu)} L_A\left[\left(1 + \frac{L_B}{L_A}\right)\ln \frac{L_B}{r_0{}^e} + 1 - (1-\ln 2)\frac{L_B}{L_A} - \frac{3}{4}\frac{L_B{}^2}{L_A{}^2}\cdots\right],$$

which in its leading term agrees with equation (8.52) (with $L_A = L$ and $L_B = \rho$; for a prismatic pair $x_2 = \rho$). The corrections are smaller than the dominant terms by a factor of order ρ/L. Similarly taking $L_A \gg L_B$ in equation (8.54) gives equation (8.52) with $x_2 = 0$ and taking $L_B \gg L_A$ gives equation (8.48). In both cases the errors are smaller than the leading terms by a factor of order ρ/L.

E *Circular loops – prismatic case*

Consider first the general case of two coaxial circles of radii a and c on planes distance d apart. Taking general points $(a\cos\theta, a\sin\theta, 0)$ on $\Gamma^{(A)}$ and $(c\cos\phi, c\sin\phi, d)$ on $\Gamma^{(B)}$, we obtain directly from equation (8.34)

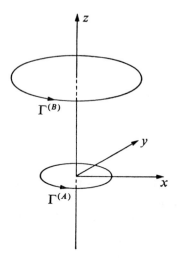

FIGURE 8.5

$$M_{33}{}^{(AB)} = \frac{\mu ac}{8\pi} \int \int \left([\sin\theta \sin\phi\, \partial_2 \partial_2 + \cos\theta \cos\phi\, \partial_1 \partial_1 \right.$$
$$+ \sin(\theta+\phi)\partial_1\partial_2]\, |\mathbf{r}-\mathbf{r}'|\, \frac{1+\nu}{1-\nu}$$
$$\left. + \cos(\theta-\phi)\,(\partial_1\partial_1+\partial_2\partial_2)\, |\mathbf{r}-\mathbf{r}'| \right) d\theta d\phi$$

$$= \frac{\mu ac}{4\pi(1-\nu)} \int \int \cos(\theta-\phi)\,(\partial_1\partial_1+\partial_2\partial_2)\, |\mathbf{r}-\mathbf{r}'|\, d\theta d\phi.$$

Here we have used $\oint \partial_k |\mathbf{r}-\mathbf{r}'| \cdot dl_k = 0$ in order to remove the mixed derivatives. Thus, changing variables to θ and $\alpha = (\theta-\phi)$ and performing the θ integration leaves

$$M_{33}{}^{(AB)} = \frac{\mu ac}{2(1-\nu)} \int_0^{2\pi} \left(\frac{1}{|\mathbf{r}-\mathbf{r}'|} + \frac{d^2}{|\mathbf{r}-\mathbf{r}'|^3} \right) \cos\alpha\, d\alpha. \tag{8.56}$$

It is then straightforward to show that

$$\int_0^{2\pi} \frac{\cos\alpha\, d\alpha}{|\mathbf{r}-\mathbf{r}'|} = \int_0^{2\pi} \frac{\cos\alpha\, d\alpha}{(a^2+c^2+d^2-2ac\cos\alpha)^{1/2}}$$
$$= 2\left(\frac{m}{ac}\right)^{1/2} \left(\frac{2-m}{m} K(m) - \frac{2}{m} E(m) \right) \tag{8.57}$$

and

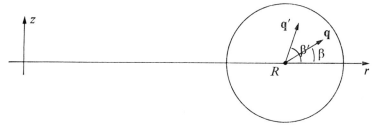

FIGURE 8.6

$$\int_0^{2\pi} \frac{\cos \alpha \, d\alpha}{|\mathbf{r}-\mathbf{r}'|^3} = \left(\frac{m}{ac}\right)^{3/2} \left(-\frac{1}{m} K(m) + \frac{2-m}{2m(1-m)} E(m)\right), \tag{8.58}$$

where $K(m)$ and $E(m)$ are the complete elliptic integrals of the first and second kinds with parameter m given by

$$m = \frac{4ac}{(a+c)^2+d^2}. \tag{8.59}$$

Combining equations (8.56)–(8.58) then gives the M_{33} component of mutual inductance.

In this case of coaxial prismatic loops, the case $a = c$ is particularly important, corresponding to successive loops gliding on the same cylinder. Then $m = 4a^2/(4a^2+d^2)$ and the mutual inductance takes the simple form

$$M_{33}^{(AB)} = \frac{\mu a}{(1-\nu)} m^{1/2} [K(m)-E(m)]. \tag{8.60}$$

These results can be used to obtain the self-energy of a single prismatic loop, by considering the two loops $\Gamma^{(A)}$ and $\Gamma^{(B)}$ to be two fibres within the toroidal core filament of the main loop. Let the central fibre have radius R and let the two fibres be at positions $\mathbf{q}$ and $\mathbf{q}'$ in the r–z plane relative to the central one (Fig. 8.6). Then

$$a = R+q \cos \beta, \qquad c = R+q' \cos \beta', \qquad d = q' \sin \beta' -q \sin \beta, \tag{8.61}$$

and for $q, q' \ll R$ we have

$$m \approx 1 - |\mathbf{q}-\mathbf{q}'|^2/4R^2. \tag{8.62}$$

The behaviour of the elliptic integrals for m near unity is

$$E(m) \sim 1 \quad \text{and} \quad K(m) \sim \ln \frac{4}{(1-m)^{1/2}} \sim \ln \frac{8R}{|\mathbf{q}-\mathbf{q}'|}. \tag{8.63}$$

Hence substituting equation (8.56) into equation (8.40) and keeping only the leading terms gives

$$M_{33}{}^{(AA)} = \frac{\mu R}{1-\nu} \int \gamma(\mathbf{q})d^2\mathbf{q} \int \gamma(\mathbf{q}')d^2\mathbf{q}' \left(K(m) - 2E(m) + \frac{d^2}{4R^2(1-m)} \right)$$

$$= \frac{\mu R}{1-\nu} \left(\ln \frac{8R}{r_0} - 1 - \kappa \right). \tag{8.64}$$

Here the definition (8.46) of r_0 is used, and κ is used in the same sense as before, which in the present notation means that

$$\kappa = \int \gamma(\mathbf{q})d^2\mathbf{q} \int \gamma(\mathbf{q}')d^2\mathbf{q}' \frac{(q'\cos\beta' - q\cos\beta)^2}{|\mathbf{q}-\mathbf{q}'|^2}. \tag{8.65}$$

F *Circular glide-loops*

With the same geometry as in Figure 8.5, for glide-loops we can take the Burgers vectors in the x_1-direction. Then from equation (8.34)

$$M_{11}{}^{(AB)} = \frac{\mu a c}{8\pi} \int\int \left[\left(\frac{1+\nu}{1-\nu} \cos\theta\cos\phi + \cos(\theta-\phi) \right) \partial_3\partial_3 |r-r'| \right. $$
$$\left. + \cos(\theta-\phi)\, \partial_2\partial_2 |r-r'| \right] d\theta d\phi,$$

with the same notation as in section E. Again transforming to θ and $\alpha = (\theta-\phi)$ as variables and performing the θ-integration gives

$$M_{11}{}^{(AB)} = \frac{\mu a c}{8\pi} \int\int \left[\left(\frac{1+\nu}{1-\nu} \cos\theta\cos(\alpha+\theta) + \cos\alpha \right) \left(\frac{1}{|r-r'|} - \frac{d^2}{|r-r'|^3} \right) \right.$$
$$\left. + \cos\alpha \left(\frac{1}{|r-r'|} - \frac{(a\sin\theta - c\sin(\alpha+\theta))^2}{|r-r'|^3} \right) \right] d\alpha d\theta$$

$$= \frac{\mu a c}{4(1-\nu)} \int_0^{2\pi} \left((2-\nu) \frac{\cos\alpha}{|r-r'|} - d^2 \frac{\cos\alpha}{|r-r'|^3} \right) d\alpha. \tag{8.66}$$

The two terms here are again given in equations (8.57) and (8.58).

The most important case this time is when the two loops are concentric on the same glide-plane ($d = 0$). Such a situation would arise when successive loops are emitted from a Frank-Read source. Then $m = 4ac/(a+c)^2$ and

$$M_{11}{}^{(AB)} = \frac{\mu(a+c)(2-\nu)}{4(1-\nu)} [(2-m)K(m) - 2E(m)]. \tag{8.67}$$

For the self-energy of a single loop, we proceed as in the last section (Fig. 8.6 and eqns. (8.61)–(8.63)) and obtain from equation (8.66)

$$M_{11}{}^{(AA)} = \frac{\mu R(2-\nu)}{2(1-\nu)} \left(\ln \frac{8R}{r_0} - 2 - \frac{\kappa}{2-\nu} \right). \tag{8.68}$$

This time, however, κ is not given by equation (8.65) since the glide-planes are now $z = $ const. The numerator in the integrand is now $(q_2' - q_2)^2 = (q' \sin \beta' - q \sin \beta)^2$.

The present results for a single glide-loop are based on the assumption that the structure function $\gamma(\mathbf{q})$ is the same at all points of the loop. Although this would be reasonable for many prismatic loops, it has dubious validity for a glide-loop where the edge/screw character of the dislocation changes around the loop. For example the Peierls-Nabarro model predicts different widths for the two dislocation types, and presumably the widths of mixed dislocations are somewhere between those for the pure types. Thus equation (8.68) is likely to be somewhat in error in practical applications. (Note that in equation (8.55) for a rectangular prismatic loop the change of character is properly accounted for.)

8.4
EQUILIBRIUM CONFIGURATIONS OF DISLOCATION LINES

A *Introductory remarks*

Often we are concerned with the problem of finding the equilibrium position of a dislocation line or of a group of lines in a medium which is subjected to certain external stresses. We have in fact already encountered a number of problems of this type in discussing the equilibrium of planar arrays of parallel straight dislocations in Chapter 5. These are of importance, for example, in situations in which slip is blocked by extended barriers, such as grain boundaries, and we are interested in the conditions under which microcracks can form or the slip can propagate through the barrier. However, not all questions of interest concern straight dislocation lines, the simplest example to the contrary being the Frank-Read source (see §1.3D). Here a dislocation segment whose ends are pinned down is forced to bow out as in Figures 1.18 and 1.19 under the action of an external shear stress until eventually a position of instability is reached. Earlier we treated this source on the basis of the line-tension model according to which a curved dislocation line is analogous to a stretched string with a certain equivalent tension T. The 'string' is acted on by a normal force per unit length of magnitude $b\sigma$ where σ is the appropriate resolved shear stress (cf. eq. (3.34)). Our main aim in the following sections will be to substantiate this model, to discuss its inadequacies, and to calculate values for T.

Before proceeding, however, we shall mention very briefly some further examples of the application of the line-tension model to complex equilibrium problems, beginning with the theory of precipitation-hardening as given initially by Mott and Nabarro[7,8] and by Orowan.[9] Many materials contain, often by design, a sizable proportion of foreign atoms which are distributed throughout the parent medium as precipitate particles. The average size and spacing of these precipitates is determined by the prior heat treatment which the material has been given. The presence of precipitate particles intersecting its glide-plane can provide a barrier to the motion of a dislocation, which differs from that of a grain boundary in being localized around the precipitates. Thus a dislocation will be pinned by precipitates at various points along its line, and will bow out in between, like a row of Frank-Read sources. The stress needed to propagate the slip through the precipitates will then roughly equal that necessary to bend the free segments into semicircular shapes, and will be inversely proportional to the distance Λ between the precipitates. (It will in fact equal $2T/b\Lambda$, using the line-tension model.)

This theory becomes inapplicable when the distance Λ becomes so small that free segments cannot be regarded as independent straight-line sources. This is most apparent in the extreme case of *solid solutions* in which the impurity particles are very small and close together, Λ being only a few lattice spacings. A dislocation line is not sufficiently flexible to be able to avoid all of the precipitates when they are as close together as this, and it will in fact adopt an almost straight configuration. Then the effects of the stress-fields of the precipitate particles, which will be randomly distributed along the dislocation line, will almost cancel one another out – they would do so exactly if the line did not deviate slightly from a straight line. Thus very little external stress is necessary in this case to drive the dislocation through the field of precipitates.

As the distance between the precipitates increases, the dislocation is able to deviate more and more from a straight line so as to take advantage of the positions of minimum energy between the particles. With this change the stress necessary to drive the dislocation over the barriers increases. The optimum condition occurs when the stresses due to the precipitates are just sufficient to bend the dislocation to a radius of curvature equal to $\Lambda/2$. Then the dislocation can avoid almost all of the precipitates, and in order to propagate the slip each of the barriers has to be overcome in full by the external stress. But as Λ is further increased, the dislocation becomes able to loop through between the precipitates, as we discussed first. Thus the increase in the yield stress produced by precipitates reaches a maximum at a certain intermediate value of the precipitate spacing (typically of the order of 50 lattice spacings).

A somewhat similar analysis can be applied to estimate the hardening effect of immobile dislocations in a material. All crystalline materials contain dislocations, and in the unstrained state these will tend to arrange themselves in a characteristic network so as to minimize their strain energy. Such networks will intersect the slip-plane of any mobile dislocation in a number of points, thus presenting barriers to the slip through the action of their stress-fields on that dislocation. The forces of interaction between the mobile dislocation and any 'tree' in the 'forest' of intersecting network dislocations may be attractive or repulsive, but in any case they tend to inhibit the progress of slip.

Using Λ again to denote the average spacing within the slip-plane of the intersections with trees, then the average stresses at points mid-way between the trees will be of order $\sigma_i \sim \mu b/2\pi\Lambda$. These will be capable of bending a mobile dislocation to a radius of curvature equal to $T/b\sigma_i \sim 2\pi\Lambda$, since, as we shall show later, $T \sim \mu b^2$. Such a curvature is sufficient to enable the dislocation to avoid most of the places of high energy and to lie along the floor of the energy valleys. The stress needed to move a length Λ of dislocation line from one floor to the next, and hence to propagate the slip, will be of the same order as σ_i, that is $\sim \mu b/2\pi\Lambda$. Thus, as expected, the yield stress increases with the density of network dislocations.

A more detailed investigation of network hardening has been given by Saada[10] and Carrington, Hale, and McLean.[11] For a complete discussion the reader is referred, for example, to the book by Friedel.[12]

The line-tension analogy can also be used to determine the equilibrium configurations of intersecting dislocations, such as the node in Figure 1.15(b) (see Probs. 8.17 and 8.18). From this it is possible to investigate the configurations of Frank networks.[13] For example if the analogy were exact, with all dislocations having the same line tension, then clearly one possible configuration would be a regular plane hexagonal net. In actual fact this is not exactly correct, because of material anisotropy, and because even in an isotropic material the line tension depends on the relative orientation of the line direction and the Burgers vector (not forgetting also that there are intrinsic errors in the line-tension model). Furthermore in practice the network would be a more general three-dimensional one, but nevertheless similar considerations would apply.

B *Line-tension model*[14]

Consider a dislocation line segment, pinned at its two end-points A and B, and lying entirely in its own glide-plane. We take this plane to be the xz-plane with the x-axis parallel to the Burgers vector direction. A parameter t is chosen along the segment in such a way that $t = 0$ corresponds to A and $t = 1$ to B. We denote by $x(t)$ and $z(t)$ the coordinates of a general point on the segment and by $\theta(t)$ the angle between the tangent at t and the Burgers vector (i.e. the x-axis). Thus $\tan \theta(t) = z'(t)/x'(t)$. The basic assumption of the line-tension model is that the contribution to the energy from the dislocation segment has the form

$$U^d = \int_A^B U[\theta(t)]\, ds. \tag{8.69}$$

Here s is the arc-length along the segment, so that $U[\theta(t)]$ can be described as the energy per unit length at t. It is assumed to depend on the angle θ at t and on nothing else. Comparing with the complete expression (8.40), we see that the model is ignoring at least some of the coupling between different parts of the segment.

If the dislocation is in equilibrium, then the change δU_{tot} in the total energy of the system that would result from an arbitrary small change in the dislocation position must be zero to first order. If we suppose that in such a virtual change the point t is displaced by an amount $\mathbf{\delta r}(t) = (\delta x(t), \delta z(t))$, then comparing with equation (3.32) gives the equilibrium condition

$$\delta U^d = \int_A^B \mathbf{f}^0(t) \cdot \mathbf{\delta r}(t) ds, \tag{8.70}$$

where $\mathbf{f}^0(t) = \mathbf{b} \cdot \mathbf{\sigma}^0 \wedge \mathbf{\xi}$. In equation (3.34) this quantity was interpreted as the force per unit length on the dislocation line due to an external stress-field $\mathbf{\sigma}^0$. In view of our remarks in §3.3c, $\mathbf{\sigma}^0$ can include the effects of other dislocations, and in particular of the rest of the loop or loops to which the segment AB is attached at its end-points. $\mathbf{\xi}$ is a unit vector, tangent to the segment at t. Since we are not considering climb motions, we can ignore the component of $\mathbf{f}^0$ out of the slip-plane, with the result that, effectively,

$$\mathbf{f}^0(t) = b\sigma_{xy}^0(\sin\theta(t), 0, -\cos\theta(t)). \tag{8.71}$$

Now, from equation (8.69),

$$\delta U^d = \int_0^1 \{U'(\theta)\delta\theta(t)\,[x'(t)^2+z'(t)^2]^{1/2} + U(\theta)\delta[x'(t)^2+z'(t)^2]^{1/2}\}dt$$

$$= \int_0^1 \{[U(\theta)\cos\theta - U'(\theta)\sin\theta]\delta x' + [U(\theta)\sin\theta + U'(\theta)\cos\theta]\delta z'\}\,dt.$$

Integrating by parts and using the fact that δx and δz vanish at $t = 0$ and 1, since the end-points are pinned, we obtain

$$\delta U^d = \int_0^1 [U(\theta)+U''(\theta)]\,[\sin\theta\,\delta x - \cos\theta\,\delta z]\,(d\theta/dt)dt. \tag{8.72}$$

If we now substitute this into equation (8.70), and use the fact that the variation $(\delta x, \delta z)$ is an arbitrary function of t, we can set the integrand equal to zero, giving

$$\mathbf{f}^0(t) \cdot \mathbf{\delta r}(t) = [U(\theta)+U''(\theta)]\,[\sin\theta\,\delta x - \cos\theta\,\delta z]\,(d\theta/ds).$$

Finally, substituting from equation (8.71) gives the condition

$$b\sigma_{xy}^0 = [U(\theta)+U''(\theta)]\,(d\theta/ds). \tag{8.73}$$

This equation is identical with the equilibrium equation of a stretched string subject to a transverse force of magnitude $b\sigma_{xy}^0$ when the tension in the string is

$$T = U(\theta)+U''(\theta). \tag{8.74}$$

Thus we arrive at the stretched-string analogy. There is, however, one difference, namely that T depends on the orientation of the dislocation line, and so in particular varies from one point to another on the segment.

In evaluating T, it is usual to take $U(\theta)$ to be the energy per unit length of a long mixed dislocation, given for example by the right-hand side of equation (8.53) with ψ replaced by θ. From this, equation (8.74) then gives

$$T = \frac{\mu b^2}{4\pi(1-\nu)}\left[[(1+\nu)\cos^2\theta + (1+2\nu)\sin^2\theta]\left(\ln\frac{2L}{r_0} - 1\right)\right.$$
$$\left. + \frac{3}{2}(1-\nu)\cos 2\theta - \kappa(2\cos^2\theta - \sin^2\theta)\right]. \quad (8.75)$$

This result illustrates a difficulty inherent in the line-tension model, in that a quantity L has appeared which has no immediate interpretation. It is the length of straight dislocation which is equivalent to our present curved line in the sense of having the same energy per unit length. If the segment AB is almost straight, as would be the case for small shear stresses, then L should approximately equal AB itself. If the segment is highly curved, this may not be true, and the radius of curvature would seem a more reasonable value for L, but the model is ambiguous on this point.

For almost straight dislocations, where the model appears to be satisfactory, an additional simplification can be made in that T is roughly constant along the segment. Thus, for example, under a uniform shear stress σ_{xy}, according to equation (8.73), the line would be approximately a circular arc of radius $T/b\sigma_{xy}$. We used this result in §1.4 when discussing the Frank-Read source.

c Tension of a circular loop

We can assess the line-tension model to some extent by examining the changes in energy of particular dislocation configurations using the full result equation (8.40).[15] The simplest geometry to consider is the circular glide-loop. If the radius is R, the strain energy is $\frac{1}{2}M_{11}b^2$ with M_{11} given in equation (8.68). When R changes to $R+\delta R$, the change in energy is then given by

$$\frac{\delta U}{2\pi\delta R} = \frac{\mu b^2(2-\nu)}{8\pi(1-\nu)}\left(\ln\frac{8R}{r_0} - 1 - \frac{\kappa}{2-\nu}\right). \quad (8.76)$$

If the dislocation were a stretched string of tension T, the corresponding energy change would be $\delta U = T.2\pi\delta R$. Hence the line tension of the loop is given by the expression on the right-hand side of equation (8.76).

Comparing this with equation (8.75) we see that the two results are very similar. If we set $L = 4R$ in the earlier equation and average over θ, the two agree exactly. Thus the circular loop provides some support for the line-tension model with T given by equation (8.75).

Unfortunately the same result does not follow if an effective line tension is calculated by considering geometries other than the circular loop.[15,16] For example the change in energy divided by change in length can be calculated for an expanding rectangular or hexagonal loop, rather than a circular one. Or, alternatively, the change in energy can be found when an initially straight dislocation changes its shape by bowing out in the middle in some particular way. In each of these cases a somewhat different value of the effective line tension is found: the logarithmic term in equation (8.75) is reproduced in every case, but the other terms change from one geometry to another. Thus the effective line tension depends on the over-all configuration of the dislocation line. For $L \gg r_0$, the logarithmic term is dominant, so that the line-tension model will in such cases be approximately correct. However, in practical applications, the additional terms will be by no means insignificant, perhaps representing a 20 per cent contribution to T.

D *Equilibrium of a line segment*

In this section, we shall examine more closely the equations of equilibrium of a dislocation line segment deforming in its glide-plane under the influence of an external stress-field or the stress-field of other dislocations. Rather than starting with the simplified dislocation strain energy of equation (8.69), we shall use an energy more directly related to the full expression of equation (8.40), and thus we shall be led to something more than a simple line-tension model.

We consider the same geometry as in §B, with a general point on the dislocation segment AB being $\mathbf{l}(t) = [x(t), 0, z(t)]$ $(0 \le t \le 1)$ and the Burgers vector being in the x-direction. The dislocation energy is then $\frac{1}{2}M_{11}b^2$, with M_{11} being given by equation (8.40). For our present geometry, this takes the approximate form

$$U^d = \frac{\mu b^2}{16\pi} \int\int [Fx'(t)x'(t') + Gz'(t)z'(t')] \, dt \, dt', \tag{8.77}$$

where

$$F = (2X^2 + Z^2)/(X^2 + Z^2)^{3/2},$$
$$G = [(3-\nu)X^2 + 2Z^2]/(1-\nu)(X^2 + Z^2)^{3/2}, \tag{8.78}$$

$$X = x(t) - x(t'), \qquad Z = z(t) - z(t'). \tag{8.79}$$

The approximation we have made in writing equation (8.77) is to drop the core variables $\mathbf{q}$ and $\mathbf{q}'$. In order to have a convergent integral, we shall adopt the procedure of omitting a small length of dislocation arc from the neighbourhood of the point $t' = t$ when carrying out the first of the two integrations. The length of the omitted piece will be taken as equal to the core radius. This method is actually often used in evaluating energies in cases such as those considered in §8.3, and leads to identical results to the ones we obtained there. It is, however, a rather unsatisfactory expedient, even in energy calculations, but particularly in the force calculations we are about to give. It is by no means obvious that the dislocation core does not give a rigidity to the dislocation line which will not show up in the present analysis. There would be considerable advantage in extending the work of this and the following sections to include the core structure, but although it is possible to obtain the equations of equilibrium with this generalization the evaluation of the additional effects presents great algebraic difficulties, even with the simplest geometry. Thus our results, while lending some support to the line-tension model, will not be conclusive.

We have omitted the limits from the t and t' integrations in equation (8.77) since, as we have seen on several occasions in §8.3, equation (8.40) can only be applied to closed loops, and if these two integrations were to be restricted to the interval (0, 1), equation (8.77) would be incorrect. We must include integrations over the other part of the closed loop of which AB forms one segment. Formally we can do this by using the same parametrization, $[x(t), 0, z(t)]$, for points on this other part, with t running, say, between -1 and 0. The value $t = -1$ corresponds again to B, as does $t = +1$. Then if we take the t and t' integrations in equation (8.77) over $(-1, 1)$ we obtain the correct contribution to the energy from the whole loop.

Our procedure will be very similar to that in §B. We shall consider a small change in the dislocation position, and require that in equilibrium the energies balance to first order, as in equation (8.70). Since we are interested only in the equilibrium of AB, we allow the changes $\delta x(t)$ and $\delta z(t)$ to be non-zero only for $0 < t < 1$. They must vanish at $t = 0$ and $t = 1$ since A and B are supposed to be held fixed. In evaluating the variation δU^d, because of the symmetry between t and t' in equation (8.77) we need only count variations at one of these values of the parameter, and afterwards multiply by 2. Thus

$$\delta U^d = \frac{\mu b^2}{8\pi} \int_0^1 dt \int_{-1}^1 dt' \left[\delta x'(t) x'(t') F + \delta z'(t) z'(t') G \right.$$

$$+ \delta x(t) \left(x'(t) x'(t') \frac{\partial F}{\partial X} + z'(t) z'(t') \frac{\partial G}{\partial X} \right)$$

$$\left. + \delta z(t) \left(x'(t) x'(t') \frac{\partial F}{\partial Z} + z'(t) z'(t') \frac{\partial G}{\partial Z} \right) \right].$$

The first two terms can be integrated by parts with respect to t with no resulting contributions from the fixed end-points. The remaining terms can be simplified using

$$\frac{dF}{dt} = \frac{\partial F}{\partial X} x'(t) + \frac{\partial F}{\partial Z} z'(t), \qquad \frac{dG}{dt} = \frac{\partial G}{\partial X} x'(t) + \frac{\partial G}{\partial Z} z'(t).$$

Then δU^d becomes

$$\delta U^d = \frac{\mu b^2}{8\pi} \int_0^1 [z'(t)\delta x(t) - x'(t)\delta z(t)] \, dt \int_{-1}^1 \left(-x'(t) \frac{\partial F}{\partial Z} + z'(t) \frac{\partial G}{\partial X} \right) dt'.$$

In equilibrium, by equations (8.70) and (8.71) this variation is required to be equal to

$$\int_0^1 b\sigma_{xy}(t) \, [z'(t)\delta x(t) - x'(t)\delta z(t)] dt,$$

where $\sigma_{xy}(t)$ denotes the shear stress component at the point t on AB. It includes the stresses of other dislocation loops and those produced by external tractions, but does *not* include any stresses produced by the loop under consideration (unlike $\sigma_{xy}{}^0$ in eq. (8.71) in which it is usual to include the stresses of the segment $-1 \le t \le 0$ which closes AB into a loop). The changes $\delta x(t)$ and $\delta z(t)$ are arbitrary functions of t, so that we can compare the integrands in these two expressions, giving the equilibrium equations

$$\sigma_{xy}(t) = \frac{\mu b}{8\pi} \int_{-1}^1 \left(-x'(t') \frac{\partial F}{\partial Z} + z'(t') \frac{\partial G}{\partial X} \right) dt'$$

$$= \frac{\mu b}{8\pi} \oint \left(\frac{Z(4X^2 + Z^2)}{(X^2 + Z^2)^{5/2}} \, dx' - \frac{X[(3-\nu)X^2 + 2\nu Z^2]}{(1-\nu)\,(X^2 + Z^2)^{5/2}} \, dz' \right). \qquad (8.80)$$

In this last step, we have substituted from equations (8.78).

For a given stress-field, equation (8.80) provides an integral equation for the equilibrium curve $\{x(t), z(t)\}$ of the dislocation line. Although it is the equation we are seeking, it has the disadvantage of treating the complementary segment to AB in a special way, differently from all the other dislocation segments in the medium whose effect is to contribute to the shear stress through equation (8.30). This is particularly unfortunate when the segment AB forms part of a complex tangle of dislocation lines, and cannot be regarded as part of a unique loop. Therefore we shall now rewrite equation (8.80) in a form in which all dislocation segments are treated on the same footing.

The stress components of the dislocation loop can be obtained from equation (8.30). At a point $(x, 0, z)$ in the same plane as the loop itself, we have

$$\sigma_{xy}^{\text{loop}}(x, 0, z) = \frac{\mu b}{4\pi} \oint \left(- \frac{z - z(t')}{\{[x - x(t')]^2 + [z - z(t')]^2\}^{3/2}} \, dx' \right.$$
$$\left. + \frac{1}{1 - \nu} \frac{x - x(t')}{\{[x - x(t')]^2 + [z - z(t')]^2\}^{3/2}} \, dz' \right).$$

If we let $(x, 0, z)$ tend to the point (t) on the loop, this stress of course diverges. But let us adopt the same procedure used for the energy integrals, of omitting a small arc-length to a distance of $r_0/2$ on either side of the limit point. Then the limiting value of the stress is

$$\sigma_{xy}^{\text{loop}}(t) = \frac{\mu b}{4\pi} \oint \left(- \frac{Z}{(X^2 + Z^2)^{3/2}} \, dx' + \frac{1}{1 - \nu} \frac{X}{(X^2 + Z^2)^{3/2}} \, dz' \right), \quad (8.81)$$

which is convergent. Combining this with equation (8.80) then gives

$$\sigma_{xy}(t) + \sigma_{xy}^{\text{loop}}(t) = \frac{\mu b}{8\pi} \oint \left(\frac{(2X^2 - Z^2)Z}{(X^2 + Z^2)^{5/2}} \, dx' + \frac{(2Z^2 - X^2)X}{(X^2 + Z^2)^{5/2}} \, dz' \right)$$
$$= \frac{\mu b}{8\pi} \oint d\left(\frac{XZ}{(X^2 + Z^2)^{3/2}} \right)$$
$$= (\mu b / 8\pi) [x'' z'(2x'^2 - z'^2) + z'' x'(2z'^2 - x'^2)] / [x'^2 + z'^2]^{5/2}.$$

In evaluating the last integral, we have taken the difference between $XZ/(X^2 + Z^2)^{3/2}$ at the points $t \pm \delta t$, corresponding to the limits of the omitted segment. Now the omitted length was chosen to have equal *arc-lengths* of $r_0/2$ on each side of $x(t)$, and therefore the parameter t must be proportional to the arc-length. Then $x' = k \cos \theta$, $z' = k \sin \theta$, $x'' = k^2 \sin \theta / R$, and $z'' = k^2 \cos \theta / R$, where θ is the angle between the tangent to the curve at t and the x-axis, R is the radius of curvature at t, and k is the constant of proportionality between t and the arc-length. Thus we must have, for equilibrium,

$$\sigma_{xy}(t) + \sigma_{xy}^{\text{loop}}(t) = \mu b \cos 2\theta / R. \quad (8.82)$$

This condition can be expressed in the form of an integral equation for the location of the segment AB by splitting equation (8.81) into two parts, one from AB itself and a second from the remainder of the loop of which AB is part. If we use $\sigma_{xy}^{0}(t)$ in the same capacity as in §B, that is as the shear stress at the point t on AB arising from all sources except AB itself, then the equilibrium equation becomes

$$\sigma_{xy}^{0}(t) = \frac{\mu b}{4\pi} \int_A^B \left(\frac{Z}{(X^2 + Z^2)^{3/2}} \, dx' - \frac{1}{1 - \nu} \frac{X}{(X^2 + Z^2)^{3/2}} \, dz' \right) - \frac{\mu b \cos 2\theta}{8\pi R}. \quad (8.83)$$

Again a small element centred at the point t is to be omitted from the integration. In using this equation it must be remembered that the contributions to $\sigma_{xy}{}^0$ from any open segments must be calculated using equation (8.30) and not using any other equation which may be equivalent to it for closed loops.

The integral equations (8.80) or (8.83) are quite complex; besides the obvious non-linearity of the integral terms, an additional complication arises from the implicit dependence of the left-hand side on the solution. Thus it appears unlikely that anything approaching a general solution can be found. In the following sections we shall use an inverse method to investigate two problems. We shall assume the existence of certain dislocation configurations – in fact a straight segment and a circular arc – and use equation (8.83) to calculate the shear stress necessary to maintain equilibrium. Our main aim will be to assess the accuracy with which the line-tension model provides an equivalent solution in these cases.

E *Straight-line segment*

As a first example we consider the straight-line segment of Figure 8.7 making an angle θ with the x-axis. We denote the total length by L, and take the origin at the centre of the segment. The parameter t we take as the distance along the segment, so that

$$x(t) = t \cos \theta, \qquad z(t) = t \sin \theta,$$

$$X = (t-t') \cos \theta, \qquad Z = (t-t') \sin \theta.$$

Substituting these values in equation (8.83) we then have the equilibrium stress

$$\sigma_{xy}{}^0(t) = \frac{\nu\mu b}{4\pi(1-\nu)} \sin \theta \left[\int_{-L/2}^{t-r_0/2} + \int_{t+r_0/2}^{L/2} \right] \frac{(t'-t)dt'}{|t'-t|^3}$$

$$= -\frac{\nu\mu b}{4\pi(1-\nu)} \sin 2\theta \frac{t}{L^2/4-t^2}. \tag{8.84}$$

Thus we see that, apart from the special cases of pure edge and screw dislocations, a shear stress is necessary in order to hold a finite dislocation segment straight. Consequently, if this stress is not applied, the dislocation will distort and, because of the oddness of $\sigma_{xy}{}^0$ as a function of t, will adopt and S-shaped configuration. This is certainly a departure from the line-tension analogy, although probably not a very serious one in practice. For taking, say, $t = L/4$ the equilibrium stress is roughly $\sigma_{xy}{}^0/\mu \sim 2 \times 10^{-2} b/L$, and this is quite small for typical segment lengths ($b/L \sim 10^{-3}$ to 10^{-4}). It is in fact of the same order as the values of the Peierls resistance which we calculated in Chapter 4, indicating that the S-distortion would be at most a

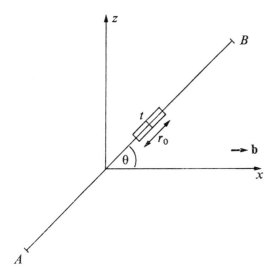

FIGURE 8.7

few lattice spacings. Furthermore, it will also be of the same order of magnitude as the shear stress contributions of the segments which are attached to AB at its ends. Thus the actual configuration of an initially straight-line segment will be influenced as much by these other neighbouring segments as by its own 'self-stress' as given by equation (8.84). In this sense the straight segment is unlike a curved segment which, as we shall see in the next section, requires, in addition to stresses similar to that in equation (8.84), a shear stress which depends on the local curvature at t.

F *Circular dislocation segment*

We wish now to examine the effect of curvature of the dislocation segment on the shear stress necessary to maintain it in equilibrium. The simplest configuration we can consider is a circular arc. To the extent that a general curved segment can be locally approximated by a circular arc, we expect our results to be indicative of more general ones.

We consider the geometry of Figure 8.8, with the origin at the centre of the arc under investigation, and take as parameter t the usual polar angle α. Then in equation (8.83) $\theta = \alpha + \pi/2$ is the angle between the tangent to AB and the x-axis. We have in equation (8.83)

$$X = R(\cos\alpha - \cos\alpha'), \qquad Z = R(\sin\alpha - \sin\alpha'),$$

so that

$$\sigma_{xy}{}^0(\alpha) = -\frac{\mu b}{32\pi R} \int_A^B \Big([\sin\alpha\,(1-\cos\phi) - \cos\alpha\,\sin\phi]\sin(\alpha+\phi)$$

$$+ \frac{1}{1-\nu}[\cos\alpha\,(1-\cos\phi) \quad +\sin\alpha\,\sin\phi]\cos(\alpha+\phi)\Big)\frac{d\phi}{|\sin^3\phi/2|}$$

$$- \frac{\mu b\cos 2\theta}{R}.$$

Here $\phi = \alpha' - \alpha$, and the integration runs over the total extent of the arc, from $-\phi_1$ to ϕ_2, excluding the interval $|\phi| < r_0/2R$. Carrying out the integration and replacing α by $\theta - \pi/2$, we obtain

$$\sigma_{xy}{}^0 = \frac{\mu b}{4\pi(1-\nu)R}\bigg\{[(1-2\nu)\sin^2\theta + (1+\nu)\cos^2\theta]$$

$$\cdot\ln\left[\frac{8R}{r_0}\left(\tan\frac{\phi_1}{4}\tan\frac{\phi_2}{4}\right)^{1/2}\right]$$

$$+\cos 2\theta\left[\frac{1}{2}(1+3\nu) - \nu\left(\cos\frac{\phi_1}{2} + \cos\frac{\phi_2}{2}\right)\right]$$

$$+\frac{1}{2}\nu\sin 2\theta\left(\sin\frac{\phi_1}{2} - \sin\frac{\phi_2}{2}\right)\left(\operatorname{cosec}\frac{\phi_1}{2}\operatorname{cosec}\frac{\phi_2}{2} - 4\right)\bigg\}. \quad (8.85)$$

Given this equilibrium shear stress, we can define an effective line tension T at the point θ by setting the transverse force $b\sigma_{xy}{}^0$ equal to T/R. The result bears considerable resemblance to that in equation (8.75). However, T is seen to depend not only on the orientation θ but also on the position of the

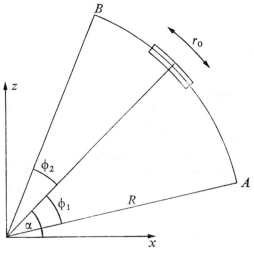

FIGURE 8.8

point relative to the ends of the arc. This is especially apparent as the point θ approaches either end of the arc, when $\sigma_{xy}{}^0$ becomes unbounded. The actual unboundedness is a result of our having ignored the core structure, but even had we kept this in, the equilibrium stress would still become very large at the ends. These large stresses are supplied by the dislocation segments attached to our particular segment at A and B. This feature of equation (8.85) illustrates the fact that it is not profitable to consider the equilibrium of dislocation segments in isolation from the neighbouring segments, at least as far as the end regions are concerned. For the central regions, however, equation (8.85) does provide a useful result.

In equation (8.85), the logarithmic term dominates the remaining terms, except near the ends of the segment. Near the centre of the arc, this term is only slowly varying, and T is given approximately by

$$T \approx \frac{\mu b^2}{4\pi(1-\nu)} [(1+\nu) \cos^2 \theta + (1-2\nu) \sin^2 \theta] \ln \left(\frac{8R}{r_0} \tan \frac{\phi}{4} \right), \qquad (8.86)$$

where $2\phi = \phi_1 + \phi_2$ is the angular extent of the arc. This is identical with the logarithmic term in equation (8.75) provided we identify $L = 4R \tan \frac{1}{4}\phi \approx R\phi$, that is approximately half the arc-length.

To summarize our findings, we can conclude that the equilibrium equation (8.83) does not lead exactly to the line-tension model, either for a straight segment or for a circular segment. The equilibrium stress at a particular element of a dislocation line depends not only on the curvature and orientation of that element, but on the whole configuration. Near the ends of a segment of dislocation, use of the line-tension model becomes particularly dubious – in fact to discuss the end regions correctly we need to consider the neighbouring dislocation segments as well. Nevertheless, the dominant term in the stress for regions near the centre of the segment is correctly given by the line-tension model. Thus we can expect that model to provide a first approximation to a problem of dislocation equilibrium, at least for regions away from the ends of segments and for segments whose curvature is not too great. In such cases, the inaccuracy introduced by neglecting the additional terms in equation (8.85) will be of the order of 20 per cent. In addition, we have been able to remove the ambiguous quantity L which appears in the line-tension model. However, as mentioned in §D, these results are predicated on a particular treatment of the dislocation core, and further analysis is needed before it can be regarded as substantiated.

PROBLEMS

1 Write the following in components and establish its validity:

$$\text{inc inc } \mathbf{T} = \nabla\nabla(\nabla\cdot\mathbf{T}\cdot\nabla) - \nabla[\nabla\cdot(\nabla^2\mathbf{T})] - [(\nabla^2\mathbf{T})\cdot\nabla]\nabla + \nabla^4\mathbf{T}.$$

2 Prove equation (8.18) [Hint: use the identity

$$\varepsilon_{ijk}\varepsilon_{lmn} = \begin{vmatrix} \delta_{il} & \delta_{im} & \delta_{in} \\ \delta_{jl} & \delta_{jm} & \delta_{jn} \\ \delta_{kl} & \delta_{km} & \delta_{kn} \end{vmatrix}$$

and expand the determinant.]

3 Obtain the special forms of the representation (8.8) when

(a) $\mathbf{\nabla}\cdot\mathbf{T} = \mathbf{0}$ or (b) $\mathbf{\nabla}\wedge\mathbf{T} = \mathbf{0}$ or (c) $\mathbf{\nabla}\wedge\mathbf{T} = \mathbf{0}$ and $\mathbf{T} = \mathbf{T}^T$.

4 Generalize equations (8.8) and (8.9) to third-order tensors (perhaps inventing your own notation, since the one we have used becomes ambiguous when the tensor has three indices). What replaces equations (8.13) in this case? [Note: the Green's function for the operator ∇^6 is $-|\mathbf{r}-\mathbf{r}'|^3/96\pi$.]

5 Using Stokes' theorem show that the integral in equation (8.17) is path-independent if and only if inc $\mathbf{T} = \mathbf{0}$. Verify that def $\mathbf{v} = \mathbf{T}$.

6 Use Stokes' theorem to obtain from equation (8.30) an expression, for the stress-field of a dislocation loop as an integral over any surface spanning the loop. The result is

$$\sigma_{il}(\mathbf{r}) = \frac{\mu}{4\pi}\int\left(-\frac{1}{2}b_n(\varepsilon_{lmn}\varepsilon_{ipq}+\varepsilon_{imn}\varepsilon_{lpq})\partial_{mq}\nabla^2\,|\mathbf{r}-\mathbf{r}'|\right.$$
$$\left.+\frac{1}{1-\nu}\,(\delta_{il}b_n\partial_{np}\nabla^2-b_n\partial_{nipl}-\delta_{il}b_p\nabla^4+b_p\partial_{il}\nabla^2)\,|\mathbf{r}-\mathbf{r}'|\right)n_p dS'.$$

Using the identity in Problem 2 obtain an equivalent result.

7 Obtain the stress-field of an infinitesimal planar glide-loop using the results of the previous question.[17,18] Take the loop to be in the plane $y = 0$ with its centre at the origin and Burgers vector in the x-direction. In particular the shear stress in the glide-plane itself, at the point $(x, 0, z)$, has one component

$$\sigma_{xy}(x, 0, z) = -[\mu b dS/4\pi(1-\nu)]\,[(1+\nu)x^2+(1-2\nu)z^2]/(x^2+z^2)^{5/2},$$

where dS is the area of the loop. Show that all stress components are of order r^{-3} at large distances from the centre of the loop. [These results for an infinitesimal loop also give the asymptotic stresses for a finite loop which is plane, or approximately plane, at distances much greater than the loop diameter.]

8 In a similar way to the previous question, derive the stress components of an infinitesimal prismatic loop. Taking the loop to be positioned as before but with Burgers vector in the y-direction, show that in particular

$$\sigma_{yy}(x, 0, z) = -[\mu b dS/4\pi(1-\nu)]/(x^2+z^2)^{3/2}.$$

9 Use equation (8.30) to calculate the stress-fields of infinite straight dislocations of edge and screw type, thus retrieving equations (3.2) and (3.19).

10 In the solution (8.27) suppose that the dislocation density tensor $a_{ni}(\mathbf{r}') = a_{ni}(x', y')$ is independent of the third coordinate. Show that

$$\partial_{pq}\chi_{ij}(x, y) = \frac{1}{4\pi}\left[\varepsilon_{Jmn}\int a_{ni}(x', y')\left(-\frac{\delta_{mp}X_q+\delta_{mq}X_p+\delta_{pq}X_m}{R^2}\right.\right.$$
$$\left.\left.+\frac{2X_mX_pX_q}{R^4}\right)dx'dy'\right]^{(s)}$$

provided $p \neq 3$ and $q \neq 3$ and with the m summation omitting the value $m = 3$; $X_1 = x-x'$, $X_2 = y-y'$, and $R^2 = X_1^2+X_2^2$.

11 A prismatic dislocation loop has the shape of a regular hexagon of side L.

Show that its energy is

$$U = [\mu b^2/4\pi(1-\nu)](6L)[\ln(L/r_0)+A],$$

where A is a constant of order unity.[19]

12 In the previous question, suppose that the loop is a glide-loop with Burgers vector parallel to one of its sides. Show that the expression for U is multiplied by $(2-\nu)/2$, with A being different. Use this result to calculate an over-all line tension for the hexagonal loop analogous to equation (8.76) for the circular loop.

13 Obtain the value for the over-all line tension of a square glide-loop, using the energy in equation (8.54). Compare the result with equation (8.76).

14 Equation (8.60) gives the interaction of two coaxial circular prismatic dislocation loops. Use this result to find the energy of a cylindrical surface dislocation whose Burgers vector is uniformly distributed along the length of a cylindrical surface of radius a and length L. The axis of the cylinder is parallel to the Burgers vector. The energy is[1,3]

$$\frac{4\mu b^2 a^3}{1-\nu}\left(\frac{E(m_0)}{\sqrt{m_0}}-1\right)$$

where the parameter m_0 is $4a^2/(4a^2+L^2)$. Recover equation (8.64) (with $\kappa = 0$) when $L \ll a$. [Note the result (8.47) for the core radius of a uniform flat dislocation.] Obtain an approximate value of the energy when $L \gg a$.

15 The dislocation line in Figure 8.9 has a right-angle kink of depth a. When the Burgers vector is along the x-direction, show that the excess of energy over a straight dislocation of the same length is equal to

$$\frac{\mu b^2}{4\pi}a\left(\ln\frac{2a}{r_0}-\frac{2-\nu}{1-\nu}\right)$$

when $L \gg a$. Obtain corresponding expressions for Burgers vectors in the z and y directions. [Note: such steps are traditionally called *jogs* when they do not lie in the slip-plane of the main dislocation line. Kinks can be removed by an appropriate amount of relative slip of the two main segments, but jogs cannot. Kinks and jogs are created when two non-parallel dislocations cross one another.[20]]

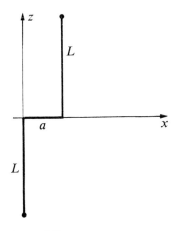

FIGURE 8.9

16 The dislocation line in Figure 8.10 has two kinks a distance r apart. If the Burgers vector is in the x-direction, show that the energy contains an additional term

$$[\mu b^2(1-2\nu)/8\pi(1-\nu)]\,(a_1 a_2/r)$$

over and above that of the two individual kinks. Hence the kinks will tend to repel one another, unless a_1 and a_2 have opposite signs in which case they attract one another. Obtain similar expressions for other directions of Burgers vector.

17 Using the line-tension model, show that the triple node in Figure 8.11, in which the three dislocations lie in a common glide-plane, is in equilibrium if

$$T_1/\sin \alpha_1 = T_2/\sin \alpha_2 = T_3/\sin \alpha_3.$$

Here T_1, T_2, T_3 are the three line tensions. Discuss the case when the three dislocations have different glide-planes.

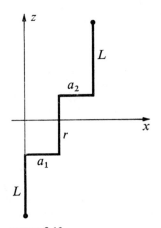

FIGURE 8.10

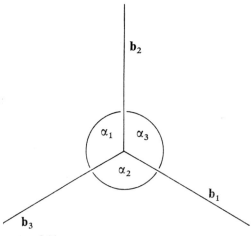

FIGURE 8.11

18 Figure 8.12(a) shows two dislocation segments intersecting at their mid-points, making an angle α with one another. Show that the configuration in Figure 8.12(b) in which the dislocations have reacted to form a short segment of Burgers vector $\mathbf{b}_2 - \mathbf{b}_1$ has lower energy than the initial configuration if

$$\sin \alpha/2 > (\mathbf{b}_2 - \mathbf{b}_1)^2 / (b_1{}^2 + b_2{}^2).$$

[Assume that any dislocation with Burgers vector $\mathbf{b}$ has a line tension equal to $\mu \mathbf{b}^2$, independent of orientation.]

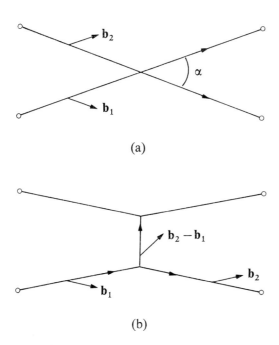

(a)

(b)

FIGURE 8.12

REFERENCES

1 Kröner, E., *Kontinuumstheorie der Versetzungen und Eigenspannungen* (Berlin: Springer, 1958)
2 Kröner, E., Arch. Rat. Mech. Anal. **4** (1960), 273
3 de Wit, R., Solid State Phys. **10** (1960), 247
4 Abraham, M., and Becker, R., *Electricity and magnetism* (London: Blackie, 1950), chap 2
5 Peach, M.O., and Koehler, J.S., Phys. Rev. **80** (1950), 436
6 Blin, J., Acta Metall. **3** (1955), 199
7 Mott, N.F., and Nabarro, F.R.N., *Report on strength of solids* (London: Physical Society, 1948), p. 1
8 Mott, N.F., Phil. Mag. **43** (1952), 1151

9 Orowan, E., *Symposium on internal stresses* (London: Institute of Metals, 1947), p. 451
10 Saada, G., Acta Metall. **8** (1960), 841
11 Carrington, W., Hale, K.F., and McLean, D., Proc. Roy. Soc. **A259** (1960), 203
12 Friedel, J., *Dislocations* (London: Pergamon, 1964), chaps. 8, 9, and 14
13 Frank, F.C., *Pittsburgh conference on plastic deformation of crystals* (Washington: Office of Naval Research, 1950), p. 100
14 de Wit, G., and Koehler, J.S., Phys. Rev. **116** (1959), 1113
15 Hirth, J.P., Jøssang, T., and Lothe, J., J. Appl. Phys. **37** (1966), 110
16 Hirth, J.P., and Lothe, J., *Theory of dislocations* (New York: McGraw-Hill, 1968), chap. 6
17 Kroupa, F., Czech. J. Phys. **10B** (1960), 284
18 Kroupa, F., Czech. J. Phys. **12B** (1962), 191
19 Yoffe, E.H., Phil. Mag. **5** (1960), 161
20 Heidenreich, R.D., and Shockley, W., *Report on the strength of solids* (London: Physical Society, 1948), p. 57

9

DYNAMICAL PROBLEMS

9.1
ANTIPLANE STRAIN: SCREW DISLOCATIONS

In this chapter we shall be concerned with situations in which dislocations move through a material, giving rise to a time-dependent elastic field and usually to a non-zero rate of plastic strain. The equations describing such a situation are, as we shall see in §9.3, in general quite complex, so that we shall begin in this and the following sections by considering the special cases of antiplane and plane deformations. As particular examples we shall derive the solutions for a single moving dislocation of screw or edge type.

A *The basic equations for dislocation motion*

We considered in §7.2A a state of deformation of a dislocated crystal in which the triad of local lattice vectors is given by

$$\mathbf{e}_1 = \mathbf{u}_1 + \alpha_1 \mathbf{u}_3, \qquad \mathbf{e}_2 = \mathbf{u}_1 + \alpha_2 \mathbf{u}_3, \qquad \mathbf{e}_3 = \mathbf{u}_3, \tag{9.1}$$

which corresponds to a shear of the $\mathbf{e}_1$ and $\mathbf{e}_2$ vectors in the third coordinate direction. For such states, when α_1 and α_2 are small, the only non-zero component of dislocation density is a_{33}, so that only screw dislocations with lines parallel to the z-axis are present. Denoting a_{33} simply by a, we found earlier that

$$a = \partial_1 \alpha_2 - \partial_2 \alpha_1. \tag{9.2}$$

The only non-zero components of the elastic strain tensor were also shown to be

$$e_{13} = \alpha_1/2, \qquad e_{23} = \alpha_2/2, \tag{9.3}$$

to first order in α_1 and α_2.

Now let us consider a situation in which the screw dislocations are moving relative to the medium. We shall suppose that they translate always as straight lines, with all points on any one line moving with the same velocity. We shall also in this section make the assumption that all the positive screw dislocations in the neighbourhood of a given point move with the same

L*

velocity $\mathbf{v} = (v_1, v_2, 0)$, while the negative dislocations all have velocity $-\mathbf{v}$. The motivation for this assumption is that the controlling factor which determines the dislocation velocity is the local resolved shear stress, which will be the same for all positive dislocations but equal and opposite for all negative ones at the point in question, However, this is certainly an over-simplification: in a crystal containing a distribution of discrete dislocation lines, there will be localized effects, such as barriers to slip, which can influence perhaps only one dislocation line, leaving its neighbouring lines unaffected. Thus in reality there will exist at any point a whole distribution of dislocation velocities, rather than a single velocity. We shall return to this question in §9.1E, showing there that our main conclusions remain valid within the more general framework of a distribution of velocities.

Consider a small rectangle $ABCD$ (Fig. 9.1) whose sides are parallel to the x and y axes and have lengths δx and δy. In a small time δt, the number of positive dislocations crossing AD in the positive x-direction is $a^+(x, y)v_1(x, y)\delta y\delta t$ and the number of negative dislocations is $-a^-(x, y)v_1(x, y)\delta y\delta t$. Here we use $a^\pm(x, y)$ to represent the contributions to the total dislocation density arising from the positive and negative dislocations, so that $a(x, y) = a^+(x, y) + a^-(x, y)$. Then a^+, say, equals the number of positive lines per unit area of the xy-plane (multiplied by the Burgers vector). Returning to Figure 9.1, the numbers of positive and negative dislocations crossing BC in time δt are equal to $\pm a^\pm(x+\delta x, y)v_1(x+\delta x, y)\delta y\delta t$ respectively. Including also the numbers of the two types of dislocation crossing AB and CD, taking the limit as δx, δy, and δt all tend to zero, we have an expression for the rate of change of density of each type of dislocation:

$$\frac{\partial a^+}{\partial t} = -\frac{\partial}{\partial x_1}(a^+v_1) - \frac{\partial}{\partial x_2}(a^+v_2) + \dot{a}_p, \tag{9.4}$$

$$\frac{\partial a^-}{\partial t} = \frac{\partial}{\partial x_1}(a^-v_1) + \frac{\partial}{\partial x_2}(a^-v_2) - \dot{a}_p. \tag{9.5}$$

In these equations we have included terms $\pm \dot{a}_p$ to allow for the creation of

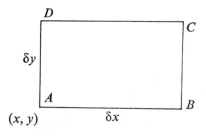

FIGURE 9.1

dislocations within the small area $ABCD$. Since the creation process always results in equal numbers of positive and negative dislocations, an equal term appears in the two equations. $\dot{a}_p$ measures the rate of production per unit area of positive-negative dislocation pairs.

Equations (9.4) and (9.5) have the standard form for the conservation law of a flow process in which $(a^+v_1, a^+v_2, 0)$ for example represents the 'current' of positive dislocations. The conservation is in fact not exact because of the pair production terms. However, if we add the two equations together, we obtain an exact conservation law for the total Burgers vector:

$$\frac{\partial a}{\partial t} = -\frac{\partial}{\partial x_1}(\Delta v_1) - \frac{\partial}{\partial x_2}(\Delta v_2), \tag{9.6}$$

where $\Delta \equiv a^+ - a^-$ is the total numerical density of the two types of dislocation. For Δ we have the conservation equation

$$\frac{\partial \Delta}{\partial t} = -\frac{\partial}{\partial x_1}(av_1) - \frac{\partial}{\partial x_2}(av_2) + 2\dot{a}_p. \tag{9.7}$$

In time δt, the total Burgers vector crossing the small element AD in Figure 9.1 is equal to $(a^+ - a^-)v_1\delta y\delta t = \Delta v_1\delta y\delta t$. If the points A and D were positioned on specific material particles during this time interval, they would be displaced relative to one another in the z-direction by an amount equal to this total Burgers vector. The Burgers vector crossing AB during δt equals $\Delta v_2\delta x\delta t$, which displaces a particle at B relative to one at A by this amount in the $-z$ direction. Besides these plastic displacements, there is a contribution to the changes in relative displacements of a purely elastic nature arising from the time dependence of the local lattice vectors. Consider, for example, a material element $\overrightarrow{AB} = \delta x \cdot \mathbf{u}_1 = \delta x \cdot (\mathbf{e}_1 - \alpha_1 \mathbf{e}_3)$, after using equations (9.1). As the lattice deforms, material vector $\overrightarrow{AB}$ maintains a constant relationship to the lattice vectors, so that $\delta\overrightarrow{AB} = \delta x(\delta\mathbf{e}_1 - \alpha_1\delta\mathbf{e}_3) = \delta x\delta\alpha_1\mathbf{u}_3$. Thus we have total changes

$$\delta\overrightarrow{AB} = [-\Delta v_2\delta x\delta t + \delta x\delta\alpha_1]\mathbf{u}_3, \qquad \delta\overrightarrow{AD} = [\Delta v_1\delta y\delta t + \delta y\delta\alpha_2]\mathbf{u}_3.$$

If now $u_3(x, y, t)$ is the over-all displacement field in the body, these total changes are $\delta\overrightarrow{AB} = \partial_1\dot{u}_3\delta x\delta t$ and $\delta\overrightarrow{AD} = \partial_2\dot{u}_3\delta y\delta t$, using dots to denote time derivatives. Thus we have

$$\partial_1\dot{u}_3 = -\Delta v_2 + \dot{\alpha}_1, \tag{9.8}$$

$$\partial_2\dot{u}_3 = \Delta v_1 + \dot{\alpha}_2. \tag{9.9}$$

In this way we have split the total rate of change of displacement gradient into an elastic part, $\dot{\alpha}_i$, and a plastic part, Δv_j.

It will be convenient later to introduce quantities β_1 and β_2, called the components of plastic distortion, by the definitions

$$\partial_1 u_3 = \alpha_1 + \beta_1, \qquad \partial_2 u_3 = \alpha_2 + \beta_2. \tag{9.10}$$

It then follows that

$$\dot{\beta}_1 = -\Delta v_2, \qquad \dot{\beta}_2 = \Delta v_1, \tag{9.11}$$

and from equation (9.2) that

$$a = \partial_2 \beta_1 - \partial_1 \beta_2. \tag{9.12}$$

Equation (9.6) can now be obtained by combining these two results, so that we have in fact only six independent equations, (9.7) and (9.10)–(9.12).

A further equation arises from the Cauchy equations of motion,

$$\rho \ddot{u}_3 = \partial_1 \sigma_{31} + \partial_2 \sigma_{32}, \tag{9.13}$$

with the stress components σ_{31} and σ_{32} being given by some constitutive relation in terms of the elastic strain components of equation (9.3). For an isotropic medium obeying Hooke's law, we would have

$$\sigma_{13} = \mu \alpha_1, \qquad \sigma_{23} = \mu \alpha_2, \tag{9.14}$$

so that the equation of motion takes the form

$$\rho \ddot{u}_3 = \mu(\partial_1 \alpha_1 + \partial_2 \alpha_2). \tag{9.15}$$

Using equation (9.10), we have

$$\frac{1}{c^2} \ddot{u}_3 = \nabla^2 u_3 - (\partial_1 \beta_1 + \partial_2 \beta_2), \tag{9.16}$$

where $c = (\mu/\rho)^{1/2}$ is the velocity of elastic shear waves in the medium.

B Solutions with prescribed plastic distortion

The most basic type of problem we can consider involving a time-dependent situation is to suppose that the dislocation densities and velocities are pre-scribed functions and to find the resulting displacement and stress-fields. In view of equations (9.11) and (9.12), if Δ, a, v_1, and v_2 are given, then the vector (β_1, β_2) is determined up to the gradient of some time-independent field, say $u_3^{(0)}$. This quantity can be included in with u_3 itself in equation (9.16) without changing the form of that equation. Thus we can take β_1 and β_2 as known, and the problem is to solve equation (9.16) for u_3. The resulting solution will differ from the 'true' displacement field by the quantity $u_3^{(0)}$; however, in a plastic problem the displacement field cannot be uniquely specified, and this arbitrariness is to be expected. Only changes in the displacement are physically meaningful, for instance between the values at some initial time t_0 and the current time, and these are unaffected by $u_3^{(0)}$, which is time-independent.

Equation (9.16) is the inhomogeneous wave equation in two dimensions, and for an infinite medium it has a well-known general solution.[1] It is, however, convenient to take the solution for three dimensions, and then use the independence of the third coordinate afterwards to simplify the solution. Thus we have[1]

$$u_3(x, y, t) = \int_{-\infty}^{t} dt' \int_{-\infty}^{\infty} G(\mathbf{r}, t, \mathbf{r}', t') (\partial_1'\beta_1 + \partial_2'\beta_2) dx'dy'dz', \qquad (9.17)$$

where G denotes the usual retarded potential,

$$G(\mathbf{r}, t, \mathbf{r}', t') = -\frac{1}{4\pi} \delta\left(t' - t + \frac{1}{c}|\mathbf{r}-\mathbf{r}'|\right)|\mathbf{r}-\mathbf{r}'|^{-1}. \qquad (9.18)$$

Integrating by parts with respect to t', we obtain

$$u_3(x, y, t) = -\frac{1}{4\pi}\int_{-\infty}^{\infty} \frac{\partial_1'\beta_1(\mathbf{r}', -\infty) + \partial_2'\beta_2(\mathbf{r}', -\infty)}{|\mathbf{r}-\mathbf{r}'|} dx'dy'dz'$$

$$- \int_{-\infty}^{t} dt' \int_{-\infty}^{\infty} K(\mathbf{r}, t, \mathbf{r}', t') (\partial_1'\beta_1 + \partial_2'\beta_2) dx'dy'dz', \qquad (9.19)$$

where

$$K(\mathbf{r}, t, \mathbf{r}', t') = \left[1 - H\left(t' - t + \frac{1}{c}|\mathbf{r}-\mathbf{r}'|\right)\right]\Big/ 4\pi |\mathbf{r}-\mathbf{r}'|.$$

Here $H(\tau)$ denotes the Heaviside step function. The first term in u_3 is a time-independent displacement arising from the initial values of the plastic distortion at $t = -\infty$. It can be included in with the unknown $u_3^{(0)}$. The second integral vanishes as $t \to -\infty$, and so represents the change in displacement between the times $-\infty$ and t. This is the quantity in which we are usually interested, and we shall from now on use u_3 to denote this change only. Then we can integrate by parts with respect to the relevant positional coordinate to give the final expression

$$u_3(x, y, t) = \int_{-\infty}^{t} dt' \int K_l(\mathbf{r}, t, \mathbf{r}', t')\dot{\beta}_l(\mathbf{r}', t') dx'dy'dz', \qquad (9.20)$$

where

$$K_l(\mathbf{r}, t, \mathbf{r}', t') = \partial_l K(\mathbf{r}, t, \mathbf{r}', t') = \left[\frac{|\mathbf{r}-\mathbf{r}'|}{c} \delta\left(t' - t + \frac{|\mathbf{r}-\mathbf{r}'|}{c}\right)\right.$$

$$\left. + 1 - H\left(t' - t + \frac{|\mathbf{r}-\mathbf{r}'|}{c}\right)\right]\frac{x_l - x_l'}{4\pi|\mathbf{r}-\mathbf{r}'|^3}, \qquad (9.21)$$

If we now substitute from equation (9.11) for $\dot{\beta}_1$ and $\dot{\beta}_2$ we obtain the solution in terms of directly prescribed quantities.

c *Single moving dislocation*

The solution (9.20), holding for any time-dependent distribution of dislocations, can be applied to the case of a single screw by allowing the dislocation density to have a δ-function behaviour. If the position of the dislocation at time t is $(\xi(t), \eta(t))$, then we take

$$a = \Delta = b\delta[x - \xi(t)]\delta[y - \eta(t)]. \tag{9.22}$$

The dislocation velocity has components $v_1 = \dot{\xi}(t)$, $v_2 = \dot{\eta}(t)$, and using these values with equations (9.11) and (9.20) gives

$$u_3(x, y, t) = b \int_{-\infty}^{t} dt' \int_{-\infty}^{\infty} [-K_1 \dot{\eta}(t') + K_2 \dot{\xi}(t')]$$
$$\cdot \delta[x' - \xi(t')]\delta[y' - \eta(t')]dx'dy'dz'.$$

Performing in turn the x', y', and z' integrations, we obtain the result[2, 3]

$$u_3(x, y, t) =$$

$$\frac{bc}{2\pi} \int_{-\infty}^{t} \frac{(t-t')\ \{[y-\eta(t')]\dot{\xi}(t') - [x - \xi(t')]\dot{\eta}(t')\}dt'}{\{[x-\xi(t')]^2 + [y-\eta(t')]^2\}\ \{c^2(t'-t)^2 - [x-\xi(t')]^2 - [y-\eta(t')]^2\}^{1/2}}. \tag{9.23}$$

Although we have written the upper limit here as t, it must in fact be taken as a smaller value, t_0, at which

$$c^2(t-t_0)^2 - [x - \xi(t_0)]^2 - [y - \eta(t_0)]^2 = 0.$$

This time t_0 is the latest instant at which the shear waves emanating from the dislocation at position $(\xi(t_0), \eta(t_0))$ can reach the field point (x, y) before the current time t. At times t' later than t_0, the dislocation does not influence the situation at (x, y) until after time t. We also remark that equation (9.23) only holds provided $\dot{\xi}^2 + \dot{\eta}^2 < c^2$ at all times – that is, provided the dislocation velocity remains subsonic.

As a particular example we can consider a screw dislocation moving along the x-axis with constant velocity v. Then $\xi(t) = vt$, $\eta(t) = 0$, enabling the t' integration to be performed. The result is

$$u_3(x, y, t) = \frac{b}{2\pi} \tan^{-1} \frac{\beta y}{x - vt}, \tag{9.24}$$

where $\beta = (1 - v^2/c^2)^{1/2}$. The corresponding stresses can be obtained from equations (9.14) and (9.10), noting that β_1 and β_2 are zero except on the dislocation path itself, and they are given by

$$\sigma_{xz} = -\frac{\mu b}{2\pi} \frac{\beta y}{(x-vt)^2 + \beta^2 y^2}, \qquad \sigma_{yz} = \frac{\mu b}{2\pi} \frac{\beta(x-vt)}{(x-vt)^2 + \beta^2 y^2}. \tag{9.25}$$

For this case of uniform motion, the solution can be obtained much more directly than through the general solution (9.23) by introducing the Lorentz transformation[4]

$$x' = \beta^{-1}(x-vt), \qquad y' = y, \qquad t' = \beta^{-1}(t-vx/c^2). \tag{9.26}$$

The wave equation (9.15) remains invariant when expressed in terms of the primed coordinates, and the dislocation becomes fixed at the new origin, $x' = y' = 0$. Thus the solution $u_3(x', y', t')$ must be independent of the new time t', and is given in fact by the static solution which we found in §3.1.

The elastic strain energy in a body containing a steadily moving screw dislocation is obtained by combining equations (2.30) and (9.3) as

$$U/L = \int \tfrac{1}{2}\mu[(\partial_1 u_3)^2 + (\partial_2 u_3)^2]dxdy$$

per unit length in the z-direction. In addition, there is a kinetic energy

$$T/L = \int \tfrac{1}{2}\rho[\dot{u}_3]^2 dxdy.$$

It is convenient to take polar coordinates centred on the current position of the dislocation, so that $x - vt = r \cos \theta$ and $y = r \sin \theta$. The region of integration is then taken to be $r_0 < r < R$. The lower limit $r = r_0$ is plausible since we expect the dislocation to carry its core along with it. However, the outer limit can only be regarded as an approximation, valid when the dislocation lies within distances much smaller than R of the centre of the actual cylindrical region. With these limits, the total energy becomes

$$(U+T)/L = \beta^{-1}(\mu b^2/4\pi) \ln(R/r_0) = \beta^{-1}U_0/L$$

and so is changed from the static value U_0/L of equation (3.4) by the relativistic factor β^{-1}.

This result enables us to associate an inertia with a dislocation line.[4] In equation (3.33) we defined the force on a dislocation in terms of the change in the total static energy of the system when the dislocation undergoes a small virtual displacement. (The system here consists of the deforming body together with those external agents which are supplying tractions on its boundaries.) During an actual motion, we must have conservation of energy, so that any decrease in the energy of the external agents must be balanced by a corresponding increase in the internal energy of the medium. This could take any of a number of forms, such as dissipation into heat energy, possibly extra surface energy, elastic strain energy, or kinetic energy. However, if we assume that there is no dissipation and that no new surface is being formed, only the last two of these energies increase, and this occurs via an increase in the velocity of the dislocation. If we ignore any radiation of energy caused by the dislocation acceleration, we can set the force F/L per unit length of dislocation line equal to

$$\frac{F}{L} = \frac{\partial}{\partial x}\frac{U+T}{L} = \frac{1}{v}\frac{\partial}{\partial t}\beta^{-1}(U_0/L) = \frac{m\dot{v}}{\beta^3}, \tag{9.27}$$

where $mc^2 = U_0/L$ is the rest energy. m is called the effective mass of the dislocation line. For $v \ll c$ we can put $\beta \approx 1$ and the dislocation obeys Newton's second law, but for higher velocities effects of a relativistic nature are present.

We shall return to the question of the equation of motion of a dislocation for fuller discussion in §9.5, examining in particular the effects of energy dissipation and radiation which we have ignored here.

D *The electromagnetic analogy*

The general solution (9.23) involves quite difficult integrals when applied even to the simplest cases. It was shown by Eshelby[5] that there is a complete correspondence between the equations of antiplane dislocation and Maxwell's equations for a two-dimensional electromagnetic field, and using this he was able to express the solution in terms of simpler integrals. The correspondence is obtained by defining electric and magnetic fields **E** and **H** by

$$E_1 = \sigma_{23}/(\mu\varepsilon_0)^{1/2}, \qquad E_2 = -\sigma_{13}/(\mu\varepsilon_0)^{1/2}, \qquad E_3 = H_1 = H_2 = 0,$$
$$H_3 = \dot{u}_3(\rho/\mu_0)^{1/2},$$

where we use MKS units with ε_0 and μ_0 the permittivity and permeability of the vacuum. They satisfy $c^2 = 1/(\varepsilon_0\mu_0)^{1/2}$, where we identify the velocity of light with that of the shear waves. Then using $\mathbf{D} = \varepsilon_0\mathbf{E}$ and $\mathbf{B} = \mu_0\mathbf{H}$, we can verify Maxwell's equations in turn: div $\mathbf{B} = 0$ is satisfied identically; curl $\mathbf{E}+\dot{\mathbf{B}} = \mathbf{0}$ becomes equation (9.16); div $\mathbf{D} = \rho_c$ becomes equation (9.2) provided we identify the electric charge density ρ_c with $(\mu\varepsilon_0)^{1/2}a$; and curl $\mathbf{H}-\dot{\mathbf{D}} = \mathbf{j}$ becomes equations (9.8) and (9.9) if we set the electric current density $\mathbf{j}$ equal to $(\rho/\mu_0)^{1/2}\Delta v$.

One of the most fruitful techniques of electromagnetic theory is the introduction of scalar and vector potentials, which we write in the form $H = \mu_0^{1/2}$ curl $A, E = -\varepsilon_0^{1/2}(\nabla\phi+c^{-1}\dot{A})$. These are related to the mechanical variables by

$$\dot{u}_3 = \rho^{-1/2}(\partial_1 A_2 - \partial_2 A_1),$$
$$\sigma_{13} = \mu^{1/2}(\partial_2\phi + c^{-1}\dot{A}_2), \tag{9.28}$$
$$\sigma_{23} = -\mu^{1/2}(\partial_1\phi + c^{-1}\dot{A}_1).$$

Then if **A** and ϕ are chosen to satisfy the subsidiary condition $c^{-1}\dot{\phi}+$ div **A** $= 0$, the field equations become equivalent to the two inhomogeneous wave equations

$$c^{-2}\ddot{\mathbf{A}}-\nabla^2\mathbf{A} = \mu_0^{1/2}\mathbf{j} = \rho^{1/2}\Delta\mathbf{v},$$
$$c^{-2}\ddot{\phi}-\nabla^2\phi = \varepsilon_0^{-1/2}\rho_c = \mu^{1/2}a.$$

The solution of these equations for an infinite medium is

$$\mathbf{A}(\mathbf{r}, t) = -\rho^{1/2} \int_{-\infty}^{t} dt' \int G(\mathbf{r}, t, \mathbf{r}', t') \Delta\mathbf{v}(\mathbf{r}', t') dx' dy' dz',$$

$$\phi(\mathbf{r}, t) = -\mu^{1/2} \int_{-\infty}^{t} dt' \int G(\mathbf{r}, t, \mathbf{r}', t') a(\mathbf{r}', t') dx' dy' dz', \tag{9.29}$$

where the Green's function is given by equation (9.18).

Substituting from equation (9.22) for the case of a single dislocation, the three spatial integrations can be performed to give the result

$$\{\phi, A_1, A_2\} = \frac{b\mu^{1/2}}{2\pi} \int_{-\infty}^{t} \frac{\{c, \dot{\xi}(t'), \dot{\eta}(t')\} dt'}{\{c^2(t-t')^2 - [x - \xi(t')]^2 - [y - \eta(t')]^2\}^{1/2}}, \tag{9.30}$$

where we have combined the three formulae. The solution in terms of equations (9.30) and (9.28) involves much simpler integrals than equation (9.23).

It is an easy matter to verify, by differentiating under the integral sign, that the solution contained in equations (9.29) and (9.28) is equivalent to that given in equation (9.20). It is not as simple to demonstrate so directly the equivalence of equations (9.23) and (9.30) for the single dislocation case, since differentiation under the integral sign yields an infinite boundary term and a divergent integral.

Eshelby[5] evaluated explicitly the potentials for a motion in which $\eta(t) = 0$ and $\xi(t) = x_0$ for $t < 0$, $\xi(t) = (x_0^2 + c^2 t^2)^{1/2}$ for $t > 0$, and also calculated the resulting shear stress σ_{yz} in the neighbourhood of the dislocation. The significance of this particular motion is that, relative to a Lorentz frame (cf. eqs. (9.26)) in which the dislocation is momentarily at rest at some instant $t = \tau$, its acceleration at this instant τ has the constant value c^2/x_0. The dislocation velocity approaches c asymptotically.

We mention finally that the analogy between electromagnetic and mechanical quantities has been chosen to give a partial correspondence between the energy-momentum tensors. The energy density $\frac{1}{2}(\mathbf{D}\cdot\mathbf{E} + \mathbf{B}\cdot\mathbf{H})$ of the electromagnetic field equals $\frac{1}{2}\rho\dot{u}_3^2 + (e_{13}\sigma_{13} + e_{23}\sigma_{23})$, the kinetic plus elastic energy densities. The Poynting vector given by $\mathbf{E}\wedge\mathbf{H} = -(\sigma_{13}\dot{u}_3, \sigma_{23}\dot{u}_3, 0)$ is the vector giving the rate of elastic energy flow. The momentum densities do not, however, precisely correspond.[5]

E *Distributions of dislocation velocities*

We mentioned in §A that in a real crystal the velocity of the dislocations in any small region will not be uniquely defined, but rather a distribution of velocities will occur. Thus for completeness we need to introduce density functions in velocity space as well as in ordinary space. We let $d^+(\mathbf{r}, \mathbf{v})dv_1 dv_2$ denote the total Burgers vector per unit area of positive dislocation lines in the neighbourhood of the point $\mathbf{r}$ and with velocities between $\mathbf{v}$ and $\mathbf{v}+d\mathbf{v}$; $d^-(\mathbf{r}, \mathbf{v})$ denotes the corresponding density for negative dislocations. Then, returning to Figure 9.1, the Burgers vector of positive dislocations crossing, say, AB in a time δt is $\delta t \int d^+(x, y, \mathbf{v})v_1 dv_1 dv_2$. There are similar expressions for the other three sides of $ABCD$, and for the negative dislocations, and if we combine them and take the limits as δt and the sides of the rectangle tend to zero, we obtain the conservation law in place of equations (9.4) and (9.5):

$$\partial a^\pm / \partial t = -\partial_1 f_1{}^\pm - \partial_2 f_2{}^\pm \pm \dot{a}_p. \tag{9.31}$$

Here the spatial density and currents are defined by

$$a^\pm(\mathbf{r}) = \int d^\pm(\mathbf{r}, \mathbf{v})dv_1 dv_2, \qquad f_i{}^\pm(\mathbf{r}) = \int v_i d^\pm(\mathbf{r}, \mathbf{v})dv_1 dv_2.$$

If we combine the equations for a^+ and a^- we again obtain an exact conservation equation for a, similar to equation (9.6) except that each Δv_i is replaced by $f_i \equiv f_i{}^+ + f_i{}^-$.

The analysis of the plastic strain rate leading to equations (9.8) and (9.9) can similarly be extended, and leads to the same replacement $\Delta v_i \rightarrow f_i$. Then equations (9.10) and (9.12)–(9.16) are still valid and equation (9.11) becomes

$$\dot{\beta}_1 = -f_2, \qquad \dot{\beta}_2 = f_1. \tag{9.32}$$

Consequently the solutions (9.20) and (9.29) remain valid in this case provided we replace Δv_i throughout.

9.2
PLANE STRAIN: EDGE DISLOCATIONS

A *The basic equations*

In terms of the lattice vectors, we can define a state of plane strain as one in which

$$\mathbf{e}_1 = \mathbf{u}_1 + \alpha_{11}\mathbf{u}_1 + \alpha_{12}\mathbf{u}_2, \qquad \mathbf{e}_2 = \mathbf{u}_2 + \alpha_{21}\mathbf{u}_1 + \alpha_{22}\mathbf{u}_2, \qquad \mathbf{e}_3 = u_3, \tag{9.33}$$

where (α_{ij}) are functions of x_1 and x_2 which are restricted to be small. For such a lattice distortion, we showed in §7.2B that the non-zero strains and the rotation are given by

$$e_{11} = \alpha_{11}, \quad e_{22} = \alpha_{22}, \quad e_{12} = \tfrac{1}{2}(\alpha_{12} + \alpha_{21}), \quad \theta = \tfrac{1}{2}(\alpha_{12} - \alpha_{21}). \tag{9.34}$$

We found earlier that the non-zero dislocation density components are given by equation (7.25), which becomes, when expressed in terms of the α_{ij},

$$a_{13} = \partial_1\alpha_{21} - \partial_2\alpha_{11}, \qquad a_{23} = \partial_1\alpha_{22} - \partial_2\alpha_{12}. \tag{9.35}$$

As pointed out in §7.2B, such densities can be realized by having two sets of edge dislocations with Burgers vectors in the x_1 and x_2 directions and line directions along x_3.

In the following we shall set $a_1 = a_{13}$ and $a_2 = a_{23}$ and we shall refer to dislocations as being of the four types $1\pm$ and $2\pm$ with the $\pm$ denoting whether they are positive or negative dislocations. (We use the convention that the dislocation line is directed along the positive x_3-direction always, and then a dislocation is referred to as positive if its Burgers vector points in the positive direction of the relevant coordinate axis.)

If we consider glide motions only, then dislocations of type $1+$ will have a velocity in the x_1-direction, and we suppose as before that they all have the same velocity, v_1, while dislocations of type $1-$ have a velocity $-v_1$. Similarly dislocations of types $2\pm$ can glide only in the x_2-direction, say with velocities $\pm v_2$. Then we can derive conservation equations for each of the four dislocation densities, $a_1{}^\pm$ and $a_2{}^\pm$, similar to equations (9.4) and (9.5), except that just a single velocity component enters each equation. Expressing these in terms of the total density components, $a_i = a_i{}^+ + a_i{}^-$, and the quantities $\Delta_i = a_i{}^+ - a_i{}^-$, we have the analogues of equations (9.6) and (9.7):

$$\dot{a}_1 = -\partial_1(\Delta_1 v_1), \qquad \dot{\Delta}_1 = -\partial_1(a_1 v_1) + 2\dot{a}_{p1},$$

$$\dot{a}_2 = -\partial_2(\Delta_2 v_2), \qquad \dot{\Delta}_2 = -\partial_2(a_2 v_2) + 2\dot{a}_{p2}. \tag{9.36}$$

To relate the dislocation flow to the plastic strain rate, we again consider the rate of change of a material element, say $\overrightarrow{AD}$ in Figure 9.1. We have

$$\partial \overrightarrow{AD}/\partial t = (\partial_2 \dot{u}_1 \mathbf{u}_1 + \partial_2 \dot{u}_2 \mathbf{u}_2)\delta y$$

$$= \Delta_1 v_1 \mathbf{u}_1 \delta y + (\dot{\alpha}_{21} \mathbf{u}_1 + \dot{\alpha}_{22} \mathbf{u}_2)\delta y.$$

In the first line we have expressed the change in $\overrightarrow{AD}$ in terms of the total displacement field (u_1, u_2) in the body, and in the second we have broken this up into the two separate contributions, one from the flow of type 1 dislocations through $\overrightarrow{AD}$ and the other from the change in the local lattice vectors. Comparing coefficients of the two basis vectors, we therefore have

$$\partial_2 \dot{u}_1 = \Delta_1 v_1 + \dot{\alpha}_{21}, \qquad \partial_2 \dot{u}_2 = \dot{\alpha}_{22}. \tag{9.37}$$

Similar considerations applied to a small material element in the x_1-direction give

$$\partial_1 \dot{u}_2 = -\Delta_2 v_2 + \dot{\alpha}_{12}, \qquad \partial_1 \dot{u}_1 = \dot{\alpha}_{11}. \tag{9.38}$$

Finally we have the equations of motion

$$\rho \ddot{u}_1 = \partial_1 \sigma_{11} + \partial_2 \sigma_{12}, \qquad \rho \ddot{u}_2 = \partial_1 \sigma_{12} + \partial_2 \sigma_{22}, \tag{9.39}$$

and the stress-strain relations, which for an isotropic medium in plane strain are

$$\sigma_{11} = (\lambda+2\mu)e_{11}+\lambda e_{22}, \quad \sigma_{22} = \lambda e_{11}+(\lambda+2\mu)e_{22}, \quad \sigma_{12} = 2\mu e_{12}. \quad (9.40)$$

As was the case with screw dislocations, the above set of equations constitutes a complete system for problems in which the dislocation densities and velocities are prescribed. In fact, for such problems equations (9.36) are not needed: the first pair can be deduced from equations (9.35), (9.37), and (9.38) and the second pair merely provides the values of the production rates, $\dot{a}_{p1}$ and $\dot{a}_{p2}$, which are necessary in order to maintain the prescribed velocities and densities. The remaining equations can again be simplified by introducing components (β_{ij}) of plastic distortion defined by

$$\beta_{ij} = \partial_i u_j - \alpha_{ij}. \tag{9.41}$$

Then equations (9.37) and (9.38) give

$$\dot{\beta}_{12} = -\Delta_2 v_2, \quad \dot{\beta}_{21} = \Delta_1 v_1, \quad \dot{\beta}_{11} = \dot{\beta}_{22} = 0, \tag{9.42}$$

and from equations (9.35) we obtain the dislocation density as

$$a_1 = -\partial_1\beta_{21}+\partial_2\beta_{11}, \quad a_2 = -\partial_1\beta_{22}+\partial_2\beta_{12}. \tag{9.43}$$

The field equations are derived from equations (9.39) by substituting from equations (9.40) for the σ_{ij}, from equations (9.34) for the e_{ij}, and finally from equations (9.41) for the α_{ij}, giving

$$\rho\ddot{u}_i-(\lambda+\mu)\partial_i\Delta-\mu\nabla^2 u_i = -\lambda\partial_i(\beta_{11}+\beta_{22})-\mu\partial_j(\beta_{ij}+\beta_{ji}), \tag{9.44}$$

where $\Delta = \partial_i u_i$ is the dilatation.

B Uniformly moving edge dislocation

We can obtain the general solution of equations (9.44) in a similar way to the antiplane case by introducing the appropriate Green's function for the operator on the left-hand side. We shall not pursue this here, however, since in §9.4 we shall derive the general solution for three-dimensional states of strain, which contain the present states as a subclass. In this section we shall examine the simplest plane problem of all, that of a single edge dislocation moving with uniform velocity.

The plane strain equations (9.44) are not invariant under the Lorentz transformation (9.26) since they give rise to two types of wave, dilatational and shear, which have different velocities, $c_1 = [(\lambda+2\mu)/\rho]^{1/2}$ and $c_2 = (\mu/\rho)^{1/2}$ respectively. Thus it is not possible to solve the moving dislocation problem in terms of the stationary one in the trivial way we could for the screw dislocation. (See, however, Prob. 3.) We shall follow Eshelby[6] and use Fourier transforms to find the solution.

We consider a positive edge dislocation with Burgers vector b in the x-direction gliding with constant velocity v on the plane $y = 0$. The displacement fields depend on the variables x, y, t only in the combinations $x - vt$ and y, and we shall temporarily use the notation $x_1 \equiv x - vt$, $x_2 \equiv y$. Then we introduce the Fourier transforms

$$\bar{u}_i(s, x_2) = \int_{-\infty}^{\infty} u_i(x_1, x_2) e^{isx_1} dx_1. \tag{9.45}$$

Since the u_i are real functions, the transforms satisfy $\bar{u}_i^*(s, x_2) = \bar{u}_i(-s, x_2)$, where * denotes complex conjugation. Because of this, the inversion formulae can be written[7]

$$u_i(x_1, x_2) = \frac{1}{2\pi} \int_{-\infty}^{\infty} \bar{u}_i(s, x_2) e^{-isx_1} ds \tag{9.46}$$

$$= \frac{1}{\pi} \int_0^{\infty} \{\mathscr{R}e[\bar{u}_i(s, x_2)] \cos sx_1 + \mathscr{I}m[\bar{u}_i(s, x_2)] \sin sx_1\} ds. \tag{9.47}$$

The field equations (9.44) have zero right-hand sides except on the $x_2 = 0$ plane. Thus we can approach the problem by solving the homogeneous equations in each of the half-spaces $x_2 > 0$ and $x_2 < 0$ and then matching the two solutions appropriately across the interface $x_2 = 0$. The homogeneous equations become, after substituting from equation (9.46),

$$(c_1{}^2 - v^2)s^2 \bar{u}_1 - c_2{}^2 \partial_{22} \bar{u}_1 + is(c_1{}^2 - c_2{}^2)\partial_2 \bar{u}_2 = 0,$$
$$(c_2{}^2 - v^2)s^2 \bar{u}_2 - c_1{}^2 \partial_{22} \bar{u}_2 + is(c_1{}^2 - c_2{}^2)\partial_2 \bar{u}_1 = 0. \tag{9.48}$$

There are four independent solutions of this pair of equations, with $u_i \propto e^{\pm \beta_1 sx_2}$ and $u_i \propto e^{\pm \beta_2 sx_2}$, where $\beta_i{}^2 = 1 - v^2/c_i{}^2$. The inversion formula in the form (9.47) involves only positive values of the transform variable s, and since the solutions must remain bounded as $x_2 \to \pm \infty$, only the negative exponents can appear for $x_2 > 0$ and the positive exponents for $x_2 < 0$. An additional simplication arises from the fact that u_1 is an odd function of x_1 and u_2 an even function, so that $\mathscr{R}e[\bar{u}_1] = \mathscr{I}m[\bar{u}_2] = 0$. Then solving equations (9.48) and substituting into equation (9.47) we obtain for $x_2 > 0$

$$u_1(x_1, x_2) = \frac{1}{2\pi} \int_0^{\infty} [I(s)e^{-\beta_1 sx_2} - \beta_2 R(s)e^{-\beta_2 sx_2}] \sin sx_1 \, ds,$$

$$u_2(x_1, x_2) = \frac{1}{2\pi} \int_0^{\infty} [\beta_1 I(s)e^{-\beta_1 sx_2} - R(s)e^{-\beta_2 sx_2}] \cos sx_1 \, ds. \tag{9.49}$$

A similar pair of representations holds for the displacements in $x_2 < 0$ except that the β_i are everywhere replaced by $-\beta_i$, and the functions $I(s)$ and $R(s)$ are different (we denote them by I' and R').

The continuity condition $u_2(x_1, 0+) = u_2(x_1, 0-)$ across the interface

$x_2 = 0$, when combined with the continuity of σ_{xy}, leads to the result that $R' = R$ and $I' = -I$. The condition that $\sigma_{yy}(x_1, 0+) = \sigma_{yy}(x_1, 0-)$ then gives

$$I(1 - v^2/2c_2{}^2) = \beta_2 R.$$

Finally we have

$$u_1(x_1, 0+) - u_1(x_1, 0-) = \tfrac{1}{2}b[H(-x_1) - H(x_1)]$$

$$= -\frac{b}{2\pi} \int_0^\infty \left(\frac{1}{s - i\varepsilon} + \frac{1}{s + i\varepsilon}\right) \sin sx_1 \, ds,$$

where $H(x)$ is the Heaviside step function and ε is any positive number. The equality of the last two lines can easily be verified by contour integration. Substituting from equation (9.49_1) and the corresponding equation for $x_2 < 0$ then leads to the result

$$I(s) = -\frac{bc_2{}^2}{v^2}\left(\frac{1}{s - i\varepsilon} + \frac{1}{s + i\varepsilon}\right).$$

It then remains to evaluate the integrals in equations (9.49), giving

$$u_1 = \frac{bc_2{}^2}{\pi v^2}\left[-\tan^{-1}\frac{x - vt}{\beta_1 y} + \left(1 - \frac{v^2}{2c_2{}^2}\right)\tan^{-1}\frac{x - vt}{\beta_2 y}\right],$$

$$\tag{9.50}$$

$$u_2 = \frac{bc_2{}^2}{2\pi v^2}\left[\beta_1 \ln[(x - vt)^2 + \beta_1{}^2 y^2] - \beta_2{}^{-1}\left(1 - \frac{v^2}{2c_2{}^2}\right)\ln[(x - vt)^2 + \beta_2{}^2 y^2]\right]$$

It can be seen from this result how Lorentz transformations with respect to both velocities c_1 and c_2 are combined in relating the dynamic and static solutions.

We remark in passing that this same method can be used to derive the dynamic fields of an edge dislocation which is climbing with uniform velocity. If we consider the Burgers vector to be in the y-direction with the climb occurring in the positive x-direction on $y = 0$, the same transforms as in equations (9.45)–(9.47) can be used. This time, however, u_1 is even and u_2 is odd in x_1. The quantities u_1, σ_{xy}, and σ_{yy} are continuous across $x_2 = 0$, and $u_2(x_1, 0+) - u_2(x_1, 0-) = \tfrac{1}{2}b[H(x_1) - H(-x_1)]$. The resulting solutions are quite similar to equations (9.50), and can be obtained from them by replacing u_1 by $-u_2$, u_2 by u_1, and interchanging β_1, and β_2 everywhere that they occur.

Returning to the glide case, the displacements in equations (9.50) an be used to calculate the stress-field and the energy. The results have been given by Weertman[8] and show some interesting properties, especially regarding the behaviour of the shear stress. From equations (3.16), the shear stress σ_{xy} around a stationary positive edge dislocation is positive in the quadrant $-\pi/4 < \theta < \pi/4$ (where θ is the usual polar angle) and in the two octants $\pi/2 < \theta < 3\pi/4$ and $5\pi/4 < \theta < 3\pi/2$ and is negative in the quadrant $3\pi/4 < \theta < 5\pi/4$ and in the remaining two octants. For a moving dislocation the above two quadrants become squeezed down towards the x-axis, the boundaries becoming the lines of slope $\pm \tan \phi$, where

$$\tan^2 \phi = [\beta_1\beta_2 - (1 - v^2/2c_2^2)^2]/[\beta_1^2(1 - v^2/2c_2^2)^2 - \beta_1\beta_2^3].$$

As v increases to the value at which $\beta_1\beta_2 = (1 - v^2/2c_2^2)^2$, the sectors between these lines tend to zero. This condition is satisfied when the velocity $v = c_r$, the velocity of Rayleigh waves, and at velocities $v > c_r$ the value of σ_{xy} on the slip-plane $y = 0$, $x_1 > 0$ becomes negative. This has the somewhat surprising implication that a second positive edge dislocation on the same slip-plane would be attracted towards the moving edge when its velocity exceeds c_r while a negative dislocation would be repelled.

Weertman[8] discusses this result at some length and shows that it is consistent with the fact that when $v > c_r$ the kinetic energy of a moving edge exceeds the elastic energy. Such a situation does not occur for screw dislocations. Weertman also concludes that groups of edge dislocations of the same sign moving on the same plane will coalesce if their velocities are sufficiently high and produce a single dislocation of large Burgers vector. Such an occurrence could provide a mechanism for microcrack formation under high-speed deformation without the necessity for dislocation pile-ups at obstacles.

The kinetic and elastic strain energies for a uniformly moving edge are found by integrating $\frac{1}{2}\rho(\dot{u}_1^2 + \dot{u}_2^2)$ and $W(e_{ij})$ (cf. eq. (2.29)) throughout the body. Weertman evaluates these two quantities for a hollow cylindrical body of inner and outer radii r_0 and R, centered on the moving dislocation line, and the total energy per unit length turns out to be

$$\frac{U}{L} = \frac{U_0}{L} \frac{2(1-v)}{1-\beta_2^2} \left(\frac{2}{\beta_1} + 4\beta_1 + \frac{1}{2\beta_2^3} (1+\beta_2^2)(1-7\beta_2^2) \right). \tag{9.51}$$

Here $U_0/L = [\mu b^2/4\pi(1-v)] \ln(R/r_0)$ is the energy per unit length of a stationary dislocation.

At high velocities, approaching the velocity c_2 of the shear waves, the expansion of U/L in powers of β_2 gives

$$U/L \sim (U_0/L)(1-v)(\beta_2^{-3} - 5\beta_2^{-1} + \text{bounded terms}),$$

after correcting the result in the literature. This has a stronger divergence, as β_2^{-3}, than the corresponding energy which we found earlier for the screw case, which only diverged as β_2^{-1} when $v \to c_2$.

At low velocities, the expansion of equation (9.51) in a power series in the velocity gives

$$U/L = U_0/L + \tfrac{1}{2}mv^2,$$

where $mc_2^2 = (U_0/L)(1-\nu)(1+c_2^4/c_1^4)$. Thus again we have an energy analogous to that of a Newtonian particle of mass m. However, this result holds only in the classical mechanical approximation of $v \ll c_2$, and it is not possible for an edge dislocation to establish the correspondence with relativistic particle mechanics which we could do for a moving screw for all velocities.

9.3
GENERAL DISLOCATION MOTIONS

A *Non-linear plastic strains*

The theory of the previous sections, applicable so far only to plane or anti-plane states of infinitesimal strain, can be extended to arbitrary states of finite strain and dislocation density. In the present section we shall derive the generalized equations and in §9.4 we shall obtain their solution for isotropic materials in the case of small deformations. The present section is not a prerequisite for the rest of this chapter.

When discussing the dislocation density tensor in §7.3A we introduced three configurations of the body (see Fig. 7.2): the current configuration, with positions denoted by $\mathbf{x}$; the global reference configuration, with positions $\mathbf{X}$; and the local reference configurations of each neighbourhood of the body, with positions $\boldsymbol{\xi}$. The chain-rule relationship between the various deformation gradients can be written, in view of equation (7.34a), as

$$\frac{\partial x_i}{\partial X_a} = \frac{\partial \xi_k}{\partial X_a} (e|u)_{ki}. \tag{9.52}$$

In this section we shall use the same coordinate system, and latin indices, for each of the three configurations. This equation expresses the total deformation gradient as the product of $(\partial \xi_k/\partial X_a)$ with the matrix of elastic distortion, and so is the non-linear analogue of equation (9.41). For this reason we shall term

$$p_{ak} \equiv \partial \xi_k/\partial X_a \tag{9.53}$$

the (non-linear) plastic distortion.[9] For small strains, we will have $x_i = X_i + u_i$, so that $\partial x_i/\partial X_a = \delta_{ia} + \partial_a u_i$. Then if we set

$$(e|u)_{ki} = \delta_{ki} + \alpha_{ki}, \qquad p_{ak} = \delta_{ak} + \beta_{ak} \tag{9.54}$$

(cf. eq. (9.33)), we shall obtain the form of equation (9.52) relevant in such a case to be

$$\partial_a u_i = \alpha_{ai} + \beta_{ai}. \tag{9.55}$$

In the previous sections we have examined the problem of finding the displacement field when (β_{ij}) is given. The non-linear analogue of this is then to solve for the motion $x_i = x_i(X_a, t)$ when the new plastic distortion p_{ak} is a prescribed function of position and time. The equations describing this situation are then the equations of motion (2.73) and the stress-strain relations (2.84), which can be combined to give

$$\rho \ddot{x}_i = \frac{\partial}{\partial x_j} \left(\frac{\rho}{\rho_0} \frac{\partial x_i}{\partial \xi_k} \frac{\partial x_j}{\partial \xi_l} \Sigma_{kl}(E_{pq}) \right) \tag{9.56}$$

after using equation (2.53). In using equation (2.84), we have replaced the (X) reference configuration in the stress-strain relations by the local reference configuration (ξ). Thus E_{pq} is the strain tensor given in equations (7.37), and Σ_{kl} denotes the symmetrized derivative of the strain energy density, $\partial W / \partial E_{(kl)}$.

Using equations (9.52) and (9.53), we obtain

$$E_{pq} = \frac{1}{2} \left(p_{pa}^{-1} p_{qb}^{-1} \frac{\partial x_l}{\partial X_a} \frac{\partial x_l}{\partial X_b} - \delta_{pq} \right), \tag{9.57}$$

which allows us to express Σ_{kl} as a function of the deformation gradients $\partial x_l / \partial X_a$. Equation (9.56) itself can also be written in the form

$$\rho \ddot{x}_i = \frac{\partial}{\partial x_j} \left(\frac{\rho}{\rho_0} \frac{\partial x_i}{\partial X_a} \frac{\partial x_j}{\partial X_b} p_{ka}^{-1} p_{lb}^{-1} \Sigma_{kl}(E_{pq}) \right). \tag{9.58}$$

Thus when the (p_{ak}) are prescribed, equation (9.58) becomes the required system of partial differential equations for the components $x_i(X_a, t)$ of the motion.

For small deformations, the approximations of equation (9.54) can be used. Then equation (9.57) gives, to first order,

$$E_{ik} = \tfrac{1}{2}(\partial_i u_k + \partial_k u_i - \beta_{ik} - \beta_{ki}). \tag{9.59}$$

Using the strain energy function of equation (2.92) we can now put the equations of motion (9.56) in the first-order form

$$\rho \ddot{u}_i = c_{ijkl} \partial_j (\partial_k u_l - \beta_{kl}). \tag{9.60}$$

So, as in the previous simple deformations (cf. eqs. (9.16) and (9.44)), the presence and motion of dislocations in the body add an extra term in the Navier equations proportional to the derivative of the plastic distortion. We shall discuss the solution of these equations in §9.4, but first we shall relate the plastic distortion to the dislocation flow field.

B The dislocation flux tensor

The dislocation density tensor defined by equations (7.1) and (7.2) in terms of dislocation lines or by equation (7.3) in terms of Burgers circuits is not ideally suited to treating non-linear flow situations. It is more appropriate to use quantities referring to the reference configuration ($\mathbf{X}$) than to the current one. Thus if the Burgers circuit $\mathscr{C}$ in equation (7.3) corresponds to an element $N_a dS_0$ of surface in the global reference state, then we write the *true* Burgers vector of $\mathscr{C}$ as

$$\mathbf{B}(\mathscr{C}) = A_{ka}N_a dS_0 \mathbf{u}_k. \tag{9.61}$$

Thus $A_{ka}\mathbf{u}_k$ is the total true Burgers vector of all the dislocation lines in the current configuration which cross a unit area of the reference configuration normal to the ath direction. The relationship between A_{ka} and a_{kl} can be obtained by using equation (2.54), which relates elements of area in the two states together with the transformation between local and true Burgers vectors. It is

$$A_{lc} = (u|e)_{ml}a_{mj} \frac{\partial X_c}{\partial x_j} \det\left(\frac{\partial x_i}{\partial X_a}\right). \tag{9.62}$$

From equation (7.38) we then obtain

$$A_{lc} = -\varepsilon_{cab}\, \partial p_{bl}/\partial X_a. \tag{9.63}$$

The advantage of this dislocation density is now immediately apparent, since upon taking the time derivative of this equation with $\mathbf{X}$ held fixed, we obtain a conservation equation of the usual type:

$$\dot{A}_{lc} = -\partial(\varepsilon_{cab}\dot{p}_{bl})/\partial X_a. \tag{9.64}$$

Thus it is clear that if we seek to generalize the conservation equations (9.36) to the non-linear case, it will be simplest to deal with these new densities.

Each type, i, of dislocation will have its own associated density tensor, which can be written

$$A_{lc}^{(i)} = N^{(i)}B_l^{(i)}T_c^{(i)}, \tag{9.65}$$

where $N^{(i)}$ is the number of i lines per unit reference area, $\mathbf{B}^{(i)}$ is the relevant true Burgers vector, and $\mathbf{T}^{(i)}$ is a unit vector along the i lines in the positions they would occupy if mapped onto the global reference configuration. The total density is

$$A_{lc} = \sum_{(i)}A_{lc}^{(i)}.$$

Now in the dynamic situation each line will have a certain velocity, and we suppose that all lines of type i have the same velocity $\mathbf{V}^{(i)}$ relative to the $\mathbf{X}$-configuration. Then we define the dislocation flux tensor for dislocations of type i as

$$F_{ldc}^{(i)} = A_{lc}^{(i)}V_d^{(i)} \tag{9.66}$$

and the total flux to be the sum of these over all dislocation types:

$$F_{ldc} = \sum_{(i)} F_{ldc}^{(i)}. \tag{9.67}$$

Consider now a material element $\delta \mathbf{X}$. During a small time interval δt, all those dislocations of type i lying within an area $\mathbf{V}^{(i)} \delta t \wedge \delta \mathbf{X}$ cross the element. Therefore the total amount of true Burgers vector crossing $\delta \mathbf{X}$ is

$$\sum_{(i)} A_{lc}^{(i)} \mathbf{u}_l \varepsilon_{cda} \delta X_a V_d^{(i)} \delta t = \varepsilon_{cda} F_{ldc} \delta X_a \delta t \, \mathbf{u}_l. \tag{9.68}$$

This flow produces a plastic change in the corresponding element $\delta \mathbf{x}$ in the current configuration equal to the local Burgers vector crossing it, so that

$$\delta \dot{\mathbf{x}}^{(p)} = \varepsilon_{cda} F_{ldc} \delta X_a \mathbf{e}_l. \tag{9.69}$$

Now

$$\delta \mathbf{x} = \frac{\partial x_i}{\partial X_a} \mathbf{u}_i \delta X_a = p_{ak} \mathbf{e}_k \delta X_a.$$

Thus the total rate of change of $\delta \mathbf{x}$ is

$$\delta \dot{\mathbf{x}} = \dot{p}_{ak} \mathbf{e}_k \delta X_a + p_{ak} \dot{\mathbf{e}}_k \delta X_a. \tag{9.70}$$

The first term here corresponds to the plastic change $\delta \dot{\mathbf{x}}^{(p)}$ calculated above, and the second gives the change in $\delta \mathbf{x}$ induced by any change in the lattice vectors. The latter will in general contain contributions from both elastic strain and rotation. However, we are at the moment concerned only with the plastic term, and comparing it with equation (9.69) gives

$$\dot{p}_{ak} = \varepsilon_{cda} F_{kdc}. \tag{9.71}$$

Combining equations (9.64) and (9.71) gives the dislocation conservation equation in the almost familiar form:

$$\dot{A}_{lc} = -\partial(F_{lbc} - F_{lcb})/\partial X_b. \tag{9.72}$$

Recalling the definition (9.66) of the flux tensor, we see that the first term on the right is the usual convective term. The second term arises from other processes which we shall discuss to some extent in §c.

For small deformations, the distinction between the different density tensors becomes unimportant. To first order we have, from equations (7.38) and (9.63),

$$a_{mj} = A_{mj} = -\varepsilon_{jpl} \partial_p \beta_{lm}. \tag{9.73}$$

Equation (9.71) gives

$$\dot{\beta}_{ak} = \varepsilon_{cda} F_{kdc}, \tag{9.74}$$

and in calculating F_{kdc} we can, to this approximation, use the dislocation velocities relative to the material in the current configuration. We can also treat $\dot{\beta}_{ak}$ as a spatial time derivative to first order.

c Dislocation conservation equations

Consider a small surface element S with boundary $\mathscr{C}$, drawn in the reference configuration $(\mathbf{X})$. During a time δt the total Burgers vector which crosses into the corresponding element in the current configuration due to the motion of i-type dislocations is obtained from equation (9.68) as

$$-\oint_{\mathscr{C}} \epsilon_{cda} F^{(i)}_{ldc} dX_a \delta t \, \mathbf{u}_l.$$

Using Stokes' theorem, we can transform this into a surface integral over S to obtain

$$-\int_S \epsilon_{eba} \frac{\partial}{\partial X_b} (\epsilon_{cda} F^{(i)}_{ldc}) N_e \, dS \, \delta t \, \mathbf{u}_l = -\int_S \frac{\partial}{\partial X_b} (F^{(i)}_{lbc} - F^{(i)}_{lcb}) N_c \, dS \, \delta t \, \mathbf{u}_l.$$

Now we would like to equate this with the change in the Burgers vectors of type i lines which thread the circuit S, namely

$$\delta \int_S A^{(i)}_{lc} N_c \, dS \, \mathbf{u}_l - \int_S \dot{A}^{(i)(p)}_{lc} N_c \, dS \, \mathbf{u}_l \, \delta t.$$

The last term here accounts for the possibility of creation of dislocations of the ith type. If we do this, we obtain the conservation law

$$\dot{A}^{(i)}_{lc} = -\partial (F^{(i)}_{lbc} - F^{(i)}_{lcb})/\partial X_b + \dot{A}^{(i)(p)}_{lc}. \tag{9.75}$$

This is the direct extension of equations such as (9.4) and (9.5). What is more, by summing over all types and noting that the sum of the production terms must vanish, we obtain a conservation law for the total dislocation density which is identical with equation (9.72). Nevertheless equation (9.75) will not in the general case be valid for each type i.

The reason for this is that there are other mechanisms in a three-dimensional state of strain which can change the number of dislocations of a particular type threading a surface S. There will in general exist in such a state dislocation lines with corners on them at which the type of dislocation changes. The passage of such corners through the element of surface S will not change the total Burgers vector of the corresponding circuit, but it will change the contributions to that Burgers vector from the two types of dislocation whose intersection forms the corner. A similar result occurs if dislocation nodes pass through S. A second mechanism leading to a violation of equation (9.75) is the possibility that the velocity of a dislocation line may not be constant along the line, so that its motion either produces a change in type to a mixed dislocation or creates one or more kinks of some other dislocation type along the length of the line. These questions have been investigated[10] for simple cubic crystals, and the correct conservation laws have been written down when mixed dislocations do not occur. As might be expected, new factors arise, and in particular the conservation laws are found to contain the densities of various dislocation corners. Because of this, it is uncertain as yet whether much use can be made of them.

Dislocation corners and nodes will be absent (in an average sense) if each dislocation type satisfies the condition (cf. eq. (7.8))

$$\partial A_{lb}^{(i)}/\partial X_b = 0.$$

The velocity of dislocations of type i will be uniform along the line direction, $\mathbf{T}^{(i)}$, if $T_b^{(i)}\partial V_c^{(i)}/\partial X_b = 0$, and using equation (9.65) this implies that

$$A_{lb}^{(i)}\partial V_c^{(i)}/\partial X_b = 0.$$

If these two conditions hold, then the conservation law (9.75) is valid for each type of dislocation. Furthermore, $\partial F_{lcb}^{(i)}/\partial X_b = 0$ in this case, and the conservation equation takes the simpler form

$$\dot{A}_{lc}^{(i)} = \partial F_{lbc}^{(i)}/\partial X_b + \dot{A}_{lc}^{(i)(p)}. \tag{9.76}$$

By adding and subtracting the equations corresponding to positive and negative dislocations of each type, we can obtain pairs of equations similar to equations (9.6) and (9.7). These will provide a valid description of the dislocation flow process in cases where the plasticity is dominated by the uniform motion of long straight dislocation segments. They will not be applicable to plastic flows involving substantial numbers of non-elongated dislocation loops. Although this appears to make equation (9.76) of only rather limited interest, this is in fact nor entirely the case, since there is substantial experimental evidence that in a number of materials edge dislocations travel much faster than screws (see §9.6c) so that loops produced by Frank-Read sources rapidly become elongated and the plastic flow in such materials is controlled by the motion of the remaining long straight screws.

9.4
DYNAMICAL SOLUTIONS FOR INFINITESIMAL STRAINS

A *The general solution*

The system of equations describing the deformation field of a distribution of moving dislocations can be summarized for small distortions as follows:

$$\rho\ddot{u}_i = \partial_j\sigma_{ij}, \tag{9.77}$$

$$\sigma_{ij} = c_{ijkl}\alpha_{kl}, \tag{9.78}$$

M

$$\partial_i u_j = \alpha_{ij} + \beta_{ij}, \tag{9.79}$$

$$\varepsilon_{jkl}\partial_k \beta_{lm} = -a_{mj}. \tag{9.80}$$

The first of these is just the Cauchy equation of motion, and the second is the stress-strain relation expressed in terms of the lattice distortion α_{kl} (cf. eq. (7.11)). When plastic flow occurs as part of the deformation, the lattice distortion is not equal to the displacement gradient, and the difference between these two defines the plastic distortion β_{ij} in equation (9.79). Equation (9.80) is obtained by substituting from equation (9.79) into equation (7.13).

We have seen for the antiplane and plane cases that the time derivatives of the plastic distortion components are equal to appropriate dislocation fluxes (cf. eqs. (9.11), (9.32), and (9.42)). These equations may readily be generalized to arbitrary dislocation flows. The required results were in fact derived in §9.3 from the non-linear theory and are contained in equation (9.74), where the dislocation flux tensor F_{kdc} is defined in equations (9.66) and (9.67). In these definitions, $A_{lc}^{(i)}$ denotes the density tensor for dislocations of the ith type and $V_d^{(i)}$ denotes their velocity. The total flux is obtained by summing over all types of dislocations.

We wish to solve these equations in the domain $\mathscr{B}$ occupied by the body and in a certain time interval $\mathscr{T}$, with the dislocation density a_{mj} and the rate of plastic distortion $\dot\beta_{lm} = \varepsilon_{lpq}F_{mqp}$ being given functions of $\mathbf{r}$ and t. (Of course, in view of the fourth equation, only the initial values of a_{mj} need be specified if $\dot\beta_{lm}$ is also given.) We shall also need certain boundary conditions on $\partial\mathscr{B}$, which we take to be of the usual traction or displacement types on two parts of $\partial\mathscr{B}$:

$$\sigma_{ij}n_j = t_i \quad \text{on } S_1, \qquad u_i = U_i \quad \text{on } S_2. \tag{9.81}$$

We shall take the interval $\mathscr{T}$ to be $(-\infty, \infty)$, although it is a simple matter to extend the discussion to an initial-value problem in which the displacement and velocity fields, u_i and $\dot u_i$, are given at some initial instant.

In §2.4 we introduced the Green's function for static elasticity, which is chosen to satisfy equation (2.41) together with homogeneous boundary conditions on S_1 and S_2. The generalization of these equations to the dynamical Green's function, $u_{lm}(\mathbf{r}, \mathbf{r}', t-t')$, is as follows:

$$\rho\ddot u_{im} = c_{ijkl}\partial_j\partial_k u_{lm} + \delta_{im}\delta(\mathbf{r}-\mathbf{r}')\delta(t-t'), \tag{9.82}$$

$$c_{ijkl}\partial_k u_{lm}n_j = 0 \quad \text{for } \mathbf{r} \text{ on } S_1,$$
$$u_{lm} = 0 \quad \text{for } \mathbf{r} \text{ on } S_2. \tag{9.83}$$

It is also necessary to stipulate some initial conditions in order to determine the Green's function uniquely, and we take these to be that $u_{lm}(\mathbf{r}, \mathbf{r}', t-t') \equiv 0$ for $t < t'$, with suitable continuity at $t = t'$. The physical interpretation of u_{lm} is that it is the l-component of displacement at position $\mathbf{r}$ and time t caused by a unit point impulse in the m-direction acting at $\mathbf{r}'$ at time t' in a medium which is undisturbed until that instant. The static Green's function can be obtained by integrating the dynamical one over t'.

We can use the dynamical Green's function to solve the boundary-value problem (9.77)–(9.81) in much the same way as in §2.4. First of all, using the divergence theorem and the boundary conditions (9.81) and (9.83), we have

$$\int_{\mathscr{B}} c_{ijkl}\partial_j(u_i\partial_k u_{lm}-u_{im}\alpha_{kl})dV = -\int_{S_1} u_{im}t_i dS + \int_{S_2} U_i c_{ijkl}\partial_k u_{lm}n_j dS.$$

We have suppressed the arguments of the functions involved here: all functions have arguments $(\mathbf{r}, t)$ except the Green's functions, which are $u_{lm} = u_{lm}(\mathbf{r}, \mathbf{r}', t'-t)$, and integrations and differentiations are with respect to $\mathbf{r}$. Expanding the derivatives on the left, two terms cancel in view of the symmetries of the modulus tensor, and the remaining terms can be re-expressed using the field equations (9.77)–(9.80) and (9.82) to give

$$\int_{\mathscr{B}} \{\rho[u_i\ddot{u}_{im}-u_{im}\ddot{u}_i]+c_{ijkl}\beta_{ij}\partial_k u_{lm}\}dV - u_m(\mathbf{r}', t)\delta(t'-t).$$

The first term is the t-derivative of

$$\rho[u_i(\mathbf{r}, t)\dot{u}_{im}(\mathbf{r}, \mathbf{r}', t'-t)-\dot{u}_i(\mathbf{r}, t)u_{im}(\mathbf{r}, \mathbf{r}', t'-t)],$$

which gives zero when we integrate over t from $-\infty$ to $+\infty$ provided the displacement and velocity u_i and $\dot{u}_i$ approach 0 as $t \to -\infty$. Thus we have

$$u_m(\mathbf{r}', t') = \int_{-\infty}^{\infty} \int_{\mathscr{B}} c_{ijkl}\beta_{ij}(\mathbf{r}, t)\partial_k u_{lm}(\mathbf{r}, \mathbf{r}', t'-t)dVdt$$

$$- \int_{-\infty}^{\infty} \int_{S_1} t_i(\mathbf{r}, t)u_{im}(\mathbf{r}, \mathbf{r}', t'-t)dSdt$$

$$+ \int_{-\infty}^{\infty} \int_{S_2} U_i(\mathbf{r}, t)c_{ijkl}\partial_k u_{lm}(\mathbf{r}, \mathbf{r}', t'-t)n_j dSdt. \qquad (9.84)$$

This solution takes on a more familiar form if we interchange the roles of $\mathbf{r}$ and $\mathbf{r}'$ and of t and t', and use the reciprocity condition that $u_{lm}(r, r', \tau)$ $= u_{ml}(\mathbf{r}', \mathbf{r}, \tau)$. (This can be derived by a similar argument to the one above, with $u_i(\mathbf{r}, t)$ replaced by $u_{ip}(\mathbf{r}, \mathbf{r}'', t''-t)$.) We obtain

$$u_m(\mathbf{r}, t) = \int_{-\infty}^{\infty} \int_{\mathscr{B}} c_{ijkl}\beta_{ij}(\mathbf{r}', t')\partial_k'u_{ml}(\mathbf{r}, \mathbf{r}', t-t')dV'dt'$$

$$- \int_{-\infty}^{\infty} \int_{S_1} t_i(\mathbf{r}', t')u_{mi}(\mathbf{r}, \mathbf{r}', t-t')dS'dt'$$

$$+ \int_{-\infty}^{\infty} \int_{S_2} U_i(\mathbf{r}', t')c_{ijkl}\partial_k u_{ml}(\mathbf{r}, \mathbf{r}', t-t')n_j'dS'dt'. \qquad (9.85)$$

From now on we shall drop the last two boundary terms from the solution, keeping only the first one, which provides the dislocation displacement field under homogeneous boundary conditions (9.81). The omitted terms present no difficulties, and in fact simply correspond to the solution of an appropriate boundary-value problem of classical elasticity. The remaining term can be expressed explicitly in terms of given quantities by integrating by parts with respect to t':

$$u_m(\mathbf{r}, t) = - \int_{-\infty}^{\infty} \int_{\mathscr{B}} c_{ijkl}\dot{\beta}_{ij}(\mathbf{r}', t')\partial_k' h_{ml}(\mathbf{r}, \mathbf{r}', t-t')dV'dt', \qquad (9.86)$$

where

$$h_{ml}(\mathbf{r}, \mathbf{r}', t-t') = \int_t^{t'} u_{ml}(\mathbf{r}, \mathbf{r}', t-t'')dt''. \qquad (9.87)$$

In this connection it should be noted that the t' integrations in equations (9.85) and (9.86) actually terminate at the upper limit $t' = t$ since the Green's function vanishes for larger values of t'. In arriving at equation (9.86) we have also dropped the term involving the limiting plastic distortion at $t' \to -\infty$ (cf. eq. (9.19) and the succeeding discussion).

The total displacement gradient is obtained by differentiating any one of these solutions for $u_m(\mathbf{r}, t)$. Substituting into equation (9.79) gives the elastic distortion

$$\alpha_{nm}(\mathbf{r}, t) = -\beta_{nm}(\mathbf{r}, t) + \int_{-\infty}^{\infty} \int_{\mathscr{B}} c_{ijkl}\beta_{ij}(\mathbf{r}', t')\partial_n\partial_k' u_{lm}(\mathbf{r}', \mathbf{r}, t-t')dV'dt'$$

$$= \int_{-\infty}^{\infty} \int_{\mathscr{B}}\{-\beta_{nj}(\mathbf{r}', t')\,[\rho\ddot{u}_{jm}(\mathbf{r}', \mathbf{r}, t-t') - c_{ijkl}\partial_j'\,\partial_k' u_{lm}(\mathbf{r}', \mathbf{r}, t-t')]$$

$$+ c_{ijkl}\beta_{ij}(\mathbf{r}', t')\partial_n\partial_k' u_{lm}(\mathbf{r}', \mathbf{r}, t-t')\}dV'dt'.$$

The inserted terms here can be reduced to a δ-function term on using the differential equation satisfied by the Green's function, and this can be integrated to give the preceding expression. We now integrate the first term by parts with respect to t', again dropping the terms involving the limiting values as $t' \to -\infty$. Then the remaining terms are rewritten using the divergence theorem, equation (9.80), and the reciprocity conditions to give finally

$$\alpha_{nm}(\mathbf{r}, t) = \int_{-\infty}^{\infty} \int_{\mathscr{B}}[-\dot{\beta}_{nj}(\mathbf{r}', t')\rho\dot{u}_{mj}(\mathbf{r}, \mathbf{r}', t-t')$$

$$+ \varepsilon_{inh}a_{jh}(\mathbf{r}', t')c_{ijkl}\partial_k' u_{ml}(\mathbf{r}, \mathbf{r}', t-t')]dV'dt'$$

$$+ \int_{-\infty}^{\infty} \int_{\mathscr{B}} \beta_{ij}(\partial_n + \partial_n')\sigma_{ijm}dV'dt'$$

$$+ \int_{-\infty}^{\infty} \int_{\partial\mathscr{B}} [\beta_{nj}n_i - \beta_{ij}n_n]\sigma_{ijm}dS'dt'. \qquad (9.88)$$

Here we use the notation $\sigma_{ijm} = c_{ijkl}\partial_k' u_{lm}(\mathbf{r}', \mathbf{r}, t-t')$.

The stress-field is obtained by substituting this solution into equation (9.78).

The derivation presented above was first given by Mura,[11,12] and Kosevich[13,14] found the solutions using tensor potential methods. The earliest solution of problems of this kind was given by Nabarro[2] for the motion of a single dislocation loop. The use of dynamical Green's functions to solve elastodynamic initial-value problems is, of course, much older than this. (See, for example, the book by A.E.H. Love,[15] pp. 302–5.)

B *Solution for an infinite medium*

In an infinite medium the Green's function depends only on the difference between the two spatial position vectors: $u_{lm} = u_{lm}(\mathbf{r}-\mathbf{r}', t-t')$. Because of this, the second integral in equation (9.88) vanishes. The third integral, from the boundary at infinity, is also zero provided $\beta_{ij} \to 0$ sufficiently fast at large distances. Thus we are left with only the first term, which is a formula first given by Mura.[11,12]

For an infinite isotropic medium, the dynamical Green's function is given by:

$$4\pi \rho u_{km}(\mathbf{r}, t) = \frac{x_k x_m}{r^3}\left[\frac{1}{c_1^2}\delta\left(t - \frac{r}{c_1}\right) - \frac{1}{c_2^2}\delta\left(t - \frac{r}{c_2}\right)\right]$$

$$+ \frac{\delta_{km}}{r}\frac{1}{c_2^2}\delta\left(t - \frac{r}{c_2}\right) + \partial_k\partial_m(r^{-1})\, t\left[H\left(t - \frac{r}{c_1}\right) - H\left(t - \frac{r}{c_2}\right)\right]. \quad (9.89)$$

where $c_1^2 = (\lambda+2\mu)/\rho$ and $c_2^2 = \mu/\rho$. Recalling that $u_{km}(\mathbf{r}, t)$ represents the displacement at $\mathbf{r}$ caused by a certain impulse acting at the origin at time zero, we see that physically the various terms here represent two sharp signals arriving at times $t = r/c_1$ and $t = r/c_2$ together with a continuous disturbance between these two times, arising from the last term. Thus $u_{km}(\mathbf{r}, t)$ is non-zero only in between and on the two characteristic cones of slopes c_1^{-1} and c_2^{-1} emanating from the origin, that is in the shaded region of Figure 9.2.

In order to verify that equation (9.89) does indeed give the Green's function, we can first of all show that equation (9.82) is satisfied at all points except $\mathbf{r} = 0$, $t = 0$. This is most easily done by direct substitution from equation (9.89), using the elastic modulus tensor of equation (2.5). It then remains to be shown that $u_{km}(\mathbf{r}, t)$ has the correct singularity at $\mathbf{r} = 0$, $t = 0$. For this it is sufficient if we verify that

$$\iint f(\mathbf{r}, t)\,(\rho\ddot{u}_{im} - c_{ijkl}\partial_j\partial_k u_{lm})dVdt = f(0,0)\delta_{im},$$

where $f(\mathbf{r}, t)$ is any function continuous at $\mathbf{r} = 0$, $t = 0$ and where the integration is carried out over any region containing the origin. Since the integrand vanishes at all points except the origin, the shape of the region is immaterial, and we take it as the cylinder $r < \varepsilon$, $|t| < \delta$, where $\delta < \varepsilon/c_2$

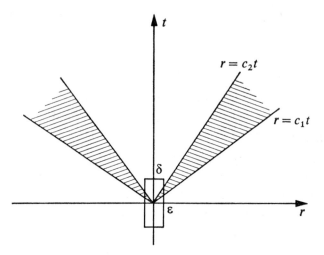

FIGURE 9.2

(see Fig. 9.2). We can also replace $f(\mathbf{r}, t)$ by $f(0, 0)$ if ε and δ are sufficiently small. Then the term $\ddot{u}_{im}$ can be integrated with respect to t, and so depends on the values of $\dot{u}_{im}$ on $t = \pm \delta$. Since the two cylinder ends do not intersect the region between the two characteristic cones, $\dot{u}_{im}$ is zero there. The second term can be transformed to a surface integral over $r = \varepsilon$ using the divergence theorem, to give

$$-f(0, 0) \int_{r=\varepsilon} c_{ijkl} \left(\partial_k \int_{-\delta}^{\delta} u_{lm}(\mathbf{r}, t)dt \right) n_j dS.$$

Now it is easily verified from equation (9.89) that the integral of $u_{lm}(\mathbf{r}, t)$ over any time interval which overlaps both characteristic cones gives the static Green's function $u_{lm}(\mathbf{r}, 0)$ of equation (2.44). But this function satisfies condition (2.36d), which gives us precisely the result that we want.

c Single dislocation lines

The general solutions of §A can readily be applied to give the disturbance produced by the motion of a single curved dislocation line. We suppose that the line has Burgers vector b_i, and that at time t it occupies a curve $\Gamma(t)$. Then the integration of the dislocation density across the filament cross-section is given by equation (8.28), and from equations (9.74) and (9.66) we have the corresponding result for the rate of plastic distortion:

$$dl' \int \dot{\beta}_{ak}(\mathbf{r}', t')dS' = \varepsilon_{cda}dl' \int F_{kdc}(\mathbf{r}')dS' = \varepsilon_{cda}b_k V_d dl_c'. \qquad (9.90)$$

Here $\mathbf{dl}'$ is a vector element along $\Gamma(t')$ and V_d is the velocity of the dislocation line at the point of interest.

Substituting these expressions into equations (9.86) and (9.88) and using the Green's function appropriate for an infinite medium, we have

$$u_m(\mathbf{r}, t) = \int_{-\infty}^{\infty} \oint_{\Gamma(t')} c_{ijkl} \varepsilon_{pqi} b_j V_q(\mathbf{r}', t') \partial_k h_{ml}(\mathbf{r} - \mathbf{r}', t - t') dl_p' dt', \qquad (9.91)$$

$$\alpha_{nm}(\mathbf{r}, t) = \int_{-\infty}^{\infty} \oint_{\Gamma(t')} \varepsilon_{ipn} b_j [\rho V_i \dot{u}_{mj}(\mathbf{r} - \mathbf{r}', t - t')$$
$$+ c_{ijkl} \partial_k u_{ml}(\mathbf{r} - \mathbf{r}', t - t')] dl_p' dt. \qquad (9.92)$$

In these integrals, $\mathbf{r}'$ denotes the position of the running point on the dislocation line in the configuration $\Gamma(t')$ that it occupies at time t'. The integral over t' extends only over the range $t' < t$ since the Green's function is zero outside this interval. In fact, as can be seen from the properties of the Green's function as demonstrated in Figure 9.2, the t' integral actually terminates at a somewhat earlier time, which is the latest instant at which dilatation waves transmitted by the moving dislocation can reach the point $\mathbf{r}$ before the current time t.

For the particular case of a screw dislocation lying parallel to the z-direction and such that it intersects the xy-plane at the point $(\xi(t), \eta(t))$ at time t, we can set $b_j = b\delta_{j3}$, $V_q = \dot{\xi}\delta_{q1} + \dot{\eta}\delta_{q2}$, and $dl_p' = dz'\delta_{p3}$. Substituting the Green's function from equation (9.89) and integrating with respect to z', equation (9.23) is retrieved.

For a straight edge dislocation, if we consider glide motions only, we can take both the Burgers vector and the velocity in the x-direction, with the dislocation line in the z-direction. Then we obtain for the velocity and elastic strain-fields[16]:

$$\dot{u}_m = -(bc_2^2/2\pi)(\partial_1 A_{2m} + \partial_2 A_{1m}),$$

$$\alpha_{1m} = (bc_2^2/2\pi)(\partial_1 \phi_{2m} + \partial_2 \phi_{1m}), \qquad (9.93)$$

$$\alpha_{2m} = -(\mu b/2\pi)\dot{A}_{2m} + (bc_2^2/\pi)\partial_2\phi_{2m} - (bc_1^2/2\pi)(\partial_1\phi_{1m} + \partial_2\phi_{2m}),$$

where

$$A_{nm}(\mathbf{r}, t) = 2\pi\rho \iint_{-\infty}^{\infty} v(t') u_{nm}(\mathbf{r} - \mathbf{r}', t - t') dz' dt',$$
$$\phi_{nm}(\mathbf{r}, t) = 2\pi\rho \iint_{-\infty}^{\infty} u_{nm}(\mathbf{r} - \mathbf{r}', t - t') dz' dt', \qquad (9.94)$$

with $v(t)$ being the dislocation velocity at time t. (We note that the velocity field is given by equation (9.91) but with h_{ml} replaced by $-u_{ml}$.) After substituting for the Green's function, the z' integrations can be performed, and we obtain

$$A_{km}(\mathbf{r}, t) = \int_{-\infty}^{\infty} \left[\left(\frac{\bar{t}^2}{S_1} - S_2 \right) \frac{\bar{x}_k \bar{x}_m}{\bar{R}^4} + \left(\frac{\bar{t}^2}{S_2} - S_1 \right) \frac{\bar{R}^2 \delta_{km} - \bar{x}_k \bar{x}_m}{\bar{R}^4} \right] v(t') dt', \qquad (9.95)$$

ϕ_{km} being given by a similar integral with $v(t')$ replaced by 1. The notation used here is $\tilde{t} = t - t'$, $\bar{x}_1 = x - \xi(t')$ with $\xi(t')$ the position of the dislocation at time t', $\bar{x}_2 = y$, $\bar{R}^2 = \bar{x}_1^2 + \bar{x}_2^2$, and $S_i^2 = \tilde{t}^2 - \bar{R}^2/c_i^2$. Although we have written the t' integral as going from $-\infty$ to ∞, it is to be understood that the upper limit is actually finite, and is, for each term in the integrand, the largest value of t' for which the corresponding quantity S_i is real.

Although these formulae do provide the solution to the general motion of an edge dislocation, their application to explicit problems, even as simple as the uniform motion case, is very tedious. It should also be pointed out that some care is necessary to avoid the occurrence of divergent integrals. For uniform motion it is possible to do this by supposing that the dislocation is stationary up to some instant $(-T)$ in the distant past. Then although the A_{nm} diverge if we let $T \to \infty$, their derivatives do not, so that the velocity field is well defined in this limit. For problems involving bounded motions of an edge dislocation these difficulties do not arise, and Kiusalaas and Mura[16] have used the results to derive the distant fields due to a small sinusoidal oscillation of an edge about a fixed position.

9.5

THE EQUATION OF MOTION OF A DISLOCATION

A *Qualitative considerations*

The derivation of equation (9.27) and the resulting value of the effective mass of a screw dislocation were based solely on the energy of a steadily moving dislocation, and ignored the effects on the energy of the acceleration itself. We can get a rough estimate of these effects in the following way. Any change dv in the dislocation velocity occurring at time t' produces a disturbance which travels outwards from the dislocation with velocity c equal to the shear wave velocity. At a later time t the disturbance fills a cylinder of radius $c(t - t')$, and so corresponds very roughly to an extra energy

$$(\mu b^2/4\pi)\ln[c(t-t')/r_0]\,d\beta^{-1}(t')$$

if we use the energy of a steadily moving dislocation with R replaced by $c(t - t')$. Integrating this over t' then gives the total energy due to a continuously changing velocity $v(t')$. If we differentiate this result with respect to t, we obtain the force on the dislocation,

$$\frac{F(t)}{L} = \frac{\mu b^2}{4\pi v(t)c^2} \int_{t-R/c}^{t-r_0/c} \frac{v(t')\dot{v}(t')}{t-t'}\,dt'.$$

We have here replaced β by 1 and have assumed that only the time taken for the cylindrical wave-front to spread from the inner core radius r_0 to the outer radius R is relevant for calculating the rate of change of energy.

To take an example, we suppose that the dislocation is at rest for $t' < 0$ and begins to move with a constant acceleration $\dot{v}$ at time $t' = 0$. Carrying out the integrations for $t > R/c$, we obtain

$$\frac{F(t)}{L} = \frac{\mu b^2}{4\pi c^2} \dot{v} \left(\ln \frac{R}{r_0} - \frac{R - r_0}{ct} \right).$$

The first term agrees with equation (9.28) and the second term gives a moderately small correction to that earlier result. On the other hand, for $t < R/c$, the lower limit in the integral must be replaced by $t' = 0$, and we have something quite different from equation (9.28):

$$\frac{F(t)}{L} = \frac{\mu b^2}{4\pi c^2} \dot{v} \left[\ln \frac{ct}{r_0} - \left(1 - \frac{r_0}{ct} \right) \right]. \tag{9.96}$$

Here the first (dominant) term increases with time, so that an increasing force must be applied to maintain a constant acceleration. We could interpret this as giving a logarithmically increasing effective mass to the dislocation. Conversely, if a constant force is applied to the dislocation, its acceleration will decrease until such a time that the zone of influence of the waves caused by the acceleration occupies the whole body.

We shall investigate these questions in more detail in §B. Meanwhile we shall describe briefly other factors affecting the equation of motion of a dislocation.

A dislocation moving through a crystalline material will undergo periodic changes as it moves from one lattice cell to the next. For example, even if its velocity is on the average constant, the dislocation will experience periodic accelerations and decelerations as it moves up and down the Peierls-Nabarro energy hills. These changes will produce a scattered elastic wave which will carry a certain amount of energy away from the dislocation. If a steady velocity is to be maintained, an external shear stress σ must be applied of such a magnitude that its rate of working, $b\sigma v$, balances the rate of radiation of energy. An approximate calculation of this effect[17] gives, for uniform velocity $v \ll c$, $\sigma/\mu \propto (v/c)^2$, with σ approaching values of the same order of magnitude as the Peierls-Nabarro resistance.

A second source of energy dissipation, relevant to edge dislocations, is thermoelastic damping. There is in most materials a coupling between the strain and temperature fields such that a rapid (and hence adiabatic) compression leads to a small increase in temperature while a rapid dilatation causes a decrease in temperature. Thus as a fast-moving edge glides past any point a temperature gradient is established across the slip-plane near the core and heat flows from the compressed to the extended side. This results in a conversion of the mechanical energy of the dislocation into internal heat energy, and again the rate of energy dissipation must be set equal to $b\sigma v$ to maintain a constant velocity.[18,19]

M*

A third mechanism arises from the interaction between the moving dislocation and sound waves (or phonons) present as thermal fluctuations in the material.[20,21]. There are two causes of such interaction. One is the direct scattering of phonons by the region of non-linear strain at the dislocation core. The other is indirect: the dislocation oscillates in the varying stress-field of the phonon and hence generates a radiating wave which carries energy away.

One final question we wish to mention is the possibility of a Lorentz force on a moving dislocation. We saw in §9.1D that there was a correspondence between the electromagnetic fields of moving line charges and the elastic fields of moving screw dislocations. Now the moving charge experiences a force per unit length equal to $e(\mathbf{E} + \mathbf{v} \wedge \mathbf{B})$, using MKS units, with e the charge per unit length. Replacing e by $(\varepsilon_0\mu)^{1/2}b$ and $\mathbf{E}$ and $\mathbf{B}$ by their analogous variables for the dislocation case, we would have a force per unit length on the dislocation line equal to $b(\sigma_{23} + \rho\dot{\eta}\dot{u}_3, -\sigma_{13} - \rho\dot{\xi}\dot{u}_3, 0)$, where $\mathbf{v} = (\dot{\xi}, \dot{\eta}, 0)$ is the velocity. The first terms in the two force components are the familiar static forces on the dislocation due to an external stress-field. The second terms, which correspond to the Lorentz force on a moving charge, are new. Much discussion has been given in the literature as to whether they should in fact be present.[21-23] They represent a force on the dislocation proportional to the absolute velocity of the medium at the dislocation position.

B *Eshelby's equation of motion*

The equation of motion of a screw dislocation was discussed extensively by Eshelby,[5] who gave an integral equation not only for the position of the centre of the dislocation but also for the distribution of Burgers vector within the core region, whose structure is allowed to vary during the motion. The basis of the equation is the same as for the Peierls-Nabarro model of a stationary screw, the only change being that the fields in the two half-spaces on either side of the slip-plane are now time-dependent. As in §4.2 we consider a screw whose glide-plane is $y = 0$, and we denote by $u_3(x, y, t)$ the displacement field of the screw alone. The external stress and displacement fields $\sigma_{ij}^{\text{ext}}(x, y, t)$ and $u_3^{\text{ext}}(x, y, t)$, which provide the driving force for the dislocation motion, will be included later. The dislocation field u_3 is an odd function of y, and we let $u(x, t) = u_3(x, 0+, t) = -u_3(x, 0-, t)$ denote its boundary values above and below the slip-plane. As in equation (5.92), the dislocation density corresponding to this smeared-out dislocation is $a_{33}(x, y, t) = -2\partial_x u(x, t)\delta(y)$, and the flux has a non-zero component $\Delta v_1 = 2\partial_t u(x, t)\delta(y)$, as can be seen from equation (9.6). Substituting these values into equations (9.29), and the resulting potentials into equations (9.28), and then integrating over y' and z' provides the stress and velocity fields. After some manipulation the shear stress σ_{yz} on $y = 0$ can be shown to be

$$\sigma_{yz}(x, 0, t) = -\frac{\mu}{\pi} \int_{-\infty}^{\infty} dx' \int_{-\infty}^{t_1} \frac{(x-x')\partial_x'u(x', t')dt'}{c(t-t')^2[c^2(t-t')^2-(x-x')^2]^{1/2}} \quad (9.97)$$

where $t_1 = t - |x - x'|/c$.

The equation of motion can now be obtained by requiring that the total stress and displacement fields satisfy the Peierls-Nabarro condition, or some generalization of it, relating the tractions and displacements across the slip-plane in a non-linear and periodic way. In doing this, it is essential to include both the external fields and the dislocation fields in the same equation. Thus equation (4.52) would be generalized to:

$$\sigma_{yz}(x, 0, t)+\sigma_{yz}^{ext}(x, 0, t) = -\frac{\mu b}{2\pi d} \sin \frac{2\pi}{b} [2u(x, t)+d\partial_y u_3^{ext}(x, 0, t)]. \quad (9.98)$$

The last term here gives the difference in u_3^{ext} between the layers of atoms on either side of the slip-plane whose separation is d. Substituting from equation (9.97) for σ_{yz} gives the required integral equation relating $u(x, t)$ to the external field.

This approach is very comprehensive, giving not only the position but also the shape of the dislocation core as a function of time. Unfortunately the resulting integral equation does not appear to be solvable. For this reason, Eshelby proposed a simpler approach which applies to rigid dislocations which maintain the same core structure during their motion. In such a case we can set

$$u(x, t) = f[x - \xi(t)],$$

where $\xi(t)$ marks the position of the dislocation centre. Furthermore, by applying this to the case $\xi(t) = 0$, the function $f(x)$ is determined by the displacement field of a stationary dislocation. If we assume the results of the Peierls-Nabarro model, equation (4.54) may be used to give

$$u(x, t) = -(b/2\pi) \tan^{-1} \{2[x - \xi(t)]/d\}. \quad (9.99)$$

In actual fact, we may expect these results to be reasonably correct for velocities small compared to c.

An equation for $\xi(t)$ may be derived from the balance of energy. The rate of working of the total stresses as the dislocation moves is

$$W(t) = \int_{-\infty}^{\infty} [\sigma_{yz}^{ext}(x, 0, t)+\sigma_{yz}(x, 0, t)]2\partial_t u(x, t)dx \quad (9.100)$$

per unit length of dislocation line. The first term can be integrated directly if we observe that $\partial_t u(x, t) = -\xi(t)\partial_x u(x, t)$. $\partial_x u(x, t)$ will be almost zero except within a few interatomic spacings of the core; in any applications

σ^{ext} will be almost constant over such a small region, so that the first integral becomes $b\dot{\xi}(t)\sigma_{yz}^{\text{ext}}[\xi(t), 0, t]$. (We have used $u(\infty)-u(-\infty) = -b/2$.) The second term in equation (9.100) could be expressed entirely in terms of $u(x, t)$ by using the shear stress of equation (9.97), in which case equation (9.100) becomes

$$b\sigma_{yz}^{\text{ext}} = \frac{W(t)}{\dot{\xi}(t)} + 2\int_{-\infty}^{\infty} \sigma_{yz}(x, 0, t)\partial_x u(x, t)dx.$$

Now, for a rigid dislocation, this provides the required equation of motion, since there is no accumulation of energy in the dislocation core, and $W(t)$ simply equals the rate of energy dissipation. Some of the physical processes giving rise to such dissipation were described in §A, and a more complete theory of these processes would provide the value of $W(t)$. Then expressing the integral on the right in terms of $u(x, t)$ provides an integral equation for the position $\xi(t)$ of the centre of the dislocation.

Eshelby[5] proceeded somewhat differently, using a Fourier transform technique to deal with this integral. After using equation (9.99) and keeping terms of order $\dot{\xi}/c$ only, he obtained the integral equation

$$b\sigma_{yz}^{\text{ext}} = \frac{W(t)}{\dot{\xi}(t)} + \frac{\rho b^2}{4\pi}\int_{-\infty}^{t}\left[\frac{\ddot{\xi}(\tau)}{[(t-\tau)^2+d^2/c^2]^{1/2}}\right.$$
$$\left. + \frac{d^2/2c^2}{[(t-\tau)^2+d^2/c^2]^{3/2}}\frac{d}{d\tau}\left(\frac{\xi(t)-\xi(\tau)}{t-\tau}\right)\right]d\tau. \quad (9.101)$$

Ignoring dissipation processes ($W = 0$) we can readily calculate the stresses necessary to force any prescribed motion. For example if the dislocation is at rest for $t < 0$ and moves with constant acceleration for $t > 0$, we obtain

$$b\sigma_{yz}^{\text{ext}} = \ddot{\xi}\frac{\rho b^2}{4\pi}\left(\ln\frac{2ct}{d} + \frac{1}{4}\right)$$

for $t \gg d/c$ (that is for times large enough for the disturbance to have propagated beyond the core region). This value is in substantial agreement with the earlier one, equation (9.96).

Eshelby[5] also gives the stresses needed to produce an impulsive change of velocity and a sinusoidal dislocation motion. In the latter case, taking $\ddot{\xi}(\tau) = \ddot{\xi}_0\cos\omega t$, the required stress is, for $\omega d \ll c$,

$$b\sigma_{yz}^{\text{ext}} = m\ddot{\xi}_0\cos(\omega t - \alpha), \quad (9.102)$$

where

$$m = \frac{\rho b^2}{4\pi}\left[\left(\frac{\pi}{2}\right)^2 + \left(\ln\frac{2c}{e^\gamma\omega d}\right)^2\right]^{1/2}$$

and

$$\tan \alpha = \frac{\pi}{2} \Big/ \ln \frac{2c}{e^\gamma \omega d}.$$

γ is Euler's constant, approximately equal to 0.577.

m gives an effective mass of the dislocation when driven by a sinusoidally varying stress, and α is the phase difference between the stress and the dislocation position. When the logarithmic term is dominant, the effective mass has a similar form to the previous case, except that the radius ct of the zone of influence of the acceleration-produced wave is replaced by c/ω, which is proportional to the wavelength of the wave produced by the oscillation. A result similar to this had previously been found by Nabarro[21] using an approximate version of the Peierls-Nabarro condition (9.98).

9.6
MACROSCOPIC PLASTICITY

A *Single dislocation velocity model*

Plastic flow on the macroscopic scale takes place through the production and motion of large numbers of dislocations, and so should be describable in terms of the formalism developed in §§9.1–9.3. There are, however, several additional complications. To take the simplest case, that of antiplane strain with all dislocations having the same velocity, the equations are the same as those of §9.1A, but in solving them we can no longer regard the four quantities a, Δ, v_1, v_2 as being prescribed. The two densities a and Δ can be obtained by solving the flow equations (9.6) and (9.7), but in order to do this we still need additional equations for the velocities and the pair production rate, $\dot{a}_p$.

By suggesting suitable equations for these quantities, Gillis and Gilman[24] were able to solve the problem of the uniform straining of a single crystal under simple tension when only a single slip system is operable. The resulting deformation solution was compared satisfactorily with observations on lithium fluoride crystals. An earlier, less complete, discussion by Johnston and Gilman[25] was applied by Li[26] to give a theory of transient creep, and a more comprehensive theory of creep was developed along these lines by Webster.[27] These solutions all refer to uniform states of plastic strain, but a similar solution to a problem of plane bending, involving a non-uniform distribution of edge dislocations, has also been given.[28]

We shall consider here the simple tension of a single crystal in which the plastic strain occurs by the motion in the x-direction of screw dislocations whose lines are in the z-direction. The tensile axis is taken to make angles α, β, and γ with the x, y, and z axes respectively (see Fig. 9.3, in which the situation is exaggerated by concentrating the slip on a single plane). If the tensile stress is T, which we assume to be prescribed, then the shear stress driving the dislocations is $\sigma_{23} = T \cos \beta \cos \gamma$. For dislocation motion in the x-direction, there is a non-zero component β_2 of plastic distortion contributing to the total deformation gradient $\partial_2 u_3$ (cf. equations (9.10) and (9.11)). Thus the plastic extensional strain in the direction of the tensile axis is $e_L = \beta_2 \cos \beta \cos \gamma$.

For a uniform deformation, equation (9.6) becomes $\partial a / \partial t = 0$, and if equal numbers of positive and negative dislocations are present in the specimen initially, then $a = 0$ during the subsequent deformation. Equation (9.7) becomes $\partial \Delta / \partial t = 2 \dot{a}_p$, and the strain-rate is $\dot{\beta}_2 = \Delta v_1$. Thus if we have available equations of the general form

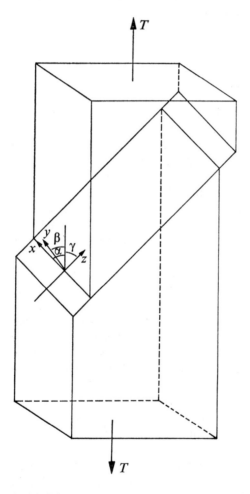

FIGURE 9.3

$$\dot{a}_p = \dot{a}_p(\Delta, a, v_1, \sigma_{23}), \qquad v_1 = v_1(\Delta, a, \sigma_{23}) \tag{9.103}$$

for the production rates and velocities, the problem reduces to solving a single first-order differential equation for Δ as a function of time.

Both Webster[27] and Gillis and Gilman[24] assume a velocity equation based on the experimental measurements of Johnston and Gilman[25] (see §c), with a friction stress which is a function of Δ, taken by Webster to be linear. For a constant stress experiment, however, the dependence of v_1 on σ_{23} need not be specified in order to determine the deformation-versus-time solution, and the assumed velocity equation can be written, for not too large dislocation densities, in the form $v_1 = v_0 - k\Delta$, where v_0 and k are certain functions of σ_{23}. The second term in v_1 accounts for the reduction in dislocation velocity caused by the interactions with other dislocations. As long as $k\Delta \ll v_0$, this linear formula for v_1 will be a plausible approximation to any reasonable relation (9.103_2) for which v_1 is an analytic function of Δ. For creep situations we can expect dislocation densities small enough to satisfy this condition,[27] but for higher plastic strain rates a more general velocity relation is necessary. We shall discuss these in more detail in §c.

The equation for $\dot{a}_p$ has contributions from a number of processes which lead to production or annihilation of dislocations.[24-28] The resulting expression in the most general case can be written as

$$\dot{a}_p = \alpha v_1 + [\beta g(v_1) - \gamma]\Delta v_1 - \delta \Delta^2 v_1 f(v_1). \tag{9.104}$$

The term αv_1 arises from the operation of Frank-Read sources, where α is proportional to the density of the sources and the rate of their operation is proportional to the velocity with which the created dislocations move away and allow the source to operate again. The term $-\gamma \Delta v_1$ accounts for dislocations colliding with fixed obstacles and being made immobile; the rate of such collisions is proportional to the dislocation flux, Δv_1. The term $+\beta \Delta v_1 g(v_1)$ allows for the creation of dislocations by moving dislocations, for example, by Frank's dynamic mechanism pictured in Figure 1.16. The factor $g(v_1)$ is the probability that this production mechanism operates at the velocity v_1. The last term accounts for the immobilization or annihilation of colliding pairs of dislocations. It is proportional to the rate of pair collisions, $\Delta^2 v_1$, with a factor $f(v_1)$ which is the probability of mutual capture of a dislocation pair moving with velocity v_1. The coefficients $\alpha, \dots, \delta$ are functions of stress, but are independent of Δ and v_1.

Substituting the linear expression for v_1 and using Taylor approximations for $g(v_1)$ and $v_1 f(v_1)$ for small dislocation densities, we obtain an equation of the form $\partial \Delta / \partial t = 2\dot{a}_p = \alpha' - \gamma' \Delta - \delta' \Delta^2 + \varepsilon' \Delta^3$ which can be easily integrated. By making appropriate choices for the coefficients, the resulting solution for the longitudinal strain, e_L, as a function of time agrees well with the observed creep behaviour of a variety of materials.[26, 27]

It is possible to draw certain qualitative conclusions about the solution obtained from this model which will hold regardless of most of the detailed properties of relations (9.103). In a freshly annealed specimen, the initial dislocation density will be low, and the production rate will be dominated by the Frank-Read source term: $\dot{d}_p \sim \alpha v_1$. Also the velocity will remain approximately constant, $v_1 \approx v_0$, so that the dislocation density Δ and plastic strain rate $\dot{e}_L$ will increase linearly with time. If dynamic production mechanisms are important, such as could occur under shock loading, $\partial \Delta / \partial t = 2\dot{d}_p$ becomes an increasing function of Δ, and the growth of Δ becomes exponential. Whether this phase occurs or not, eventually the immobilization mechanisms assume importance, and a limiting state is obtained in which creation and immobilization exactly balance, with Δ and v_1 tending to constant values. Thus the strain rate $\dot{e}_L$ also tends to a constant value, and we have an asymptotic state of steady creep.

The theory of plastic flow developed here is undoubtedly very crude, both because of the limitation to a single dislocation velocity and because of the assumptions made in arriving at explicit equations for the velocity and production rate. A more detailed theory involving a specific mechanism of dislocation capture by clusters of immobile dislocations and the creation of Frank-Read sources within the clusters has been developed by Argon and East[29] to describe easy glide in copper. In the next subsection we shall suggest some improvements that can be made in the model in the direction of allowing more than one velocity, and in the final subsection we shall describe some of the evidence, mostly experimental, in favour of certain forms for the velocity equation.

B *Models with velocity distribution*

If we do not restrict the model to a single dislocation velocity, it is no longer sufficient to deal with the spatial dislocation densities, but we must introduce densities in velocity space as well. Focusing attention on the antiplane case, as in §9.1E, we need the two densities $d^{\pm}(\mathbf{r}, \mathbf{v})$. The conservation equation for these will be a Boltzmann equation,

$$\frac{\partial}{\partial t} d^{\pm} + \frac{\partial}{\partial x_i} (v_i d^{\pm}) + \frac{\partial}{\partial v_i} (\dot{v}_i d^{\pm}) = \dot{d}_p^{\pm} + \dot{d}_c^{\pm}.$$

Here the left-hand-side gives the usual convective term in phase-space. On the right, $\dot{d}_p^{\pm}$ gives the rate of change of $d^{\pm}$ due to pair production processes and $\dot{d}_c^{\pm}$ gives the rate due to collision processes which change the dislocation velocity without creating new dislocations. Examples of such processes would be collisions between pairs of dislocations or between dislocations and fixed obstacles.

In using the Boltzmann equation we would need an expression for the dislocation acceleration $\dot{\mathbf{v}}$, which in principle we could obtain from the equation of motion, as well as some knowledge of the production and collision terms. The resulting equation would clearly be very complicated and it seems doubtful whether any progress can be made with it. A much simpler model, which still represents a great improvement over the single-velocity model, can be obtained by dividing the dislocations into two classes, those that are mobile and those that are immobile, and supposing the mobile dislocations all to have the same velocity. This amounts to assuming that the acceleration periods are short, and that most of the time mobile dislocations glide at a constant velocity which is determined by a balance of the driving force of the external stress and the retarding effect of the various dissipation mechanisms. Then we can write $a^{\pm}(\mathbf{r}) = a_m{}^{\pm}(\mathbf{r}) + a_i{}^{\pm}(\mathbf{r})$ in the antiplane case, where the subscripts m and i refer to mobile and immobile dislocations respectively. The Boltzmann equation becomes two equations holding in ordinary space only:

$$\frac{\partial}{\partial t} a_m{}^{\pm} \pm \frac{\partial}{\partial x_k}(v_k a_m{}^{\pm}) = \pm \dot{a}_p - \dot{a}_c{}^{\pm}, \qquad \frac{\partial}{\partial t} a_i{}^{\pm} = \dot{a}_c{}^{\pm},$$

where v_k is the velocity of mobile dislocations. The collision terms result in a transfer of dislocations from the mobile to the immobile category, or vice versa. The production terms are assumed to contribute only to the mobile densities.

Applying this model to the state of uniform simple tension of Figure 9.3, the numerical densities of positive and negative dislocations will again remain equal throughout ($a_m{}^{+} + a_m{}^{-} = 0$, $a_i{}^{+} + a_i{}^{-} = 0$) and it is sufficient to deal with the total numerical densities $\Delta_m = a_m{}^{+} - a_m{}^{-}$ and $\Delta_i = a_i{}^{+} - a_i{}^{-}$. These satisfy

$$\partial \Delta_m / \partial t = 2\dot{a}_p - 2\dot{a}_c{}^{+}, \qquad \partial \Delta_i / \partial t = 2\dot{a}_c{}^{+}.$$

The plastic strain rate is now $\dot{\beta}_2 = \Delta_m v_1$.

Again we need additional equations for the velocity and production rate and also for the collision rate. In the velocity equation we would expect the largest contribution to the friction stress opposing dislocation motion from the immobile dislocations, and for small densities we might have $v_1 \approx v_0 - k\Delta_i$. Assuming the same mechanisms as in equation (9.104) we would write

$$\dot{a}_p = \alpha v_1 + \beta g(v_1)\Delta_m v_1, \qquad \dot{a}_c{}^{+} = \gamma \Delta_m v_1 + \delta_1 \Delta_m{}^{2} v_1 f_1(v_1) + \delta_2 \Delta_m \Delta_i v_1 f_2(v_1).$$

It is clear from these equations that the present model is very much superior to the earlier single-velocity model, since we are able to distinguish properly between true production processes and those processes which simply immobilize dislocations.

Although no explicit solutions have been evaluated for this model, as have been obtained for the single-velocity model, it is again possible to make certain qualitative statements in general about the solution. The initial stage of deformation of an annealed specimen is again dominated by the Frank-Read sources, giving a linear increase in mobile density and in plastic strain rate. As before a period of exponential growth occurs next under rapid loading conditions, due to the dynamical sources. The asymptotic state is, however, different from before, since, at the moment when a balance between production and immobilization is achieved, which corresponds to steady-state creep, the immobile dislocation density is still increasing. This increase leads eventually to a further decrease in dislocation velocity, and to a decay of this almost-steady creep. Asymptotically the mobile density and velocity tend to zero and the immobile density increases indefinitely.

c *Dislocation velocities*

One of the components of a theory of plasticity based on microscopic considerations must be an equation for the limiting dislocation velocity under given conditions of stress and dislocation densities. Ideally this should come from an analysis of the various mechanisms of energy dissipation. However, given the current state of such analyses it is perhaps safer to base the theory on an appropriate phenomenological velocity equation. This procedure has in fact been followed by most of the authors to whom we have referred.[24–28]

Experimental measurements of dislocation velocities are quite extensive in a number of cases. For example the earliest measurements, by Johnston and Gilman[25] on lithium fluoride, cover a range of velocities for both screw and edge dislocations from 10^{-6} cm/sec to 5×10^4 cm/sec. Gilman[30] showed that over the whole range of velocities the results fitted very well the formula

$$v = v_0 \exp(-\sigma_0/\sigma) \tag{9.105}$$

where σ is the appropriate component of shear stress for the dislocations in question and v_0 and σ_0 are constants. v_0 is found to be about half the shear wave velocity, and, if equation (9.105) can be extrapolated, v_0 gives an upper limit to dislocation velocities. The constant σ_0 varies from edge to screw dislocations in such a way that edge dislocations move an order of magnitude faster than screw dislocations over most of the range of measurement. It has been suggested that this is due to the occurrence of double cross-slip[31] (see §1.3c), which produces a trail of loops behind a screw dislocation and has a retarding effect on its motion. Weertman[8] has also pointed out that the presence of jogs on moving dislocations, which would be caused by their cutting through fixed dislocations on other slip systems during their motion, would have a strong retarding effect on fast-moving screw and mixed dislocations but not on pure edges. Gilman[30] also proposed a theoretical justification for equation (9.105) based on the hypothesis that the motion of a dislocation occurs through a succession of thermally activated jumps alternating with periods in which large segments of the dislocation line are pinned by point defects.

A similarly extensive set of measurements were made on two types of sodium chloride with different impurity levels by Gutmanas et al.[32] These authors found that a formula such as equation (9.105) could fit their observations only at high velocities (> 10 cm/sec for one type) and that a better fit is given by

$$v = v_0(\sigma/\sigma_0)^n \qquad (9.106)$$

over all except very low and very high velocities. The exponent n was found to be 8 and 17 for the two purities; if the measurements on lithium fluoride are tested against equation (9.106), a reasonable fit is obtained with $n = 25$.

Earlier Stein and Low[33] had measured dislocation velocities in 3 per cent silicon-iron over the range 10^{-7} cm/sec to 10^{-2} cm/sec, and had preferred to fit their results to equation (9.106) with an exponent n ranging from 35 to 44 depending on the temperature. Their results, however, could equally well be fitted to equation (9.105) except that the resulting v_0 turned out to be quite different from the shear wave velocity, in disagreement with Gilman's theory. Measurements on other metals have been made by Greenman, Vreeland, and Wood[34] on copper (see also Suzuki[35]) and by Pope, Vreeland, and Wood[36] on zinc. Both sets of measurements favour equation (9.106) with $n \approx 1$, and this result led the authors to suggest that the important dissipative mechanisms in these metals may be associated with interactions between dislocations and phonons. Estimates of the effect of these interactions by Lothe[37] had indicated such a linear connection between v and σ. Finally measurements on tungsten by Shadler[38] again fit equation (9.106), with n being between 9 and 15 at room temperature and between 3 and 6 at 77°K.

Thus, in summary, it appears that both the Gilman formula and the Stein-Low formula are applicable to particular materials, according to the dominant dissipative processes. However the latter, equation (9.106), can also be used satisfactorily, as a purely phenomenological result, for moderate velocities even for materials which fit equation (9.105) better over the whole range. Since macroscopic plastic flow only exceptionally involves high-velocity dislocations, we could reasonably well use equation (9.106) as the velocity-stress relation for all materials.

The constant stress σ_0 specifies a reduction in effectiveness of the external shear stress as a driving force, and can be referred to as a friction stress. During plastic flow, as bigger and bigger tangles of immobile dislocations are built up, a greater resistance to dislocation motion arises, and this can be accounted for by allowing σ_0 to increase. Gillis and Gilman[24] suggest that σ_0 should be a function of the plastic strain. However, a view more consistent with our microscopic approach is to allow σ_0 to depend on the dislocation density, since it is the dislocations themselves which are the cause of the increased resistance. Thus in a completely general case with flow occurring on several slip systems, the friction stress for any one system would depend on the dislocation densities on every one of the operative systems. Webster[27] suggests a linear relation between σ_0 and the dislocation density in a situation with only one slip system operating.

The effect of strain-hardening on the velocity relation has been analysed by Argon,[39] and this discussion lends some support to such a linear dependence. Consider a dislocation moving in the x-direction, driven by an external shear stress σ, and passing through an internal stress-field $\sigma_i(x)$ caused by dislocations of the same slip system as itself but lying on parallel planes. Ignoring inertial effects, and supposing the velocity to be given by equation (9.106), we have $v(x) = v_0[(\sigma+\sigma_i(x))/\sigma_0]^n$. The average velocity over a large distance X is therefore

$$\bar{v} = v_0 \left(\frac{\sigma}{\sigma_0}\right)^n \left(\frac{1}{X}\int_0^X \frac{dx}{[1+\sigma_i(x)/\sigma]^n}\right)^{-1}.$$

At low dislocation densities, $\sigma_i(x)$ will be small, and the integrand may be expanded. The average of σ_i will be zero, and, if we denote by $\overline{\sigma_i^2}$ the mean square of σ_i, the average velocity can be set equal to $v_0(\sigma/\sigma_1)^n$, where

$$\sigma_1 \approx \sigma_0 + \frac{n+1}{2}\frac{\sigma_0\overline{\sigma_i^2}}{\sigma^2}.$$

We expect σ_i to be proportional to $\mu b/\lambda$, where λ is the mean separation of dislocations causing the internal stress-field. Since λ^{-2} equals the dislocation density, we obtain a linear relationship between the friction stress and the density. It should be noted that only the immobile dislocations contribute to σ_1.

In situations involving multiple slip, the dependence of the drag-stress on the dislocation densities on other slip systems can similarly be estimated. The intersecting dislocations provide barriers to slip which are localized around the points of intersection with the slip-plane in question, and the mobile dislocation will be temporarily pinned at these points while bowing out in between. Argon[39] argued that the crucial question as regards the dislocation velocity is the relative magnitude of the waiting time, t_w, during which the dislocation is held almost stationary by the obstacles and the average flight-time, v/λ, between obstacles. If the former dominates, then the average velocity is simply λ/t_w and hence is itself inversely proportional to the square root of the intersecting density. If the flight-time dominates, the average velocity is $v(1 + t_w v/\lambda)^{-1}$ and we obtain an additional term in the friction-stress proportional to the square root of the density.

In summary, we have seen that the application of the dislocation flow equations to macroscopic plasticity is as yet only partially understood, particular difficulties being the production and velocity equations. If a phenomenological velocity equation is taken, the correct dependence of the friction stress on the dislocation densities remains uncertain, although we have given arguments in favour of certain forms of that dependence. In spite of these uncertainties, the dislocation approach does give reasonable agreement with observations of macroplasticity for the flow of lithium fluoride and for plastic creep in a number of materials.

PROBLEMS

1 Consider a screw dislocation, parallel to the z-axis, moving in the xz-plane such that its position at time t is $\xi(t) = A \sin \omega t$. At distances r from the origin which are much larger than the amplitude A, show that the time-dependent part of the displacement field is

$$u_3 \approx -\frac{b\omega A}{4c} \frac{x_2}{r} \left[J_1\left(\frac{\omega r}{c}\right) \cos \omega t + Y_1\left(\frac{\omega r}{c}\right) \sin \omega t \right].$$

Find σ_{13} and σ_{23} at large distances. Show that the average rate of transmission of energy through a cylinder of large radius centred at the z-axis is equal to $\mu b^2 A^2 \omega^3/16c^2$.[18] [The rate of working, per unit area, of the material inside the cylinder on that outside is $-(\sigma_{13}x_1/r + \sigma_{23}x_2/r)\dot{u}_3$.]

2 In Problem 1, suppose that the dislocation is at rest at the origin for $t < 0$ and then begins to oscillate sinusoidally. Find the displacement field and show that as $t \to \infty$ it approaches the solution of the previous problem at each point. For times such that $(ct/r) - 1 \ll 1$, show that

$$u_3 \sim \frac{b\omega A}{2\pi c} \frac{x_2(c^2 t^2 - r^2)^{1/2}}{r^2}.$$

3 If $\mathbf{u}$ is a time-dependent elastic displacement field in an isotropic body, show that the quantities $\mathbf{G} = \mathbf{\nabla}_\wedge \mathbf{u}$ and $H = \mathbf{\nabla} \cdot \mathbf{u}$ both satisfying wave equations, with velocities c_2 and c_1 respectively. Find their values, $\mathbf{G}^s(x, y)$ and $H^s(x, y)$, for a stationary edge dislocation. By introducing two Lorentz transformations, to coordinates x_1' and x_2', as in equations (9.26), using the wave velocities c_1 and c_2 respectively, find the solution for a uniformly moving edge dislocation in the form $\mathbf{G}(x, y, t) = A\mathbf{G}^s(x_2', y)$ and $H(x, y, t) = B\Pi^s(x_1', y)$ where A and B are certain constants.[40]

4 Generalize equations (9.36)–(9.44) to include climb motions of the edge dislocations.

5 Show that as $v \to 0$ the displacements in equations (9.50) for the moving edge dislocation approach those of a stationary edge.

6 Find the displacement field for an edge dislocation which is climbing with constant velocity v. [See the paragraph following equation (9.50).]

7 Show that for an edge dislocation which is gliding with uniform velocity v in the x-direction, the shear stress in the glide-plane is given by

$$\sigma_{xy}(x, 0) = \frac{\mu b}{2\pi(x-vt)} \frac{c_2{}^2}{v^2\beta_2} [4\beta_1\beta_2 - (1+\beta_2{}^2)^2].$$

Find the velocity at which this stress changes sign at any value of $x - vt$.

8 Show that the total energy (kinetic plus elastic) of a uniformly moving edge dislocation is given by equation (9.51).

9 Verify that the Green's function defined by equations (9.82) and (9.83) satisfies the reciprocity condition: $u_{lm}(\mathbf{r}, \mathbf{r}', \tau) = u_{ml}(\mathbf{r}', \mathbf{r}, \tau)$. [See the remark following equation (9.84).]

10 Verify that the integral of the dynamical Green's function of equation (9.89) over all time yields the static Green's function of equation (2.44).

11 Derive equation (9.23) for a moving screw dislocation from the general result of equation (9.91).

12 Obtain from equations (9.91) and (9.92) expressions for the displacements and distortions produced by an edge dislocation which is translating with a general combination of glide and climb.

13 In order to provide a sinusoidal motion of a screw dislocation in which the acceleration is $\ddot{\xi}(t) = \dot{\xi}_0 \cos \omega t$, we have shown that a shear stress given by equation (9.102) is necessary. Calculate the average rate at which this stress does work on the dislocation and show that it equals the rate of energy radiation calculated in Problem 1. Explain why.

REFERENCES

1 Courant, R., and Hilbert D., *Methods of mathematical physics*, vol. II (New York: Interscience Publishers, 1966), p. 692

2 Nabarro, F.R.N., Phil. Mag. **42** (1951), 1244

3 Eshelby, J.D., Phil. Trans. **A244** (1951), 87

4 Frank, F.C., Proc. Phys. Soc. **A62** (1949), 131

5 Eshelby, J.D., Phys. Rev. **90** (1953), 248

6 Eshelby, J.D., Proc. Phys. Soc. **A62** (1949), 307

7 Sneddon, I.N., *Fourier transforms* (New York: McGraw-Hill, 1951)

8 Weertman, J., in *Response of metals to high velocity deformation*, ed. P.G. Shewmon and V.F. Zackay (New York: Interscience Publishers, 1961), p. 205

9 Fox, N., Quart, J. Mech. Appl. Math. **21** (1968), 67

10 Lardner, R.W., Z. angew. Math. Phys. **20** (1969), 514

11 Mura, T., Int. J. Engng. Sci. **1** (1963), 371

12 Mura, T., Phil. Mag. **8** (1963), 843

13 Kosevich, A.M., Sov. Phys. J.E.T.P. **15** (1962), 108

14 Kosevich, A.M., Sov. Phys. Uspekhi **7** (1965), 837

15 Love, A.E.H., *A treatise on the mathematical theory of elasticity* (Cambridge, 1927; New York: Dover, 1944)

16 Kiusalaas, J., and Mura, T., Phil. Mag. **9** (1964), 1

17 Hart, E.W., Phys. Rev. **98** (1955), 1775

18 Eshelby, J.D., Proc. Roy. Soc. **A197** (1949), 396

19 Weiner, J.H., J. Appl. Phys. **29** (1958), 1305

20 Leibfried, G., Z. Phys. **127** (1950), 344

21 Nabarro, F.R.N., Proc. Roy. Soc. **A209** (1951), 278

22 Lothe, J., Phys. Rev. **122** (1961), 78

23 Stroh, A.N., Phys. Rev. **128** (1962), 55

24 Gillis, P.P., and Gilman, J.J., J. Appl. Phys. **36** (1965), 3370 and 3380

25 Johnston, W.G., and Gilman, J.J., J. Appl. Phys. **30** (1959), 129
26 Li, J.C.M., Acta Metall. **11** (1963), 1269
27 Webster, G.A., Phil. Mag. **14** (1966), 775 and 1303
28 Lardner, R.W., Int. J. Engng. Sci. **7** (1969), 417
29 Argon, A.S., and East, G., Trans. Japan. Inst. Metals (Suppl., 1968)
30 Gilman, J.J., Australian J. Phys. **13** (1960), 327
31 Johnston, W.G., and Gilman, J.J., J. Appl. Phys. **31** (1960), 632
32 Gutmanas, E.Y., Nadgornyi, E.M., and Stepanov, A.V., Soviet Phys. Solid State **5** (1963), 743
33 Stein, D.F., and Low, J.R., J. Appl. Phys. **31** (1960), 362
34 Greenman, W.F., Vreeland, T., and Wood, D.S., J. Appl. Phys. **38** (1967), 3595
35 Suzuki, T., in *Dislocation dynamics*, ed. G.T. Hahn (New York: McGraw-Hill, 1969), p. 551
36 Pope, D.P., Vreeland, T., and Wood, D.S., J. Appl. Phys. **38** (1967), 4011
37 Lothe, J., J. Appl. Phys. **33** (1962), 2116
38 Schadler, H.W., Acta Metall. **12** (1964), 861
39 Argon, A.S., Materials Science and Engng. **3** (1968), 24
40 Hirth, J.P., and Lothe, J., *Theory of dislocations* (New York: McGraw-Hill, 1968), pp. 163–6

INDEX

Abraham, M., 274
Abramowitz, M., 114
accelerating dislocation, 317f., 321, 338ff.
action principle in elasticity, 57f.
Airy stress function, 44, 75f.
Amelinckx, S., 144
anisotropic materials, 42, 54, 101f.
Anderson, W.E., 233
anisotropy constant, 43
annealing, 29
antiplane deformation, 46, 177ff., 247f., 313ff.
Argon, A.S., 240, 348, 352
arrays of dislocations, 150ff., 192f.
atomistic calculations for dislocations, 144ff.
Averbach, B.L., 240

Barenblatt, G.I., 211, 213, 214
Barenblatt crack theory, 210ff.; connection with Griffith theory, 214f.
Barrett, C.S., 29
barriers to dislocation motion, 16, 18, 19f., 25ff., 295ff., 352
Bauschinger, J., 23
Bauschinger effect, 23, 28
BCS model of crack, 163ff., 193; for cyclic stressing, 234f.
Becker, R., 274
Berg, C.A., 202
Betti, E., 49
Betti's theorem, 49
Bilby, B.A., 38, 164, 169, 177, 205, 226, 234, 237, 239, 243, 251, 255, 256, 267
Blin, J., 100, 282, 283
Blin formula, 100, 283
Boas, W., 5, 29
body-centred cubic structure, dislocations in, 32f.
boundary-value problems for half-space, 177ff.

bow-out of dislocations, 20f., 295, 297ff., 300ff.
Bradshaw, F.J., 235
Bragg, W.L., 127
brittle fracture, 27, 197, 199ff.; biaxial stresses, 206ff.; crack formation, 200f., 215ff.; crack growth, 199ff., 203ff.; Griffith criterion, 199, 203ff.; pile-ups, 215ff., 220ff.; statistical theory, 200
bubble raft model, 127
bulk modulus, 43
Bullough, R., 243, 251, 256
Burgers, J.M., 4, 9, 10, 99
Burgers circuit, 10f., 244
Burgers formula for displacement, 97ff.
Burgers vector of a dislocation, 4, 10f., 12, 38; in a crystal lattice, 10f., 29ff.; in an elastic continuum, 4, 12, 244, 245, 256, 268, 330; conservation at a node, 14
Burton, W.K., 15

Cabrera, N., 15
Carrington, W., 297
Cauchy equations of motion, 42, 60, 316, 329
Cauchy stress tensor, 41f., 59f.
centre of dilatation, 68
Césaro formula, 277
Chou, Y.T., 102
circular dislocation loops, 291ff., 299f.
cleavage plane, 201, 218
climb force, 85
climb motion, 8, 326
Coffin, L.F., 231
cohesive strength, 199
Colonnetti, G., 82
combined stresses and fracture, 206ff.
compatibility equations, 45, 246, 270, 273, 274, 277